Advances in
SUBSURFACE DATA
ANALYTICS

Advances in
SUBSURFACE DATA ANALYTICS
Traditional and Physics-Based Machine Learning

Edited by

SHUVAJIT BHATTACHARYA
Research Associate, Bureau of Economic Geology,
The University of Texas at Austin, USA

HAIBIN DI
Senior Data Scientist, Schlumberger, USA

ELSEVIER

Elsevier
Radarweg 29, PO Box 211, 1000 AE Amsterdam, Netherlands
The Boulevard, Langford Lane, Kidlington, Oxford OX5 1GB, United Kingdom
50 Hampshire Street, 5th Floor, Cambridge, MA 02139, United States

Notices
Knowledge and best practice in this field are constantly changing. As new research and experience broaden
our understanding, changes in research methods, professional practices, or medical treatment may become
necessary.

Practitioners and researchers must always rely on their own experience and knowledge in evaluating and
using any information, methods, compounds, or experiments described herein. In using such information
or methods they should be mindful of their own safety and the safety of others, including parties for whom
they have a professional responsibility.

To the fullest extent of the law, neither the Publisher nor the authors, contributors, or editors, assume any
liability for any injury and/or damage to persons or property as a matter of products liability, negligence or
otherwise, or from any use or operation of any methods, products, instructions, or ideas contained in the
material herein.

ISBN: 978-0-12-822295-9

For Information on all Elsevier publications visit our website at
https://www.elsevier.com/books-and-journals

Publisher: Charlotte Kent
Acquisitions Editor: Amy Shapiro
Editorial Project Manager: Maria Elaine D Desamero
Production Project Manager: R.Vijay Bharath
Cover Designer: Mark Rogers

Typeset by Aptara, New Delhi, India

Working together
to grow libraries in
developing countries

www.elsevier.com • www.bookaid.org

CONTENTS

Contributors

Ayodeji Aboaba
West Virginia University, United States

Tariq Alkhalifah
Department of Physical Science and Engineering, King Abdullah University of Science and Technology, Thuwal, Saudi Arabia

Heather Bedle
School of Geosciences, The University of Oklahoma, Norman, OK, United States

Nasher BenHasan
EXPEC ARC, Saudi Aramco, Dhahran, Saudi Arabia

Alex Bromhead
Halliburton, United Kingdom

John P. Castagna
Department of Earth and Atmospheric Sciences, University of Houston, Houston, TX, United States

Edward Clee
Department of Computer Science, Prairie View A&M University, Prairie View, TX, United States

Kate Evans
Halliburton, United Kingdom

Hugo Garcia
Geoteric, United States

Dario Grana
Department of Geology and Geophysics, School of Energy Resources, University of Wyoming, Wyoming, United States

Chris Guenther
Department of Energy, National Energy Technology Laboratory, United States

Pablo Guillen
Department of Computer Science, University of Houston, Houston, TX, United States

Ehsan Haghighat
Department of Civil Engineering, Massachusetts Institute of Technology, MA, United States

Chris Han
Geoteric, United Kingdom

Lei Huang
Department of Computer Science, Prairie View A&M University, Prairie View, TX, United States

Lian Jiang
Department of Earth and Atmospheric Sciences, University of Houston, Houston, TX, United States

Cédric M. John
Department of Earth Science and Engineering, Imperial College, London, United Kingdom

Karelia La Marca
School of Geosciences, The University of Oklahoma, Norman, OK, United States

Mingliang Liu
Department of Geology and Geophysics, School of Energy Resources, University of Wyoming, Wyoming, United States

Yong Liu
Department of Energy, National Energy Technology Laboratory, United States; Leidos Research Support Team, United States

James Lowell
Geoteric, United Kingdom

Yvon Martinez
West Virginia University, United States

Shahab D. Mohaghegh
West Virginia University and Intelligent Solutions, Inc., United States

Philippe Nivlet
EXPEC ARC, Saudi Aramco, Dhahran, Saudi Arabia

Rafael Pires de Lima
Geological Survey of Brazil, São Paulo, Brazil

Nishath Ranasinghe
Department of Computer Science, Prairie View A&M University, Prairie View, TX, United States

Mehrdad Shahnam
Department of Energy, National Energy Technology Laboratory, United States

Robert Smith
EXPEC ARC, Saudi Aramco, Dhahran, Saudi Arabia

Chao Song
Department of Physical Science and Engineering, King Abdullah University of Science and Technology, Thuwal, Saudi Arabia

Fnu Suriamin
Oklahoma Geological Survey, Norman, OK, United States

Peter Szafian
Geoteric, United Kingdom

Miao Tian
Department of Geosciences, The University of Texas Permian Basin, Odessa, TX, USA

Sumit Verma
Department of Geosciences, The University of Texas Permian Basin, Odessa, TX, USA

Umair bin Waheed
Department of Geosciences, King Fahd University of Petroleum and Minerals, Dhahran, Saudi Arabia

Jessica Wevill
Department of Earth Science and Engineering, Imperial College London, United Kingdom

Ryan Williams
Geoteric, United Kingdom

Jeffrey Yarus
Case Western Reserve University, United States

Zhendong Zhang
The Earth Resources Laboratory, Massachusetts Institute of Technology, MA, United States

About the Editors

Dr. Shuvajit Bhattacharya is a Research Associate at the Bureau of Economic Geology, the University of Texas at Austin. He is an applied geophysicist/petrophysicist specializing in seismic interpretation, petrophysical modeling, machine learning, and integrated subsurface characterization. Prior to joining the Bureau of Economic Geology, Dr. Bhattacharya worked as an Assistant Professor at the University of Alaska Anchorage. He has completed several projects in the US, Norway, Netherlands, Australia, South Africa, and India. He has published and presented more than 70 technical articles in peer-reviewed journals, books, and conferences. His current research focuses on the pressing issues and frontier technologies in energy resources exploration, development, and subsurface storage of carbon and hydrogen. He completed his Ph.D. at West Virginia University in 2016 and an M.Sc. at the Indian Institute of Technology Bombay in 2010.

Dr. Haibin Di is a Senior Data Scientist in the Subsurface Data Intelligence team at Schlumberger. His research interest is in implementation of machine learning algorithms, particularly deep neural networks into multiple seismic applications, including stratigraphy interpretation, property estimation, denoising, and seismic-well tie. Dr. Di has published more than 70 papers in seismic interpretation and holds 7 patents on machine learning-assisted subsurface data analysis. Dr. Di received his Ph.D. degree in Geology from West Virginia University in 2016, worked as a postdoctoral researcher at Georgia Institute of Technology in 2016–18, and joined Schlumberger in 2018.

Preface

Advances in Subsurface Data Analytics: Traditional and Physics-Based Machine Learning is a compilation of selected case studies in the applications of traditional, emerging, and physics-based machine learning (ML) algorithms and approaches in subsurface imaging and characterization of heterogeneous media. We edited this book to bring together the fundamentals of several ML algorithms with their detailed applications in subsurface analysis, including geology, geophysics, petrophysics, and reservoir engineering. While we focus on traditional and physics-based ML approaches, we also shed light on how ML technologies developed in the subsurface can be used in other disciplines.

The book is intended to be used as a detailed reference book primarily by professional geoscientists and reservoir engineers working on subsurface-related research problems (oil and gas, carbon sequestration, geothermal, and hydrogen storage). It can be used by faculty, researchers, and graduate students in the geosciences/energy resources engineering departments at universities.

The edited book comprises several chapters, each focusing on a particular application of ML algorithm(s) with detailed workflow and recommendations. In addition, some of the chapters also contain a comparison of an algorithm with respect to others to better equip the readers with different strategies to implement automated workflows for case-specific subsurface analysis.

The structure of the book is easy-to-follow for the readers, with mainly four parts. The first part focuses on traditional ML algorithms with case studies, which have been widely used. This is followed by emerging and complex deep learning algorithms. The third part focuses on physics-based ML. The last chapter is on future research directions in ML to solve problems in other disciplines.

Marca and *Bedle* study deep water seismic facies classification in the Taranaki Basin of New Zealand by investigating the impact of a user-controlled (interpreter-driven) selection of seismic attributes versus machine-selected results. The user-selected attribute allows the interpreter to inspect attributes individually and select the most geologically meaningful ones for their study; however, the reliability of the exercise depends on experience of the interpreters, including bias. Authors recommend detailed human inspection and approval of ML-based results in pivotal moments, including the validation and interpretation of the ML outputs regardless of the initial approach used.

Wevill et al. discuss various ML techniques to analyze the relationships between subsurface data and production success within seven North American shale plays. They use formation depth, thickness, porosity, resource concentration, pore pressure, geothermal gradient, and gas-to-oil ratio to identify the "heavy-hitters" influencing hydrocarbon production. They conclude that a large influence on production success from these reservoirs is linked to geological factors such as depositional and tectonic history dictating play properties like mineralogy and pressure. The overall success of a hydraulically fractured well in these plays is also susceptible to the influence of completion methods.

Many geologic processes are gradational at spatial and temporal scales, which results in sequence-like features. The whole concept of sequence stratigraphy is based on sequence (and parasequence) pattern analysis. *Tian* and *Verma* demonstrate a deep learning workflow (i.e., convolutional neural network combined with long short-term memory) to classify and predict complex mudstone facies using electrical log data. Their workflow captures and employs the hidden information of data sequence and spatial dependencies, which results in a generalized and robust ML model for facies classification in multiple boreholes in the basin.

Liu et al. apply recurrent neural network (i.e., bi-directional long short-term memory) for seismic reservoir characterization. They estimate P-wave velocity, S-wave velocity, density, porosity, water saturation, and facies. Their results indicate the promising potential for the application of deep neural networks to seismic inversion. The proposed approach is a valid alternative to the conventional model-driven methods, and it avoids several data pre-processing steps, such as model calibration, and provides a robust and efficient way to efficiently automate the workflow of seismic reservoir characterization.

Core description and interpretation are vital to geologic studies but time-consuming, expensive, and subject to multiple interpretations. *Lima* and *Suriamin* show that computer vision instance segmentation models built with convolutional neural networks have the potential to greatly accelerate core interpretation in a consistent manner. They directly use standard core photographs from a 260 m (850 ft) core from a siliciclastic succession as input to their model with barely any data preprocessing for core facies classification.

Waheed et al. develop a workflow based on the emerging paradigm of a physics-informed neural network to solve various forms of the Eikonal equation in computing first-arrival travel time. This has applications in seismic velocity modeling, microseismic source localization, and seismic migration. Their workflow obtains fast and accurate travel times, compared to the conventional workflow used over decades. This workflow is applicable to both isotropic and anisotropic media.

Scientific ML brings a new research dimension to solve complex problems in geosciences that traditional ML approaches cannot solve. *Huang et al.* propose a new method to improve the performance of seismic wave simulation and inversion by integrating deep learning (recurrent neural network and AutoEncoder) and differential programming with High-Performance Computing. Their workflow improved model accuracy and efficiency.

Jiang et al. discuss a novel workflow on integrating physics-based approaches with ML into building robust rock physics models (RPM) for acoustic velocity prediction.

Building RPM is a complex multi-step process with several assumptions. Jiang et al. demonstrate the feasibility of ML-assisted RPM using synthetic and real datasets. Their RPM establishes relationships between mineral composition, porosity, fluid properties, water saturation, temperature, and pressure with density and the P-wave and S-wave velocities.

Zeng and *Alkhalifah* present a deep learning-aided elastic full-waveform inversion (FWI) strategy using observed seismic data and available well logs using a synthetic dataset and field dataset in the North Sea. They link seismic facies to the inverted P- and S-wave velocities and anisotropy using trained neural networks. The estimated facies can be used as a physical constraint for conventional elastic FWI, which can be helpful to better resolve deep-buried reservoir targets.

Mohaghegh et al. discusses new research directions where ML-based technologies, specifically Smart Proxy Modeling developed in reservoir engineering, can be used in other disciplines, such as computational fluid dynamics (CFD). CFD is widely used in gas compressors, turbines, heating, and aerodynamics, etc. Gas supplies are expected to increase in North America due to shale gas and the upcoming hydrogen economy. The effect of gas composition on combustion behavior is of interest to allow end-user equipment to accommodate the widest possible gas composition. Mohaghegh demonstrates the feasibility of highly accurate and high-speed proxy modeling of such complex systems using artificial intelligence.

We hope this book will help popularize ML in the subsurface community, providing new research directions. We hope you will enjoy reading this book and solving your own problems.

Shuvajit Bhattacharya
Haibin Di

Acknowledgments

We would like to express heartfelt thanks to all 31 authors from industry and academia who contributed to this book. This book could not have been possible without their writing efforts and patience in tolerating editorial commands and reviews. These authors have spent years studying and experimenting with advanced machine learning concepts in geosciences and reservoir engineering. We thank them for that. We would also like to express thanks to numerous individuals and professionals with whom we had an opportunity to discuss artificial intelligence and learn from their experiences.

Special thanks go to Amy Shapiro, Elaine Desamero, Vijay Bharath, and Chris Hockaday at Elsevier for making this book happen. We also acknowledge the publishers and individuals who provided permission to use figures from their technical articles and websites. We deeply appreciate all your support and encouragement.

Shuvajit Bhattacharya
Haibin Di

PART 1

Traditional machine learning approaches

CHAPTER 1

User vs. machine-based seismic attribute selection for unsupervised machine learning techniques: Does human insight provide better results than statistically chosen attributes?

Karelia La Marca, Heather Bedle
School of Geosciences, The University of Oklahoma, Norman, OK, United States

Abstract

In geosciences, a variety of machine learning (ML) algorithms are currently being employed for multiple purposes, for example, facies classification, fault prediction, and reservoir characterization. Among these are two clustering methods: principal component analysis (PCA) and self-organized maps (SOMs), which provide a fast organization of data into groups or clusters (with no geologic supervision) that aid in preliminary geological interpretation. With increasingly common usage of these techniques, the motivation of this chapter is to investigate the impact of a user-controlled selection of attributes to perform SOM for deepwater seismic facies classification versus a machine-selected result through PCA. Results reveal that whereas an appropriate combination of attributes with a clear interpretation objective enhances the SOM's results and facilitates the interpreter understanding of the output classes, PCA provides insightful information regarding the contribution of seismic attributes that may not have been initially considered. While machine learning techniques are a powerful "tool" for geological interpretation, user control on initial input attributes and validation of output using an "in-context" interpretation is necessary for an optimal elucidation, at least in unsupervised machine learning methods.

Keywords

Machine learning techniques; Principal component analysis; Self-organized maps

1.1 Introduction

Conventional seismic interpretation techniques involve seismic attribute calculations and applications to determine geometries, lithologies, and reservoir properties. A seismic attribute is any measure of seismic data that helps us visually enhance or quantify features of interest for interpretation[1]. Seismic attributes are a response to the rocks' physical properties. Therefore, the spatial distribution and relationships of these responses,

Advances in Subsurface Data Analytics
DOI: https://doi.org/10.1016/B978-0-12-822295-9.00002-9

along with geological context, help make reliable seismic interpretation whether they are used individually or combined. Some examples of the use of seismic attributes can be found in (1) Partyka et al.,[2] who used frequency attributes to study channels, (2) the use of textural attributes to recognize patterns in a turbidite system by Gao[3], (3) the application of curvature attributes for mapping faults/fractures in Chopra and Marfurt[1] and the list goes on.

In recent years, the use of machine learning (ML) has become a common practice in advanced seismic interpretation workflows. These methods are based on complex algorithms that allow quick and improved analysis of seismic data than traditional single attribute studies made by interpreters. Self-organized maps, geostatistics, and neural nets are used to extend the capability of recognizing patterns beyond the three dimensions that the interpreter is limited to[4]. With common usage of large datasets, the geophysical community has worked hard to replicate the human ability to cluster, but these methods need to be thoroughly validated before they can be relied upon.

ML applications within the geosciences extend from predicting seismic facies, to automatic fault detection to reservoir property prediction[5-7,8-10]. The choice of an unsupervised ML method over a supervised method can depend on the type of data available for the study. In exploration stages, it is often rare to find well logs or other data in the basin or the area of interest. In this scenario, unsupervised methods provide ways to find relationships among the variables in the data available. However, unsupervised methods require interpreter evaluation to approve or disapprove the output.

Methods such as Principal Component Analysis (PCA), introduced by Pearson[11] and developed by Hotelling[12], is a common technique to find patterns in data of high dimensions[13]. PCA seeks to reduce a multivariate space (dataset) down to a more manageable size of relatively independent variables. In seismic interpretation, PCA helps to determine the most meaningful seismic attributes[14,15]. Likewise, SOM is an unsupervised technique that can be used in multi-attribute analysis to extract additional information from the seismic response that is not possible with a single attribute[16]. This SOM method was first coined by Kohonen[17], but became popular recently due to the increase in computational power. PCA and SOMs have been used for many purposes. For example, in the marketing arena, Das et al.[18] applied PCA and SOM to find clusters in people's responses regarding their preference for retail store personality. In seismic data, Coleou et al.,[5] applied SOMs for seismic classification using a 1D latent space. More recently, Guo et al.,[19] computed spectral attributes and applied PCA to determine the first three principal components of the spectral variation. Later, Matos et al.,[20] Roy and Marfurt[21], and Roy[22] demonstrated the advantage of extending the SOM latent space to 2D and 3D, using 2D and 3D colorbars to delineate elements in depositional environments. More recent studies[6,15,16] integrated PCA to select attributes, and SOMs to interpret facies in various depositional environments and others specifically in deepwater deposits[8,23,24].

To better understand the effects of attribute choice, we investigated and compared a multi- attribute user-driven approach versus a machine-derived method (through PCA) to select suitable attribute combinations to use in the unsupervised self-organizing maps (SOMs) for facies prediction. We used the Pipeline 3D seismic dataset, in the southern Taranaki Basin of New Zealand, to interpret deepwater stratigraphic features based on SOMs results. We explored which attribute selection (human or machine driven) allows for a reliable classification to improve interpretation. Finally, we document the advantages and disadvantages of each method, recommending good practices to take advantage of ML tools in a suitable manner to obtain efficient and valuable geological results.

1.2 Motivation

This study's main motivation was to explore, compare, and understand the differences that machine selected attributes (via PCA) and user-selected attributes (based on interpreter experience) produce in an unsupervised ML technique classification. To compare methods, we used a seismic volume that contains a deepwater channel section to evaluate channel architecture. We evaluated how effective each combination of seismic attributes was to make an "in context" interpretation that includes a detailed characterization of architectural elements' geomorphologies. We sought to understand the best seismic attributes to characterize the deepwater setting and compare the generated clusters that helped define seismic facies and architectural elements. We then explored the advantages and disadvantages of a user-selected vs. machine-selected attribute approaches.

Geological setting

The Taranaki Basin is located in western-offshore New Zealand (Fig. 1.1). The basin deposits range from the Cretaceous to the Neogene. The basin formed and filled up due to the Tasman Sea spreading event[25,26]. The stratigraphic section studied corresponds to the Middle Miocene deposits characterized by the deposition of deepwater sequences controlled mainly by tectonic uplift in the hinterland. The high relief provided a south-east source of sediments carried and deposited in a north- northwest direction[27] (Fig. 1.1). The Miocene succession comprises intercalation of fine-grained basin floor sandstones deposited by channels and fans, in addition to silty and mudstone dominated deposits. The Moki Formation corresponds to deepwater deposits characterized by: (1) channel complex width that ranges from 600 m up to 5,000 m and ranges between 10–30 m in thickness; (2) channel sinuosity varies from low to high; and (3) a system that becomes more incised and mud-dominated during the Late Miocene[26].

1.3 Dataset characteristics

The area of study is located within the Pipeline 3D dataset (Fig. 1.1). The seismic volume data is zero phase and SEG negative polarity (a trough represents a positive change in acoustic impedance), with a sample interval of 4 ms, and bin size of 25 m by

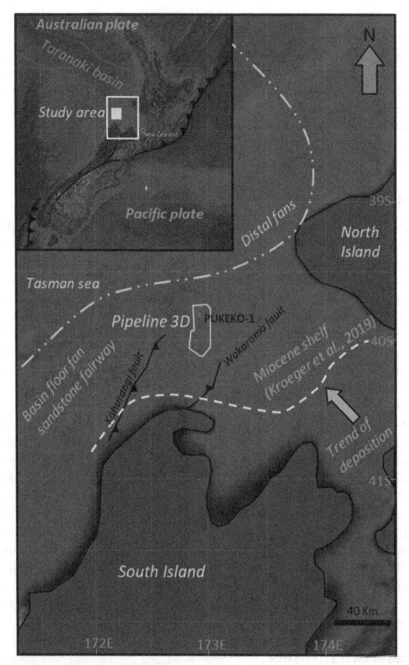

Fig. 1.1 *Map of western offshore New Zealand highlighting the Taranaki Basin and the study area. The dataset (Pipeline 3-D) location is shown in yellow and the Pukeko-1 well with a fuchsia point.* The paleo-trend of deposition (southeast-northwest) and principal faults trend (southwest-northeast) are indicated as reported in Kroeger et al.,[26]. Notice that the study location is close to the distal fans area near Miocene time, when the Moki Formation was deposited. The paleo-shelf break limit by that period is indicated using a blue dashed line after Strogen (2011).

12.5 m, with a vertical resolution of ~25 m for the interest section. Pipeline 3D seismic volume was cropped to an area of ~300 km² and a vertical window of 500 msec (1800 ms–2100 ms) to focus only on the deepwater Miocene section of interest. The projection datum used for the project was NZGD2000. The well-log data (gamma-ray) provided for the Pukeko-1 well was employed to corroborate outputs and support interpretations.

1.4 Seismic attributes

By definition, a seismic attribute is any measure of seismic data that helps us visually enhance or quantify features of interest for interpretation[28]. The seismic data, and therefore the seismic attributes are a response to the rocks' physical properties. By studying the spatial distribution and relationships of these responses, and "geological context", seismic attributes aid us in making reliable seismic interpretations. The seismic amplitude data or even a single attribute is not enough for extracting all the potential information present in the seismic data[4]. Thus, we usually combine attributes in a multiattribute analysis[29,30] or, such as the case in this study, they are used in SOMs to extract as much geological information all as possible in one analysis.

The increase in number and variety of seismic attributes in the last three decades (now we have hundreds of seismic attributes) resulted in many authors' trials of classifying them into families. Marfurt and Chopra[31] summarize the main classification attempts going from Bob Sheriff's[32] and Taner et al.[33] to convey that Liner et al.[34] classification offers a general and specific grouping of seismic attributes based on the physical or morphological character of the data related to geology for the first, and, a less well defined basis in physics or geology for the later.

In Table 1.1,[51] present a list of common seismic attributes, type, and interpretive use. Geometric attributes such as dip, coherence, and curvature help in delineating the geometry or shape of the channel, and attributes such as the instantaneous phase and sweetness are lithological indicators[24]. Grey level co–occurrence matrix (GLCM) texture attributes are useful for the determination of seismic facies analysis[4] as these attributes quantify the uniformity or disorder in the image. Frequency attributes help to interpret bed thicknesses, discontinuities, and fluids[2]. More information about attributes and their use in facies characterization is available in Taner[35], Barnes[36], Liu[37], Chopra and Marfurt[28], and Marfurt[38].

For the user-derived part in the present study, we used one of the attribute combinations chosen by La Marca[39] for use in deepwater depositional settings. Table 1.2 summarizes these five attributes (GLCM entropy, RMS, Similarity and spectral frequencies), explaining their type, underlying principles, references, and geological use/interpretation.

Table 1.1 List of common seismic attributes, their type, and interpretive use (after Roden and Sacrey, 2015).

Category	Type	Interpretive use
Instantaneous attributes	Reflection strength, instantaneous phase, instantaneous frequency, quadrature, Instantaneous Q	Lithology contrasts, bedding continuity, porosity, porosity, DHIs, stratigraphy, Thickness
Geometric attributes	Semblance and Eigen-based coherency/similarity, curvature (maximum, minimum, most positive, most negative, strike, dip)	Faults, fractures, folds, anisotropy, regional stress fields
Amplitude accentuating attributes	RMS amplitude, relative acoustic impedance, sweetness, average energy	Porosity, stratigraphic and lithologic variations, DHIs
AVO attributes	Intercept, gradient, intercept/gradient derivatives, fluid factor, Lambda-Mu-Rho	Pore fluid, Lithology, DHIs, Geomechanics
Seismic Inversion attributes	Colored inversion, sparse spike, elastic impedance, extended elastic impedance, prestack simultaneous inversion, stochastic inversion	Lithology, porosity, fluid effects, geomechanics
Spectral decomposition	Continuous wavelet transform, matching pursuit, exponential pursuit	Layer thicknesses, stratigraphic variations

Table 1.2 List of user-selected attributes used as input for SOM (adapted from[39]).

Attribute	Type	Principle	References	Geological use
GLCM (grey level co-occurrence matrix) entropy	Texture	Quantifies the lateral variation in seismic amplitude. Entropy: how smoothly varying the voxel values or seismic amplitudes are within a window	Haralick et al.[58]	Seismic facies by its textural response
RMS (root mean square amplitude)	Amplitude	Measure of reflectivity within a time window. Computes the square root of the sum of squared amplitudes divided by the number of samples within the window used	Meek[59]	Sand bodies and mud-filled channels associated with channel belts
Sobel filter (similarity)	Discontinuity	(1) Normalizes coherence (2) data to produce results between 0 and 1	1- Luo et al[60] 2- Gersztenkorn A., and K. J. Marfurt[61]	Channel edges and faults
Spectral frequency	Spectral (freq)	(1) Applies a suite of constant-bandwidth filters to the seismic data.	1-², 2-⁶²	Channels and minor architectura elements. Analyze stratigraphy and thickness changes

1.5 Principal component analysis (PCA)

When working with seismic data, we can generate dozens, if not hundreds, of attributes that ultimately amplify the "big data" issue. In machine learning, we try to avoid models that are over-fit, and for that, we need to reduce our initial variables into a more manageable set of variables. By doing this, we apply what is called "dimensionality reduction", where the variables being kept or selected are retaining the most valuable part, and having the majority of contribution, of all the possible variables.

PCA is one of the oldest and most known among multivariate analysis techniques[40]. It was first introduced by Pearson[11] and developed by Hotelling in 1933 and consists of mathematical algorithms that apply a principal axis method to transform a set of variables into a smaller number of variables called principal components (eigenvectors). The eigenvectors are vectors resulting from the PCA, where the first vector has the highest eigenvalue and represents the maximum variance in the data[4]. These eigenvectors preserve most of the variability information present in the input dataset. Therefore, PCA is primarily used as a dimensionality reduction technique. In the PCA results, the first principal component (PC1) accounts for as much variability in the data as possible (Fig. 1.2), and each succeeding component (PC2, PC3, etc.) account for as much of the remaining variability[19]. In other words, PC2 = N-1 dimensional space where the least-squares fits the residual data that was not within the PC1. Because PCA is based on Gaussian statistics, the first eigenvalues or PC preserve most of the signal and best represent the variance in the data, while the last PC has more of the uncorrelated noise[41].

In PCA, the multiattribute space will be conditioned by the N-number of attributes or variables to be used as input. We are then going to have a N-dimensional space with N-eigenvectors and N-principal components[41]. The idea of the method is to reduce dimensionality in a multiattribute data set, which in our case will provide a set of "meta seismic attributes" that allow us to reduce attribute redundancy and simplify the computation of SOM.

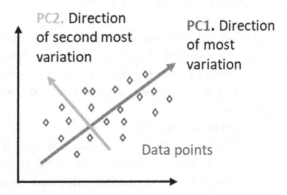

Fig. 1.2 *Explanation of the main principal components derived from PCA.*

In geoscience applications, PCA is performed on a squared symmetric matrix, usually a covariance matrix (which is the scaled sums of squares and cross products). This mathematical procedure is applied when the variances of individual variates differ, or when the units of measurement of individual variates differ, which is the case for seismic attributes of different types. After that, we compute the eigenvalues and eigenvectors of the covariance matrix. More details of the algorithm and applications can be found in Joliffe[40].

PCA has been widely used in geosciences since the 90's, providing successful results in facies prediction[19,42,43], well log correlation[44,45], fault and fracture detection[46], and lately PCA has been used as the first step in clustering techniques, such as self-organizing maps[16], generative topographic maps[22], and support vector machine analysis[41], because it reduces the number of attributes for subsequent analysis[4]. However, one of the drawbacks about PCA is that after the dimensionality reduction, there is no guarantee that the dimensions or results are interpretable (or geologically meaningful, as in our case).

1.6 Self-organizing maps (SOMs): An unsupervised technique for seismic facies identification

As mentioned previously, SOM is an unsupervised (feed-backward) machine learning technique that was first introduced by Kohonen in 1982 and is frequently used in many areas such as in technology, marketing (e.g., Hanafizadeh and Mirzazadeh, 2010[47]), medicine (e.g., Tuckova et al., 2011[48]), and environmental applications (e.g., Gibson et al., 2017[49]) to find patterns with similar characteristics within a dataset. Roy et al.[22] describe how this machine learning method has been used since the 1990s in the oil industry to resolve diverse geoscience interpretation problems. For example, SOMs have been adapted to be used in seismic facies identification as it extracts similar patterns embedded within multiple seismic attribute volumes[41]. With modern visualization capabilities and the application of 2D color maps[50], SOMs routinely produce meaningful and easily interpretable geologic patterns[15], if the input parameters are optimized.

While humans are good at pattern recognition, the task of finding the relationship between all the data points at the same time is a cumbersome analysis, if not impossible. SOMs attempt to solve this issue by reducing the number of dimensions and illustrate these similarities[51] onto a 2D lower dimensional space where the output groups (e.g., clusters) are easier to interpret.

The SOM algorithm details are presented in Kohonen[17,52]. We use a coffee bean analogy to explain how SOMs can classify the beans based on their flavor, morphology, and amount of caffeine (Fig. 1.3). We can summarize the SOM process in four steps:

1. Input data and initialization: coffee bean attributes that have a characteristic flavor, morphology, and amount of caffeine, are used as the input attributes.

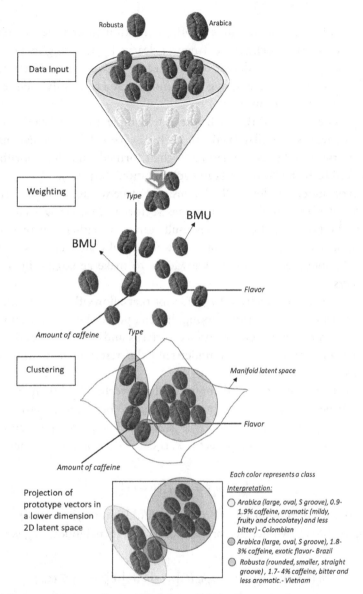

Fig. 1.3 *Example for explaining self-organizing maps (SOMs).* The input data is represented by coffee beans. When they are placed in this space, the best matching units (BMUs) will group similar beans together as they come closer to one another as a result of weighting. The classification is performed in accordance with type, flavor, and amount of caffeine. When the beans with similar characteristics are clustered together, they can be differentiated from other coffee beans in different clusters. These clusters are then projected onto the latent space and are color-coded where they can be later interpreted. In this case, the yellow class is Arabica coffee, with low caffeine content, aromatics, and less bitter in flavor. Therefore, it can be interpreted as Colombian coffee as this is characteristic of the beans that grow in that country's soils. On the other hand, a Robusta bean, with higher caffeine, high bitterness, and less aromatics is clustered in the blue color and can be interpreted as Vietnamese coffee. Even if the clustering method you decide to use is the most robust, you will need an interpreter with the knowledge that can decode the results to determine if they are meaningful.

2. Weighting and best matching unit (BMU) definition (sometimes referenced as the winning neuron) (Eq. 1.1), where a neuron learns by adjusting its position within the attribute space as it is drawn toward nearby data samples. Once the learning process has completed, the winning neuron set is used to classify each selected multiattribute sample in the survey[51].

3. Clustering: Given the BMU, each attribute sample in the dataset incrementally approaches towards a similar BMU in each case (Eq. 1.2). A SOM manifold that contains all the possible combinations is then formed, and the algorithm deforms this manifold to better fit the data in each iteration[8] (Eq. 1.3).

4. Projection of clusters in a lower 2D dimensional space where colors are assigned. Gao[50] mentions that if the number of prototype vectors is 256, we will have 256 colors. These are the potential clusters. Our results can form either 256 or considerably a smaller number of clusters (e.g., three or four). Finally, after clusters or neurons are obtained, the interpreter uses their knowledge to make geological interpretations of these clusters.

Overall, SOMs are a user-friendly ML tool that allows the seismic interpreter to cluster our data based on similarities among the data points. The organization in clusters or neurons enables the interpreter to notice patterns and similarities and enables them to analyze seismic facies and better understand the reservoir and other architecture configurations.

Consider a previously Z-scored input attribute vector k, chosen from the set of input vectors. When computing the Euclidean distance between this attribute data vector "k" and all prototype vectors PVs m_i, $i = 1,2, \ldots$ p. The prototype vector m_b, that has the minimum distance to the input vector k, is defined to be the "winner" or the Best Matching Unit

$$\|k - m_b\| = \min\{\|k - m_i\|\} \tag{1.1}$$

Where m_i = the ith grid point lying on the manifold (prototype vector).

Updating of BMU (winner prototype vector) and its neighbors is given by the following rule:

$$m_i(t+1) = m_i(t) + \alpha(t)h_{bi}(t)\big[x - m_i(t)\big]\text{if } \|r_i - r_b\| \leq \alpha(t) \tag{1.2a}$$

$$= m_i(t)) \text{ if } \|r_i - r\|_b > \alpha(t) \tag{1.2b}$$

Where (t) is the neighborhood radius, which decreases with each iteration t, and r_b, and r_i are the position vectors of the winner vectors m_b and m_i.

Definition of a neighborhood function h(t) and an exponential learning function $\alpha(t)$, where T is the training length. Both, h(t) and $\alpha(t)$ decrease with each iteration and are defined as follow:

$$h_{bi}(t) = e^{e-(\|r_b - r_i\|^2/2\alpha^2(t)} \tag{1.3a}$$

$$\alpha(t) = \alpha_0 \left(\frac{0.005}{\alpha_0} \right) t \, / \, T \qquad\qquad (1.3b)$$

Note: a subsequent iteration from (Eqs. 1.1 to 1.3) is performed until convergence.

1.7 Methodology

The study performed in the deepwater Taranaki Basin Miocene channel system consisted of six stages, shown in Fig. 1.4. We started with the dataset inspection and area of interest definition, followed by the well tie and horizon picking. We then calculated a series of geometrical, instantaneous, and spectral attributes. We used a total of 28 attributes as input in the PCA to allow for the reduction to meaningful attributes contained in the first three eigenvectors. Table 1.3 presents the 28 attributes considered for input in the PCA. We selected the attribute configuration proposed by La Marca,[39] (Table 1.2), Grey-Level-Co-ocurrence Matrix (GLCM) entropy, root mean square, sobel filter similarity, and spectral decomposition for the user inspection and selection.

Next, for the SOM, we considered the most representative attributes (the attributes that contributed the most to the total variance) in the first three principal components of the PCA results. We then performed SOM to determine the clusters (seismic facies). The selection of a window of interest allows for an optimal calculation of the preceding methods to avoid miscalculations. Details of each study phase, parametrizations and considerations taken into account are specified below:

The importance of preliminary inspection of the amplitude volume

When clustering techniques such as SOMs are thought to be applied; the first step is to quality check (QC) the input data. In our case, the seismic amplitude volume is the origin of each subsequent seismic attribute. Therefore, its inspection is necessary to identify acquisition footprint, artifacts, or noise, which can negatively influence the results. The assessment also allows for constraining and cropping the volume to the area of interest (AOI). This study comprised the Moki and Manganui formations at around 1,800–2,300 ms (Fig. 1.5), and an area of around 300 km^2.

Making the most out of available data to validate your models

The Pukeko-1 well contains check-shot data and formation tops, facilitating a depth-time conversion and a well tie. This well tie allows us to recognize the geological formations and the lithological responses present within each interval in the seismic data. Although the 25m seismic resolution of our volume is insufficient to characterize the geological sequences in detail, well logs (specifically gamma ray) are useful for output validation.

Picking horizons parallel to the reflectors of the AOI, provides a better representation of the deepwater system. We picked and flattened the seismic volumes on the Moki horizon. This horizon was determined by following a clear and continuous reflector (a peak) tied to the well log.

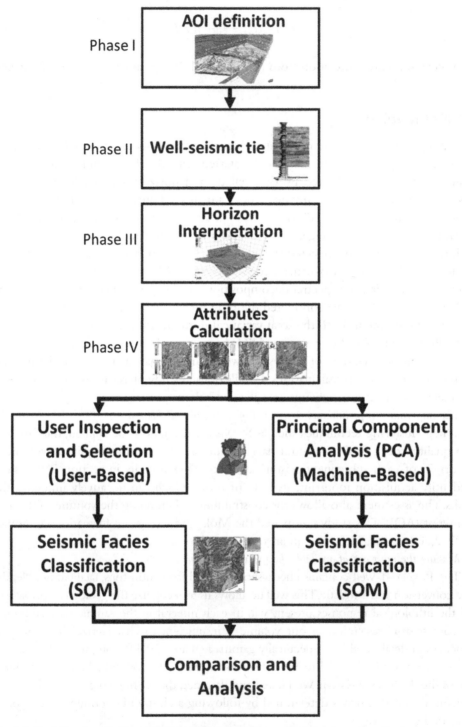

Fig. 1.4 *Workflow for machine-assisted attribute selection and comparison of methods.*

Table 1.3 List of input seismic attributes to PCA. Notice that a total of 28 attributes from different types were considered to determine the main contributors in the section of interest.

Attribute selection list

Dip magnitude	Sweetness
Chaos	Envelope
Sobel filter	Instantaneous frequency
Energy ratio similarity	Instantaneous phase
Total energy	Amplitude volume transform, AVT
K1 (most pos. curvature)	Cosine of instantaneous phase
K2 (most neg. curvature)	Peak frequency
Reflector convergence	Peak magnitude
Curvedness	Seismic Amplitude
K positive	GLCM entropy
K negative	Spect frequency 13Hz
GLCM contrast	Spect frequency 35 Hz
GLCM energy	Spect frequency 60 Hz
GLCM homogeneity	GLCM dissimilarity

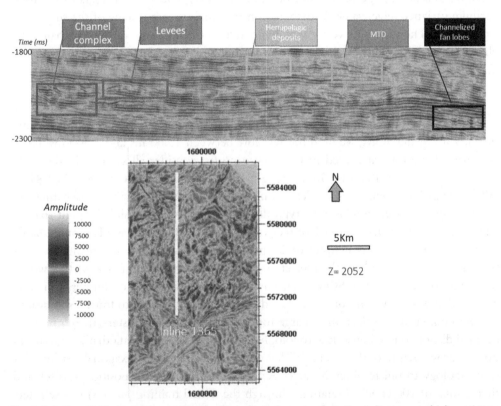

Fig. 1.5 Inline 1365 shown in amplitude. We selected a representative inline over which we interpreted the PCA results. Some deepwater architectural elements are recognizable: channel complex, levees, hemipelagic deposits, MTD, and channelized lobes. Therefore, the attributes contributing in each eigenvector should be related to these geological features.

PCA parametrizations

PCA computation was completed in a commercial software, which allows the computing of the desired seismic attributes or can import them from other platforms. We then used the 28 attributes calculated and selected the window of interest (AOI) of 500 ms, from 1,800–2,300 ms. It is essential to mention that all the seismic volumes should ideally have the same dimensions and be associated with the same original amplitude volume to be able to compare "apples to apples."

Fig. 1.6 shows the workflow used for PCA parametrization. In the workflow, we need to import or calculate the attributes, then define the window and area of interest for the PCA calculation. After getting the eigenvector results, one proceeds to evaluate each attribute's contribution to each principal component (eigenvector). This should be done by selecting from the eigenspectrum an inline or set of inlines (Fig. 1.5) that are representative of the study focus. After that, we proceed to interpret the results by using the %Max or % Total option. The %Max displays the attribute variance as a percentage of the largest attribute/eigenvector in the inline chosen (eigenvalue * attribute contribution as a percentage of the sum of all eigenvalues). The % Total refers to the total attribute variance (eigenvalue * attribute contribution as a percentage of the sum of all eigenvalues). The idea is to verify which attributes contributions are higher than 50%. However, we followed the criteria used in Leal et al.,[16] considering only the attributes that accounted for most of the variability in the data (max% > 80%).

SOM considerations

We calculated the "machine derived" SOM. For this method, we considered the attributes derived from the PCA to input in the SOM. Since we had to delimit by time or horizons in the SOM, we used the window between 1,800 ms and 2,300 ms. Then, we selected the total area used in the harvesting (to consider all samples in the same geological space), in this case Inline range: 1250-1582 and Xline range: 1179-4143. While the groups themselves are defined in the original N-dimensional space, they are mapped into a lower-dimensional, typically one- or two- dimensional, latent space[53]. This two-dimensional, hexagonal, or rectangular grid is referred to as a Kohonen map[17]. Knowing that a neuron is a cluster of datapoints, where each datapoints retains its location in 3D survey space[15], we then selected a neuron configuration of (4 × 4), that would represent 16 clusters. Since SOM is an unsupervised technique, the number of natural clusters that will occur cannot be predicted. A common practice is to force a solution to a fixed number of classifications[54]. Large topologies mean more clusters, therefore more detailed distinctions, whereas low topologies can cluster together data that may result in only a "gross picture of the reservoir"[54]. We selected a mid-size, hexagonal neuron pattern topology to isolate different facies. A random Initial Neuron position was set, and then a total of 100 epochs (iterations through the neural training process) was selected. Usually, the larger the epoch, the better the prediction, passing from a cooperative to competitive training to classify the 16 clusters in the multiattribute space.

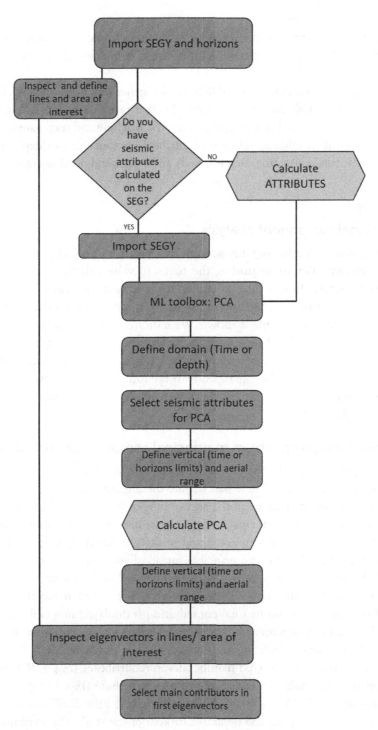

Fig. 1.6 *Principal component analysis workflow and parametrization.*

SOMs and interpretation

The interpretation process consisted of exploring the different neurons (clusters) and discriminate the ones that represented just noise from those that were geologically meaningful.

Subsequently, we compared the SOMs results generated from the machine-based selection using PCA with the SOMs results obtained using a user-based selection and determine possible biases and differences between the machine and user-based approaches. Finally, we proposed an efficient solution of attribute combinations to depict deepwater architectural elements, comparing each method's advantages and disadvantages in the study.

1.8 Results

1.8.1 Principal component analysis

The first step before conducting any seismic interpretation project is having clear the geological objective. For some studies, the focus may be structural; in others, it can be purely stratigraphic. In our case, we aimed to identify deepwater seismic facies and architectural elements in a defined area of interest in the Pipeline 3D dataset. In Fig. 1.5, Inline 1365 in seismic amplitude is shown. We indicated some of the architectures that we can identify in the vertical slice. We also present the location of the inline in a 2D amplitude map. Selecting a line or area of interest is beneficial when determining PCA since the result is shown in the eigenspectrum fashion. Also, by recognizing some elements when possible, we establish control points that can be compared to the output groups when using SOMs.

1.8.2 From the eigenspectrum to principal components and attributes

Fig. 1.7 corresponds to the results of the PCA analysis. We observe a series of blue bars, the eigenspectrum (Fig. 1.7A). These bars denote the highest eigenvalues in each inline[15]. We selected a representative inline in the area of interest (red bar) to determine the most important attributes in each principal component in that area. Remember that an eigen value shows how much variance is contained in the associated eigenvector and that an eigenvector is a vector that shows a principal spread of attribute variance in the data.

Fig. 1.7 displays the three principal eigenvectors (yellow bars within the blue bars), which contain most of the data variability. Notice that for each eigenvector $\lambda 1$, $\lambda 2$, and $\lambda 3$., a list of attributes and its total contribution is displayed in a table in the rightmost side. The first eigenvector ($\lambda 1$) has as main contributors three texture attributes, envelope, and sweetness, which together account for as much of the 60% of the data (Fig. 1.7B). Eigenvector $\lambda 2$ derived mainly curvature attributes (Fig. 1.7C), and eigenvector $\lambda 3$ resulted in amplitude and cosine of phase attribute (Fig. 1.7D).

The application of PCA to a set of 28 attributes allowed the distillation of them into nine attributes. Depending on the objective or goal of the study, the attributes derived

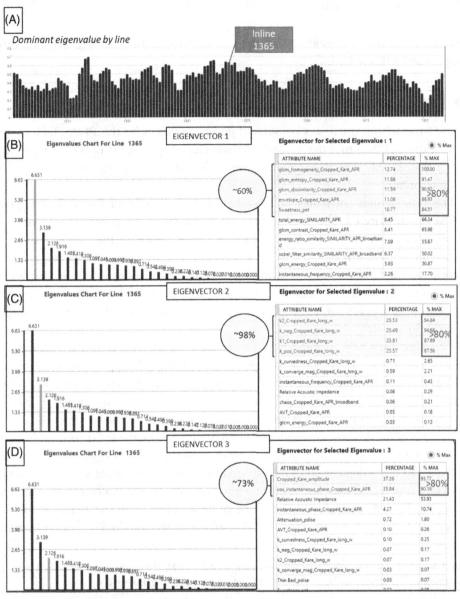

Fig. 1.7 *Principal component analysis (PCA) results from 28 attributes shown in Table 1.1.* The first three eigenvectors (also known as principal components) represent the most of the variation in the data. (A) Eigenspectrum represented by a bar chart where each bar represents the highest eigenvalue for each inline in the seismic 3D volume. Notice we selected a representative inline (1965) for the analysis presented in B, C, and D. (B) Principal component one (PC1) represented with a yellow bar. The seismic attribute contribution is presented to the right. We followed Leal et al. (2019) data analysis to select the most representative attributes. Therefore, the selected attributes were those whose maximum percentage contribution to the principle component were greater than or equal to 80%. In this case, GLCM, envelope and sweetness attributes account for around 60% of the total variability in the data. (C) Principal component two (PC2) in yellow bar, to the right we find the main seismic attribute contributors. (D) Principal component three (PC3), represented by a yellow bar in the left, and the attribute selection based on statistical contribution to the right. Depending on the interpretation objective, the highest contributors (attributes) in each PC are possible candidates as input to a SOM or other unsupervised ML technique.

from the PCA may be good candidates as inputs for self-organizing maps or other unsupervised ML method.

1.9 Self-organizing maps analysis

We used the seismic attributes considered good candidates by the PCA (Table 1.4) as input in the SOM to find the natural attribute clusters generated upon the input data. Notice that, in this case, the attributes were selected by the program and not by the user/interpreter.

The nine seismic attributes (amplitude, envelope, cosine of instantaneous phase, GLCM dissimilarity, GLCM homogeneity, GLCM entropy, K1 curvature, curvature K2 curvature, and sweetness) were inputted to calculate SOM over the area of interest. To give a sense of these attributes meaning and use, we can say that, envelope, a.k.a. instantaneous amplitude, is one of the most popular instantaneous attributes and represents the acoustic impedance contrast, can indicate bright spots and sequence boundaries[14] and even suggest porous zones. GLCM are textural attributes that operate by comparing pixels in a matrix.

Therefore, they help to distinguish facies based on differences in their texture and arrangement. Cosine of instantaneous phase is an amplitude-dependent attribute that highlights bedding or reflector configuration very well. K1, K2 are referred to as most-positive and most-negative principal curvatures; therefore, offer a measure of maximum and minimum bending of the surface at a certain point[55]. Finally, the sweetness attribute is the product of the division of the instantaneous amplitude by the square root of instantaneous frequency[56] and highlight sand-rich deposits and hydrocarbon-bearing formations.

We selected a configuration of four-by-four neurons, so we obtained 16 neurons projected on a topological map. Each neuron represents a cluster with information, geological or not, that has a distinctive color. Each one of these colors represents herein a cluster of data points. In Fig. 1.8, we show the SOM results with the topological map indicated to

Table 1.4 List of machine-selected attributes using PCA for SOM. These attributes represent the main contributors of variability for PC1, PC2, and PC3 within the section of interest.

Selected attributes	Eigenvector
Cosine of instantaneous phase	3
Amplitude (original)	3
Envelope	1
GLCM dissimilarity	1
GLCM entropy	1
GLCM homogeneity	1
K1 (most positive curvature)	2
K2 (most negative curvature)	2
Sweetness	1

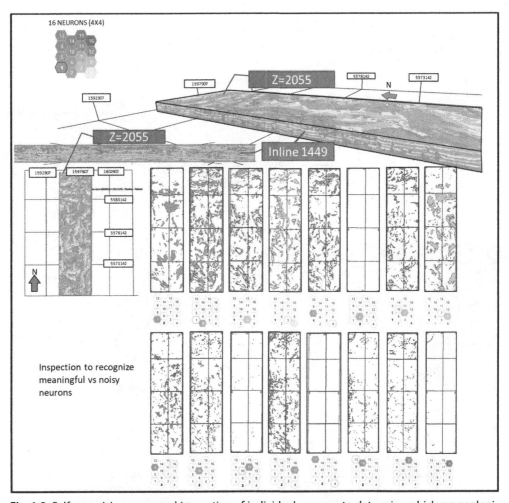

Fig. 1.8 Self-organizing maps and inspection of individual neurons to determine which are geologically meaningful and which represent random noise.

the left. In this stage, we selected the clusters (neurons) that were associated with features of interest or facies in the 3D space. We made the decomposition and inspection of each neuron and found that some of these clusters did not represent anything but random noise (e.g., Neurons 6, 9, 11, 13, and 16). On the contrary, neurons such as one, four, and eight highlighted the channel complex and associated architectural elements in the system.

Fig. 1.9 showcases the interpreted neurons and their geological significance. It is necessary to highlight that whereas ML allows for clustering, it is up to us (interpreters) the understanding and the "in context interpretation" of each of these neurons/clusters and we owe to apply seismic geomorphology principles to make sense out of the groups/clusters formed. Neurons 1 and 8 were interpreted as possible bars, whereas

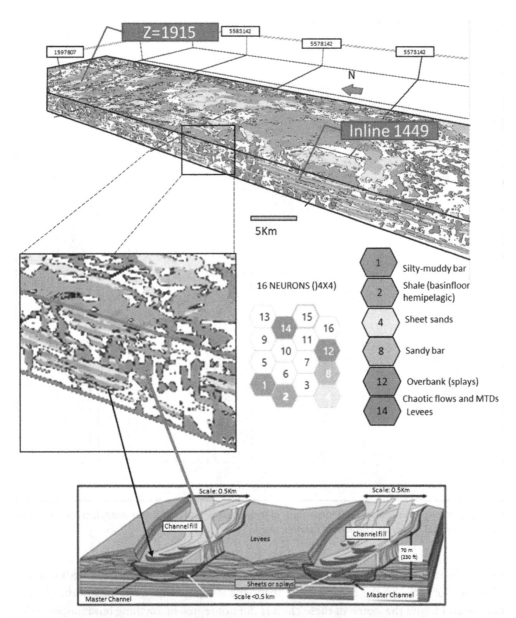

Fig. 1.9 *Self-organizing maps results from PCA is presented here, after filtering neurons considered as noise or not geologically meaningful.* Notice that out of the 16 neurons, we selected six that represent possibly architectural elements in the deepwater system evaluated. Neurons one and eight represent bars, one, more shaly, and eight sand rich. Neuron two denotes the shaley or hemipelagic background and four the sheet sands that are fan like. Neurons 12 and 14 represent minor sized features like over-banks and levees without much distinction between them. Similar clusters are located close in the topological map.

Neuron 2 represents the "background" conformed by hemipelagic facies. Neuron 4 is recognized as sheet sands (fan–like deposits) for its architecture and in context interpretation. Finally, Neurons 12 and 14 represent overbank deposits and levees.

The Neuron 14 also was identified in places where more chaotic facies such as MTD and turbiditic channels were recognized in amplitude data. Notice how similar clusters are placed nearby in the topological map (Neurons 2 and 1 are both fine-grained). A cartoon has also been included to better explain the relationship between the interpreted facies and how the system migrates to keep an equilibrium profile. This lateral channel-complex migration results from avulsion, levees' buildup and changes in energy, accommodation space and sediment supply in the system.

1.10 SOM results and discussion from user-selected attributes vs. machine-derived inputs

Now that we have seen how SOMs generated clusters based on machine-selected attributes, we compared the results with a SOM created by using attributes selected by an experienced interpreter. Fig. 1.10 depicts the comparison between the SOM derived from attributes selected in an unsupervised fashion, using PCA, the SOM results from meaningful attributes characterized by La Marca[39], and the amplitude expression. Our first impression is that when comparing the SOM results (Fig. 1.10A and B) with the seismic amplitude (Fig. 1.10C), we can differentiate with the former, architectures and different elements with particular geological meaning, that are not easily recognizable in the amplitude.

When comparing results from the SOM using the attributes selected by the PCA method (Fig. 1.10A) and the SOM derived from the user method (Fig. 1.10B), we noticed that (1) both allow the distinction of predominant architectural elements, (2) in both figures (10A and B) despite the displayed colors are different, clusters are identified to represent similar architectures. For example, in Fig. 1.10A, yellow and beige colors show sandy prone deposits (corroborated by pukeko well), whereas in Fig. 1.10B these are green and orange colored. (3) When comparing both SOMs notice that green arrows show sandy bars, whereas yellow arrows point out channel complexes. MTDs or turbiditic channels are highlighted with red arrows and marine shales (background) with dark-blue arrows. Minor features like levees (orange arrows) and splays (purple arrows) can be identified in both images A and B in Fig. 1.10. However, we found more details with the user-selected attributes. Detailed stratigraphic features, such as geomorphology of the channels and definition of subtler features within architectural elements, for example levees and sandy waves[39] allow for a better picture, hence, interpretation in the user-selected attributes SOM. For instance, the clearly depicted mud-prone channel pointed by the yellow arrow in Fig. 1.10B is not easy to interpret from Fig. 1.10A. Overall, architectures interpreted suggest that the system developed in a basin floor

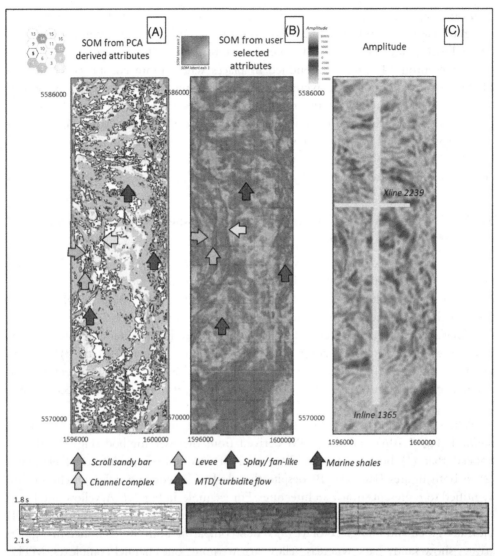

Fig. 1.10 Comparison between (A) SOM derived from the PCA-selected attributes and (B) the SOM from the user selected attributes, (C) amplitude has been provided to compare where high amplitude areas may correspond to sandstone prone facies, such as scroll bars and fans (sheet sands), but also to highlight the importance of the SOMs, since the architectural elements are not distinguishable just in amplitude expression. The position of the Inline 1365 in Fig. 1.2 and Crossline 2239 shown below the time slices, are presented as reference.

(weakly confined) area, characterized by splays, channels, and turbidites. The evaluation of these architectures' vertical succession using the clustering (SOM) results can help reconstruct the history of the system's geological evolution. The identification of multiple stratigraphic features and dominance of sandstone vs. mudstone in the architectures offers a unique opportunity to define prospective areas, isolate them and optimize volumetrics when similar workflows are applied to oil and gas producing fields. Isolation of clusters interpreted into geobodies is part of our future work.

From the comparison of both methods hereby evaluated, the user approach included attributes like spectral components, that PCA did not consider, and that may provide better detail in the classification. It is still unclear why PCA did not output/selected spectral component attributes. However, it is surprising how good results still can be with this unsupervised method, which selected attributes like envelope and sweetness that can provide insights on reservoir quality and that the user did not consider for its classification. Therefore, we can say that the PCA's use preceding a SOM can be a good practice when the interpreter's experience is limited. We would like to state that the software used for the user-selected attributes – SOM workflow is different than the one here used. Since the algorithms used for the facies classification process are the same in both cases, we concluded that our interpretation and analysis of the generated SOMs, are valid to compare clusters that were computed using the same mathematical principle. Therefore, there should not be significant differences in variables that can alter the results generated between a SOM model generated using any of the software. The visualization of the SOMs is the primary difference between the software. With the present study we wanted to compare how different the attributes selected by each method (user-driven and machine-derived) were. There are definitely advantages and disadvantages in the use of both methods, and we highlight those in the following section:

1.11 Human vs. machine comparison (upsides and downsides of PCA vs multiattribute analysis)

1.11.1 Upsides of the approaches

- The user-driven approach allows the interpreter to inspect attributes individually and select the most geologically meaningful for their interpretation purpose/goal.
- Regardless of the goal, the PCA is a dimensionality reduction technique that narrows down the data to a smaller dataset containing most of the variability in the data.
- PCA offers a quick way to explore a dataset. It is more time-efficient than a multi-attribute analysis and may consider attributes that were not previously inspected by the user. For instance, envelope attribute is cataloged as good facies distinguisher,[31] but La Marca[39] did not consider it in the user-selection for SOM analysis in the study.

- The user-selected methods allow integration of different data types that can be tested or classified to reduce redundancy.
- Both methods implemented an unsupervised machine learning technique (SOM), which allows recognizing elements that the human may not have been able to identify by himself (e.g., Fig. 1.10).

1.11.2 Downsides and limitations

- The optimal selection of attributes relies on the interpreter's experience and understanding of each attribute significance and it might be biased (not very objective).
- The PCA results may show attributes that are not the most suitable for the goal (e.g., cosine of phase), and, also requires data QC because PCA is sensitive to noise. For example, Fig. 1.8 shows clusters that only contain noise. The PCA will reduce the amount of data examined, but results may or may not related to physical features of interest.
- The multi-attribute approach may result time-consuming if started from zero, but effective if combinations used for every setting are documented and followed to apply in similar geological settings.
- The PCA method suggests attributes that are "main contributors", but that may be similar or redundant (see [57]. For example, most positive curvature and maximum curvature (Fig. 1.7C).
- SOMs do not provide a quantitative confidence measure in the facies classification[8].
- Both methods, applied to SOMs, require human inspection and approval in pivotal moments. The most straightforward example is the validation and interpretation of the SOMs outputs regardless of the initial approach used. Interpreters must apply seismic geomorphology principles to define data points with "similar" characteristics that belong to each cluster.

1.12 Recommendations

After using two unsupervised machine learning techniques and compared with a multi-attribute human approach, we can suggest the user to consider the following:

- Definition of the objective or study goal is the first step before employing any ML approach.
- The QC of data is paramount to avoid misclassifications. The idea is to be aware of the noise or artifacts within your seismic as much as possible. This step applies to any other discipline.
- Depending on the goal, predefine the suitable type of seismic attributes to use.
- Work with a small window of interest (we suggest equal or less than 500 ms) or focus on the area/features of interest. This avoids misclassifications of facies that are not proper of the depositional environment studied.

- Interpreters should be skeptical of ML methods. They should recognize that while ML are useful tools to save time, they can also lead to results associated with noise (yes, noise is also clustered), or not practical for your interest (geologically representative) in our case. Therefore, validate when possible, especially for unsupervised methods like SOMs.

1.13 Conclusion

We compared user-based (interpreter) and machine-based (PCA) attribute selections to implement in a self-organizing map (SOM) technique for deepwater seismic facies interpretation. The PCA allowed the reduction of a large dataset of 28 attributes to only nine attributes contained in the principal components. The first, second, and third principal components contained most of the variability within the data. The machine-based attribute selection included amplitude, cosine of instantaneous phase, envelope, most positive and most negative curvature, sweetness, and textural attributes such as GLCM, dissimilarity, entropy, and homogeneity to use in the self- organizing maps (SOMs).

However, we concluded that the machine-based attribute selection is composed of redundant attributes such as GLCM homogeneity and GLCM entropy, or amplitude and envelope. Such redundancy may have contributed to the lesser SOM results, when compared to the user-selected attributes. We found that the user-based approach offers a better picture and geological significance to the interpreter than the machine-based approach (PCA). This result may be related to the use of meaningful and non-redundant seismic attributes. Therefore, although the use of PCA is encouraged to diminish the amount of input attributes, an interpreter evaluation of the significance of output attributes is highly encouraged.

Since the user-based method helped to better illustrate the architecture and geomorphology of the elements and define subtle differences within them. We suggest a list of attributes provided to use in unsupervised machine learning methods in similar geological settings (Table 1.2). However, if the geological setting and objective differ of the one in this research, the user should define the most suitable attributes for their geological goal.

The use of PCA requires user control in the evaluation process to assess the input data quality, the algorithm sensitivity to noise, and the window of analysis. It is up to the interpreter to decide if results are meaningful depending on the study's scope. SOMs can depict elements that are not identifiable using single attributes and help to corroborate previous insights of the interpreter. However, the interpreter's knowledge and experience, and the use of seismic geomorphology principles and context to the interpretation are important to give meaning to each cluster.

The major difference between the compared methods is the time efficiency and accuracy: A user-based approach produces better results but is not as time-efficient as the machine-based method. Nonetheless, machine learning methods have to be treated as a tool to work efficiently to extract information from multiple datasets at the same time to overcome the interpreter's limitations. In the future, may machines try to supplant humans. But we underpin the idea that just now, geoscientists interpreters are irreplaceable.

1.14 Acknowledgments

We thank firstly, New Zealand Petroleum and Minerals for the Pipeline 3D dataset. Thanks to AASPI (used for attribute calculation and SOM analysis) for the license provided to the University of Oklahoma. Similarly, we extend our appreciation to Geophysical insights for their license on the Paradise software, used for PCA and SOM calculation. We also thank the editor and reviewers for their valuable insights.

References

1. Chopra S, Marfurt KJ. Seismic attributes for prospect identification and reservoir characterization. *Society of exploration geophysicists* 2007;ISBN 978-1-56080-141-2(SEG Geophysical Developments Series): 1–457. doi:10.1190/1.9781560801900.fm.
2. Partyka G, Gridley J, Lopez JA. Interpretational applications of spectral decomposition in reservoir characterization. *The leading Edge* 1999;**18**(3):353–60.
3. Gao D. Texture model regression for effective feature discrimination: application to seismic facies visualization and interpretation. *Geophysics* 2004;**69**(4):958–67.
4. Chopra S, Marfurt KJ. Churning seismic attributes with principal component analysis. 86th annual international meeting, SEG, Expanded abstracts 2014; 2672–2676.
5. Coleou T, Poupon M, Azbel K. Unsupervised seismic facies classification: a review and comparison of techniques and implementation. *The Leading Edge* 2003;**22**:942–53.
6. Sacrey D, Roden R. Understanding attributes and their use in the application of neural analysis- case histories both conventional and unconventional. *Search and discovery* 2014;41473:1–54. article.
7. Chaki S. Reservoir characterizationa machine learning approach: Ms. C. thesis. Indian institute of technology, 2015.
8. Zhao T, Zhang J, Li F, Marfurt KJ. Characterizing a turbidite system in Canterbury Basin, New Zealand, using seismic attributes and distance-preserving self-organizing maps. *Interpretation* 2016;**4**(1):SB79–89.
9. Liu Z, Song C, Cai H, Yao X, Hu G. Enhanced coherence using principal component analysis. *Interpretation* 2017;**5**(3):1A–T449. https://doi.org/10.1190/INT-2016-0194.1.
10. Pires de Lima R. Machine learning applications for geoscience problems: Ph.D. thesis, University of Oklahoma, 2019.
11. Pearson. K. On lines and planes of closest fit to systems of points in space. *Philos Mag* 1901;**6**(11):559–72.
12. Hotelling H. Analysis of a complex of statistical variables into principal components. *J Educ Psychol* 1933;**24**(6 and 7):417–41.
13. Smith L. A tutorial on principal component analysis. *Statistics* 2002;2:1–27.
14. Taner. M. Seismic attributes. *RECORDER* 2001;**26**:07.
15. Roden R, Sacrey D. Seismic interpretation with machine learning. *GeoExpro.* 2016;**13**(6):50–3. https://www.geoexpro.com/articles/2017/01/seismic-interpretation-with-machine-learning.
16. Leal J, Jeronimo R, Rada F,Viloria R, Roden R. Net reservoir discrimination through multi-attribute analysis. *First Break* 2019;**37**(9):77–86.
17. Kohonen T. Self-organized formation of topologically correct feature maps. *Biol Cybern* 1982;**43**:59–69.
18. Das G, Chattopadhyay m, Gupta S. A comparison of self-organising maps and principal component analysis. *Int J Mark Res* 2016;**58**(6):815–34. http://doi.org/10.2501/IJMR-2016-000.

19. Guo H, Marfurt KJ, Liu J. Principal component spectral analysis. *Geophysics* 2009;**74**(4):35–43. http://doi.org/10.1190/1.3119264.
20. Matos MC, Osorio PL, Johann P. Unsupervised seismic facies analysis using wavelet transform and self-organizing maps. *Geophysics* 2007;**72**:P9–21.
21. Roy A, Marfurt KJ. Applying self-organizing maps of multiattributes, an example from the Red-Fork Formation, Anadarko Basin. 81st Annual International Meeting Society of Exploration Geophysicists, Expanded Abstracts 2010:1591–5.
22. Roy A. Latent space classification of seismic facies: Ph.D. dissertation, University of Oklahoma, 2013.
23. Tellez JJ, Slatt R, Marfurt K. Seismic Facies Classification and Characterization of Deepwater Architectural Elements. A Case of Study, North Carnarvon Basin Australia, AAPG Annual Convention & Exhibition, Houston, TX 2016.
24. La Marca-Molina K, Silver C, Bedle H, Slatt R. Seismic facies identification in a deepwater channel complex applying seismic attributes and unsupervised machine learning techniques. A case study in the Taranaki Basin, New Zealand. 89th annual international meeting, SEG, Expanded abstracts 2019:2059–63.
25. Strogen DP, Bland N, King PR. Paleogeography of the Taranaki Basin region during the latest Eocene-early Miocene and implications for the 'total drowning' of Zealandia. *N. Z. J. Geol. Geophys.* 2014;**57**(2):110–27.
26. Kroeger KF, Thrasher GP, Sarma M. The evolution of a Middle Miocene deep- water sedimentary system in northwestern New Zealand (Taranaki Basin) depositional controls and mechanisms. *Mar Pet Geol* 2019;**101**:355–72.
27. Bull S, Nicol A, Strogen DP, Kroeger KF, Seebeck H. Tectonic controls on sedimentation in the Southern Taranaki Basin during the Miocene and implications for New Zealand plate boundary deformation. *Basin Res* 2018;**31**:2.
28. Chopra S, Marfurt KJ. Seismic curvature attributes for mapping faults/fractures, and other stratigraphic features. *Recorder* 2007;**32**:9.
29. Taner MT, Sheriff RE. Application of amplitude, frequency, and other attributes to stratigraphic and hydrocarbon determination: seismic stratigraphy- applications to hydrocarbon exploration. *Charles E. Payton* 1977. https://doi.org/10.1306/M26490C17.
30. Russell B, Hampson D. Multiattribute seismic analysis. *The Leading Edge* 1997;**16**(10):1361–552. https://doi.org/10.1190/1.1437486.
31. Chopra S, Marfurt KJ. Seismic attributes- Aa historical perspective. *Interpretation* 2005;**70**(5):IS0–Z82.
32. Sheriff RE, comp., 1984, Encyclopedic dictionary of exploration geophysics, 2nd ed.: SEG
33. Taner MT, Schuelke JS, O'Doherty R, Baysal E. Seismic attributes revisited. 64th Annual International Meeting, SEG, Expanded abstracts 1994:1104–6.
34. Liner C, Li C-F, Gerztenkorn A, Smythe J. SPICE: A new general seismic attribute. 72nd Annual International Meeting, SEG Expanded abstracts 2004:433–6.
35. Taner MT, Koehler F, Sheriff EE. Complex seismic trace analysis. *Geophysics* 1979;**44**:1041–63.
36. Barnes A. Seismic attributes in your facies. *CSEG recorder* 2001;**26**(7):41–7.
37. Liu J, Marfurt KJ. Instantaneous spectral attributes to detect channels. *Geophysics* 2007;**72**(2):P23–31.
38. Marfurt KJ. Seismic attributes as the framework for data integration throughout the oilfield life cycle. 2018 Distinguished instructor short course. *Society of Exploration Geophysicists* 2018;**21**(Distinguished instructor series):1–479. doi:https://doi.org/10.1190/1.9781560803522.
39. La Marca K, Bedle H. Seismic attribute optimization with unsupervised machine learning techniques for deepwater seismic facies interpretationusers vs machines2020The University of Oklahoma.
40. Jolliffe IT, 1986, Principal Component Analysis. Springer. ISBN 978-1-4757-1906-2; ISBN 978-1-4757-1904-8 (eBook). DOI: http://doi.org/10.1007/978-1-4757-1904-8.
41. Zhao T, Jayaram V, Roy A, Marfurt K. A comparison of classification techniques for seismic facies recognition. *Interpretation* 2015;**3**(4):SAE29–58.
42. Doveton JH. Geological log analysis using computer methods. *AAPG Computer Applications in Geology* 1994;**2**:1–169.
43. Hu. S, Zhao W, Xu Z, Zeng H, Fu Q, Jiang L et al. Applying principal component analysis to seismic attributes for interpretation of evaporite facies: Lower Triassic Jialingjiang Formation, Sichuan Basin, China. *Interpretation* 2017;**5**(4). https://doi.org/10.1190/INT-2017-0004.1.

44. Jong-Se L. Multivariate statistical techniques including PCA and rule based systems for well log correlation: developments in petroleum science. *Elsevier* 2003;**51**:673–88. https://doi.org/10.1016/S0376-7361(03)80034-1.

45. Zee Ma Y. Lithofacies clustering using principal component analysis and neural network: applications to wireline logs. *Math Geosci* 2011;**43**:401–19. http://doi.org/10.1007/s11004-011-9335-8.

46. Jahan I, Castagna J, Murphy M, Kayali MA. Fault detection using principal component analysis of seismic attributes in the Bakken Formation, Williston Basin, North Dakota, USA. *Interpretation* 2017;**5**(3):1A–T449. https://doi.org/10.1190/INT-2016-0209.1.

47. Hanafizadeh P, Mirzazadeh M. Visualizing market segmentation using self-organizing maps and fuzzy delphi method- ADSL market of a telecommunication company. *Expert Syst Appl* 2010;**38**:198–205.

48. Tuckova J, Bartu M, Zetocha P, Grill P. Self-organizing maps in medical applications. proceedings of the international conference on neural computation theory and application 1 2011:422–9.

49. Gibson PB, Perkins-Kirkpatrick SE, Uotila P, Pepler AS, Alexander LV. On the use of self-organizing maps for studying climate extremes. *Geophys. Res. Atomos.* 2017;**122**:3891–903. http://doi.org/10.1002/2016JD0262256.

50. Gao D. Application of three-dimensional seismic texture analysis with special reference to deep-marine facies discrimination and interpretation: An example from offshore Angola, West Africa. *AAPG Bull* 2007;**91**:1665–83.

51. Roden R, Smith T, Sacrey D. Geologic pattern recognition from seismic attributes: principal component analysis and self-organizing map. *Interpretation* 2015;**3**:SAE59–83.

52. Kohonen T. Self-organizing Maps, 3rd ed.: Springer-Verlag 2001.

53. Wallet BC, de Matos MC, Kwiatkowski JT, Suarez Y. Latent space modeling of seismic data: an overview. *The Leading Edge* 2009;**28**(12):1454–9.

54. Sacrey D, Sierra C. Systematic workflow for reservoir characterization in northwestern Colombia using multi-attribute classification. *First Break* 2020;**38**:77–82. http://doi.org/10.3997/1365-2397.fb2020022.

55. Roberts A. Curvature attributes and their application to 3D interpreted horizons. *First Break* 2001;**19**(2):85–100.

56. Radovich BJ, Oliveros RB. 3-D sequence interpretation of seismic instantaneous attributes from the Gorgon field. *The Leading Edge* 1998;**17**:1286–93.

57. Barnes A. Redundant and useless seismic attributes. *Geophysics* 2007;**72**(3):33–8.

58. Haralick RM, Shanmugam K, Dinstein I. Textural features for image classification: Institute of electrical and electronic engineers' transactions on systems, man and cybernetics. *SMC* 1973;**3**:610–21.

59. Meek T. Applications of 3D seismic attribute analysis workflowsa case study from Ness County, Kansas, USA2013Kansas State University.

60. Luo Y, Higgs WG, Kowalik S. Edge detection and stratigraphic analysis using 3-D seismic data. 66th annual meeting, SEG, Expanded abstracts 1996:324–7.

61. Gersztenkorn A, Marfurt KJ. Eigenstructure based coherence computations as an aid to 3D structural and stratigraphic mapping. *Geophysics* 1999;**64**(5):1468–79.

62. Castagna JP, Sun S. Comparison of spectral decomposition methods. *First Break* 2006;**24**:75–9.

CHAPTER 2

Relative performance of support vector machine, decision trees, and random forest classifiers for predicting production success in US unconventional shale plays

Jessica Wevill[a], Alex Bromhead[b], Kate Evans[b], Jeffrey Yarus[c], Cédric M. John[a]
[a]Department of Earth Science and Engineering, Imperial College London, United Kingdom
[b]Halliburton, United Kingdom
[c]Case Western Reserve University, United States

Abstract

Unconventional shale reservoirs have revolutionized the energy industry. However, the prediction of production based on reservoir geology characterization has largely focused on sweet spot definition rather than on over-arching production trends across multiple plays. This study uses machine learning (ML) techniques to analyze the relationships between well log data and production success within seven North American shale plays. Three ML algorithms were evaluated: stochastic gradient descent kernel trained support-vector machine (SGD-SVM), decision tree (DT), and random forest (RF) classifier. Accuracy of predictions using the SGD-SVM and DT classifiers did not exceed 55%. A fine-tuned RF classifier is the most successful method at predicting well success based on normalized initial production, with an accuracy of 97%. To achieve this result, the RF is trained on the following input features: average play thickness, pore pressure, TVD, and resource concentration. The main factors impacting performance of our algorithm when trying to predict success in unconventional plays are previous understanding of heterogeneities in individual formations, and consistency of data availability across multiple wells. Despite challenges, ML and the RF method in particular show promising applications in the unconventional petroleum industry as a means to streamline production and data collection.

Keywords

Decision tree; Machine learning; Random forest; SGD-SVM; Shale plays

2.1 Introduction

Signaling the start of the "golden age of gas"[1] the hydrocarbon production from economically viable tight shale in the late 2000's[2] has changed the future of the US natural gas industry in terms of energy prices, security, and CO_2 emissions. This has prompted a global shift in the petroleum industry toward exploration and production of shale reservoirs. In this paper, the term "shale gas technology" is used to describe the combination

Advances in Subsurface Data Analytics
DOI: https://doi.org/10.1016/B978-0-12-822295-9.00007-8

of horizontal drilling and high pressure, low-viscosity fluid multiple stages fracturing through low permeability organic shale formations. An increase in the availability of these techniques has enabled a transition from coal-fired to gas-fired power generation, producing approximately half the volume of CO_2 compared to using coal for the same heat output[3]. By 2012, shale gas accounted for more than half of US gas production, surpassing production from conventional reservoirs[4]. The success of the North American shale revolution can therefore be used to help inform more productive exploration and efficient production from unconventional plays globally.

However, due to the underlying complexity of fracture networks and geological heterogeneities within shales, prediction of production between two adjacent wells within the same play is challenging[5]. Well success cannot be easily predicted based on the well's landing zone or stratigraphic position and burial depth because of the complex relationships between subsurface environment and well completion methods. Understanding of these relationships exists for individual plays, but no overarching trends or models have been built so far for a range of shale plays. Due to the interconnectivity of a large range of variables influencing production, the volume of data needed to understand the different relationships can far exceed a human ability to analyse them. Data is often compared to a natural resource, and "data is the new oil" is a common motto for many industries as the value of big data rapidly increases. Comparable with hydrocarbons, data must be broken down, analyzed and refined to have any commercial or scientific value. Unlike oil however, data is a resource constantly being created[6]. Machine Learning algorithms provide the means to harness this data because of their ability to recognize patterns within high dimensionality data once trained[7]. Since unconventional shale plays are increasingly exploited globally, data from unconventional wells drilled increases as well.

Here, we propose to explore several machine learning algorithms to build an automated approach to train a model on a large range of data rather than base predictions on a prior understanding of a reservoir model or fluid flow characteristics, which only considers a limited set of features influencing production. Our aim is to determine what geological and engineering factors, based on shale reservoir characteristics, most influence production success in North American shale plays, and whether ML can identify and correctly predict the success of wells. This study provides a rigorous comparison of three different ML algorithms based on their sensitivity to the structure of the well data provided[7]. We use a proprietary dataset provided for this study by Halliburton and containing well log data obtained by different producers drilling seven different shale plays. Using this unique dataset, we could test the relative efficiency of three main algorithms for our production scenarios: Support Vector Machine (SVM), Decision Trees (DT), and Random Forest (RF). This approach allowed to understand the factors that are either limiting or improving production. It appears that the RF algorithm trained using data for average play thickness, pore pressure, True Vertical Depth (TVD), and resource concentration is the most appropriate for our use case, with 97% accuracy in predicting well production.

2.2 Methods

2.2.1 Dataset available for the North American shale plays

Geological data for the shale plays accessed within this study is provided by Neftex Insights – Halliburton for use in a project conducted with the STEPS program, and constitutes interpolated maps for various geological properties derived from public domain data sources. Intrinsic internal heterogeneities stemming from unique geological and tectonic evolutions of each play can lead to quantitative variation of the geological features in the dataset. This thus needs to be accounted for in order to predict production and provides insight for error analysis. We use a dataset of well log data from 7 major US plays, identified in Fig. 2.1 and summarized in Table 2.1.

Production data used within this study is from ShaleWellCube (version 13-06-2019) and is provided courtesy of Rystad Energy. Subsurface properties of each of the key shale plays were defined per each lateral well as continuous values. This includes both absolute and modeled data based on extrapolation of the properties in nearby wells depending on the measurements taken while drilling, and determined by the company completing the well.

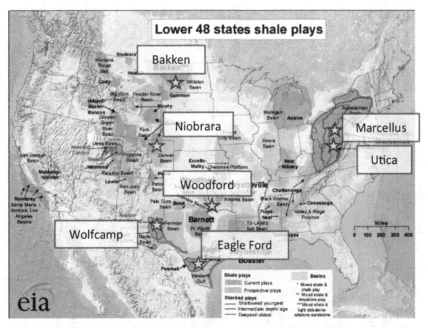

Fig. 2.1 Geographical locations of all US shale plays studied (modified from[30]).

Table 2.1 Geological summary of all US shale plays, specifically focussed on geological features likely to have a significant influence on production metrics.

Play	Target formation (*most prolific)	Dominant lithology	Depositional setting/conditions	Key heterogeneities and limiting factors to production	Hydrocarbon Phase/s
Wolfcamp shale	Wolfcamp A*, B*, C, D units	Siliceous marl	Early Permian carbonate shelf in a shallow marine setting[32].	Differences in geomechanical properties of discreet carbonate facies depositional stacking[33].	Oil and gas
Woodford shale	Woodford Formation	Siliceous mudstone	Marine and terrigenous organics deposited due to chemocline fluctuation and water density stratification with sea level change.	Lateral variation in organic content variation is a limiting factor in gas production. Lithological variations, such as clay content and calcite concretions exist[34].	Oil and gas
Niobrara shale	Niobrara Formation	Calcareous dolomitic mudstone	Cyclical sediment starvation and aeolian deposition of volcanic ash during transgression to continental runoff during regression[35].	High clay content with subtle variation in lithology and grain size due to irregular current reworking. Evaporite layers increase heterogeneity with salt movement and dissolution[36].	Oil and gas
Marcellus shale	Marcellus Formation	Siliceous mudstone	Deposition in deep anoxic water with sedimentation controlled by the Arcadian orogeny	SW and NE cores of gas production are limited by play thickness and clay content, respectively[37].	Gas
Utica shale	Utica–Point Pleasant formations	Calcareous mudstone	Deposition during a period of major transgression with restricted inner platform circulation with five cyclical episodes of sediment starvation and supply from calcareous turbidites[38]	Heterogeneity is due to diagenetic and depositional changes resulting in interfingering of the nonreservoir Point Pleasant Formation with the organic-rich Utica Formation.	Gas
Eagle Ford shale	Lower and Upper Eagle Ford formations	Calcareous dolomitic mudstone	Sediment contribution from prograding river deltas with the eagle ford representing transition between siliciclastics and carbonates in a transgressive epieric sea environment.	The calcite-rich upper Eagle Ford is more brittle than the lower Eagle Ford, though this has more capacity for hydrocarbon storage. Stratigraphy varies as clastic content increases with proximity to the laramide orogeny source. Limiting factors include smectite content with high swell potential. Production is greatest with increased play thickness.	Oil and gas

| Bakken Formation | Middle Bakken member | Mixed lithology, dolomitic siltstone | Deposition occurred during tectonic induced marine transgression. The Upper and Lower Bakken members were deposited in an offshore marine environment within a temperate climate. The Middle Bakken was deposited in a prograding shallow coastal regime during sea-level fall. Production data is supplied from the Middle Bakken where the majority of completion landing zones are. Kumar et al.[22] estimate a 12–52% contribution from the Lower/Upper Bakken to cumulative Middle Bakken contributions, with a mean contribution of 40%. Therefore, the geological properties of these formations are taken into account as the prolific Middle Bakken is often thin, with fracs extending out from the formation limits, and therefore can be sensitive to the adjacent lithology. | High lithological variation from oolitic dolostones to interbedded siltstones between the thermally mature Bakken members with irregular contacts. The Lower Bakken is kerogen-rich and thermally mature with natural fracturing resulting from variation in hydrostatic pressure during hydrocarbon generation[39]. | Oil |

Production rates and completion metrics of each well were provided for each play. In total 94,887 of the 139,199 total wells (68%) drilled in all shale plays had production rate values. Data from each play was aggregated into a common dataset of production data in order to increase sample size and to assess the overall relationships between production and geological properties of hydrocarbon-bearing shales. The origin of each well was preserved during data analysis to contrast heterogeneities between specific play.

2.2.2 Machine learning approach

We used a supervised classification approach: the training data contains samples attached to a specific well, which are fed into the algorithm with labels of the desired output. Supervised learning refers to data fed into a model where the desired solutions are present. Output values previously assigned to the data provide labels learned by the algorithm to make predictions of future instances when the model is built. Here, the output is defined as initial production normalized over the length of the producing well. The desired outputs can be either regressive, resulting in a continuous output, or a discrete class with a classification algorithm[8].

Classification algorithms are built based on multilinear regression and aim to map the separation between one domain of input data to another target class, through the use of a discrimination function $y = f(x)$. Input feature vectors, in this study representing geological factors, have the form of vectors of dimension "d" (x1, x2,... xd) with the y vector defined with a finite set of 'c' class labels (y1, y2,... yc). Instances of x and y are learned by the classification model (f') which maps the input data into target classes in order to learn from the data using the discrimination function. The trained ML model is then capable of class prediction. The task of building a supervised ML classification algorithm can be split into three stages[7], workflow design in Fig. 2.2: (A) data cleaning and preprocessing; (B) classification model training; and (C) prediction evaluation.

2.2.3 Data cleaning and preprocessing
2.2.3.1 Input feature cleaning

The initial data cleaning was completed with the aim of including only data (or features) relevant to the accurate prediction of production success. This is necessary to reduce noise and bias in extracting significant relationships within the data. Missing data was defined per play, and features where more than 50% of wells were missing data were removed to prevent increasing bias through decreased feature sample size. For the features retained but where data was missing, the median of a given feature for each individual play was calculated and imputed for all missing data points. This was done on a play-by-play basis to retain as much of the data quality and variation between shale plays as possible. We filtered out feature data present in less than 50% of the wells as it was demonstrated by Schafer and Graham[9] that RMSE increases progressively for imputation above 50%, thus increasing bias. The final geological inputs

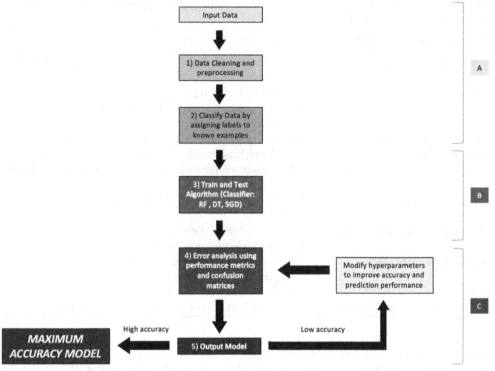

Fig. 2.2 Workflow design for this study showing the stages of ML model building as part (A) data cleaning, (B) model training, and (C) prediction evaluation.

used for ML model construction that are not taken from well log data directly are those that can be modeled at the scale of the play from a regional geological model, derived as shown in Table 2.2. These are geological features that are either commonly linked to production performance, or could perceivably be linked to production performance in shale plays.

The production metric of success was calculated as the initial production within the first year of well completion normalized over 1,000 ft (304.8 m) of lateral well length. Production classes were obtained by subdividing the entire dataset into quintiles of normalized production success, yielding five ordered classes each containing 20% of the data. The quintiles classes form the label for the data used to train the algorithm, and subsequent future case predictions.

2.2.3.2 Dimensionality reduction

Due to the large number of features in the dataset, dimensionality reduction is necessary. Correlated features are assigned to a common vector resulting in a lower number of planes of projection, while retaining most of the variance of the data[10]. This step begins with feature extraction on the cleaned data to assess significant relationships between

Table 2.2 Summary of geological features used as ML model input and their method of calculation from regional geological model data.

Geological feature for model input	Method of calculation
Average thickness (m)	Calculated from basin scale isopach maps of each shale play
Porosity (%)	Calculated from typical porosity depth trends and porosity-maturity trends. Depth and maturity are calculated from basin-scale geological model
Resource concentration (MMBOE/km²)	Resource in-place = (Net thickness x Porosity x (1–Water Saturation)) / Formation Volume Factor
TVD (m)	Depth surface for each shale play taken from a regional geological model
Pore Pressure (psi)	Calculated from basin-scale pore pressure gradient maps
Maximum burial temperature (°C)	Calculated from the regional geological model using Depth x Geothermal gradient
Reservoir pressure (MPa)	Calculated from the regional geological model using Depth x pressure gradient
GOR	Calculated from thermal maturity trends
Geothermal gradient (°C/km)	Derived from global geothermal gradient dataset

normalized production and individual attributes, followed by applying a principal component analysis (PCA) to reduce the feature dimensions. The principle of PCA is that in a dataset of dimension n there exist m principal components, that is, tensors of dimension $m < n$ that still capture most of the variability of the dataset. For the problem at hand, the data matrix (X) contains N rows of objects corresponding to geological factors influencing production with K columns of variables defining the measurements made of the objects[11]. PCA acts as a noise filter as the sample is represented by a smaller number of components that can be used to visually assess the structure of the dataset and enable the selection of variables based on residual variance. PCA offers a means to compare which principal components are of interest to the problem of production success, and how many features can be kept in order to improve prediction but without introducing noise. The linear combination of each correlated features yields the principle components, which form a subset of geological features describing 95% of the variability of the dataset. This residual variance of each input feature can be determined through factor analysis[12].

2.2.4 Training machine learning algorithms

The ML models used focused on examining the data set from a classification point of view which was necessary due to the vast range of factors known to impact production, from geology to completion metrics influenced by each operator. Therefore, the relationships between the parameters available within the dataset may be influenced by factors not accounted for only by geology, so a categorical approach was taken to make the model as predictive as possible. The main algorithms used were stochastic gradient descent kernel-trained support-vector machine (SGD-SVM) classifier, decision tree (DT) classifier, and random forest (RF) classifier.

2.2.4.1 Stochastic gradient descent kernel trained support vector machine classifier

SVM is a classification algorithm widely used for large training data sets with high dimensionality thanks to its versatility when optimized with kernels[13]. SVM defines discrimination boundaries in higher-dimensionality space through multilinear regression. Decision boundaries are determined through finding the maximal margin of hyperdimensional distance between two classes (i.e., the smallest dimensional space between the two nearest datapoints in each class) through solving a quadratic optimization problem. Expansion of the solution is in terms of training patterns, or support vectors, that lie on a margin containing all the relevant information of the classification problem. Cost parameters (C) are generated to assign penalties for misclassification of support vectors, thus high C values define complex decision boundaries responsible for fewer misclassifications.

Kernel optimization is a function that returns a mappable product between data points in the feature space where learning occurs. This is a means of increasing the simplicity of projecting data points into a feature space where non-linear decision boundaries are defined by the implicit transformation of input variables through the kernel function[14]. This allows SVM to classify non-linearly separable support vectors using linear decision boundaries. SGD Classifier is an estimator employed to regularise SVM using Stochastic Gradient Descent learning. SGD-SVM is a binary classifier that can be applied to the multiple classes of production success problem using a one-versus-all strategy. For K classes, the binary classifier discriminates between this and all other K-1 classes. When data is input for classification, a decision score is determined from each classifier and the highest confidence score is output, updating the model to reflect a decreasing learning rate[8]. The dividing decision boundaries for our dataset of 5 classes were optimized to best separate classes based on the principle components of the PCA results.

2.2.4.2 Decision tree classifier

Decision tree (DT) algorithms are powerful and commonly applied to large datasets. These are hierarchical models formed by consecutively splitting independent variables based on optimal criteria into homogeneous zones until the statistical majority of the data belonging to each of the leaf nodes, or predicted classes, has the same label[15]. This is done through defining decision rules that can be used to predict the outcome of a new instance of variables[16]. Fig. 2.3 shows a schematic of a DT model.

A node defines a point at which a question is asked, and the data split depending on the answer. The root node (1) represents the data which is input into a model, where in this case a cut-off of TVD < 250 m (~820.21 ft) defines the path followed to the next node. At the second level the child of the root node, a split node, is the point at

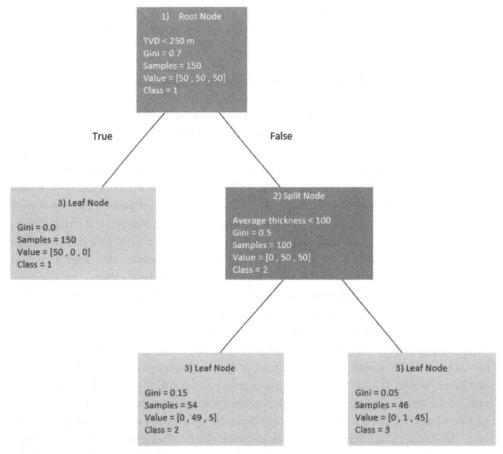

Fig. 2.3 Schematic of the anatomy of a decision tree classifier model with an example of input data based on the geological attributes used in this study.

which the algorithm chooses another feature (formation thickness in our example) and applies a second cut-off. This is done for multiple thresholds to find the purest subset through what is known as the "Gini score" or the measure of purity of the node, calculated as:

$$Gini = 1 - \sum_{i=1}^{C} (p_i)^2$$

Where, p_i is the ratio of class instances (c) among the training instances at the ith node. Therefore, each level of node in the tree is determined by the feature producing the purest subset allowing growth from this. At the leaf node (3), a Gini score approaching 0 implies a pure node, thus within this node only a single class exists. Conversely, a higher

Gini score implies a large number of heterogeneous samples and the need to split the node into different classes[17].

2.2.4.3 Random forest classifier

Random forest (RF) is an extension of the Decision Tree algorithm that produces a collection of DTs and aggregates the final prediction results through majority voting, limiting variance by training on different data samples and introducing randomness into the DTs[18]. The number of possible variables randomly selected for node splitting in each of the trees in the forest is determined through parameterisation in order for each split to select a different set of variables within which the best is chosen. This creates greater tree diversity. The Gini Index is used as a loss function to determine a best split threshold of input variables through calculation of impurity. Through this an implicit feature selection is created as a by-product based on the subset of 'strong variables' used for the classification with the lowest Gini score[19]. By including many trees within the forest, eventually all features will be included, limiting bias error and reducing variance. RF builds upon the premise that a set of aggregated DT classifiers produces a more robust RF model with a reduced risk of overfitting[20]. In addition to the predictive function of RF, it can be used to analyse the importance of input predictive variables based on the impurity determined by the Gini score of a variable.

2.2.4.4 Sample splitting and hyperparameter tuning

For all training, a randomly generated split of 80:20 training:testing data was used to prevent a 'snooping' bias occurring when the algorithm is exposed to new instance of training data. With this bias, the algorithm would learn from the data points it aims to predict, thus invalidating the accuracy of the test model result. Parameterization of all algorithms was conducted to optimize prediction based on the nature of the relevant input data. We used grid search cross-validation method to assess the best hyperparameters to use for each ML algorithm. Optimal hyperparameters were selected based on variance scores resulting from 10-fold cross-validation. When running each cross validation, stratified sampling was conducted in order to maintain an even distribution of each class label so as not to skew the data[8].

2.2.4.5 Evaluating performance of trained algorithms

The various metrics used to assess the success of each trained algorithm can be found in Table 2.3. Accuracy is used to estimate the performance of a classifier, and is calculated based on the relative numbers of correct or incorrect test sample predictions divided by the total number of samples. Accuracy is typically plotted in a confusion matrix where "true" classes are plotted against "predicted" classes, and samples correctly classified plot on the 1:1 diagonal. The confusion matrix helps to visually display the density of error and determine further hyperparameters tuning to improve the algorithm.

Table 2.3 Summary of performance metrics used to measure the capacity of the ML algorithm to correctly predict production success.

Performance measure	Definition	Calculation
Accuracy	In binary classification tasks, accuracy is calculated as a ratio of the number of correct predictions (true positives and true negatives) to the total population of input samples	$Accuracy = \dfrac{[TP + TN]}{Total\ Population}$
Precision	Precision is the number of times the algorithm correctly predicts the number of times a true positive is correctly predicted over the total number of times a negative class is predicted. This is a measure of the probability the ML model has of making a correct classification.	$Precision = \dfrac{TP}{TP + FP}$
Recall	Recall is a measure of how sensitive the model is towards identifying a positive class, taking into account the number of false negatives identified by the ML algorithm. Recall deals with how many relevant items are selected, therefore is an indicator of the quality of the measurement.	$Recall = \dfrac{TP}{TP + FN}$
F1	The F1 measure is a harmonic mean combining precision and recall values, weighting both equally to account for the trade-off between precision and recall.	$F1 = 2 * \dfrac{Precision * Recall}{Precision + Recall}$
R^2	R^2 measures how close the data points are to the regression line of a plot, defining the percentage of the response variable variation that is explained by the linear model. The higher the R^2 value, the better the fit of the model to the data, and therefore the more replicable the model to predict future outcomes[8].	$R^2 = \dfrac{Explained\ Variation}{Total\ Variation}$

2.3 Results

2.3.1 Comparative performance of our algorithms

Results from the PCA showed that over 95% of the explained variance within the data can be maintained when the data is projected onto 6 vectors. The performance of the three algorithms run using the same input features extracted from the PCA reveal (Fig. 2.4) that the RF algorithm performs the best, with an RMSE of 0.319, and the DT Classifier comes second with a similar RMSE of 0.383, likely due to the inherent similarities between the two algorithms. However, the standard deviation in DT is larger than in RF suggesting greater error and lower predictability. SVM-SGD classification is

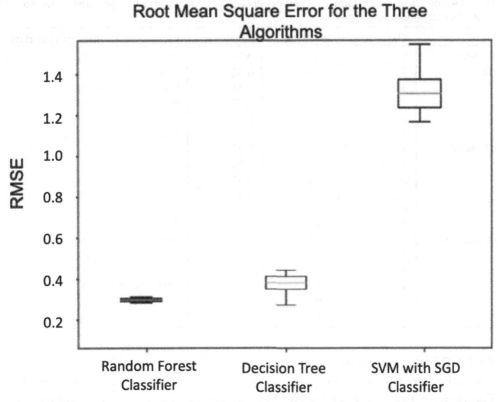

Fig. 2.4 Box plot comparison of the three ML algorithms used in this study and this study, showing the relative accuracies of these based on root mean square error (RMSE) values and standard deviation values. A lower absolute value of RMSE and a smaller error bars mean better performance.

worst fitted to the data with an RMSE of 1.31. This algorithm performance is reflected in the performance metrics per class within each algorithm (Table 2.4) with RF consistently producing the most accurate results and SVM–SGD performing worst.

2.3.1.1 SVM-SGD classifier

The confusion matrix of the SVM classifier with SGD kernel optimization (Fig. 2.5) shows this model has an accuracy score of 46%. While correct prediction of the 1st and 4th quintiles reached accuracies of 74% and 67%, only 6% of 5th quintile wells were accurately predicted by this algorithm. This unpredictability of Class 5 is also clear from Table 2.4 where large disparity between precision (50%) and recall (6%) is visible. Fig. 2.5 reveals an uneven distribution of random prediction error, suggesting limited abilities to correct for them and thus limited use of SVM–SGD classifier for our use case.

Table 2.4 Comparison of performance metrics calculated per class for the best performing trained model for each algorithm.

	Class	Precision	Recall	F1	Instances per class
SVM-SGD classifier	1	0.58	0.74	0.65	3799
	2	0.45	0.32	0.38	3777
	3	0.42	0.51	0.46	3808
	4	0.41	0.67	0.51	3828
	5	0.5	0.06	0.11	3766
	Avg/total	0.47	0.46	0.42	18978
DT classifier	1	0.75	0.7	0.73	3799
	2	0.62	0.59	0.61	3777
	3	0.64	0.41	0.5	3808
	4	0.42	0.23	0.3	3828
	5	0.45	0.86	0.59	3766
	Avg/total	0.58	0.56	0.55	18978
RF classifier	1	0.95	0.98	0.97	3799
	2	0.96	0.96	0.96	3777
	3	0.95	0.91	0.93	3808
	4	0.91	0.93	0.92	3828
	5	0.95	0.95	0.95	3766
	Avg/total	0.94	0.94	0.94	18978

2.3.1.2 Decision tree classifier

The highest prediction accuracy achieved using this algorithm remains low at 55.4%. The decision tree confusion matrix (Fig. 2.6) shows significant misclassification in prediction of the mid-success ranges of production, with prediction of the 4th quintile

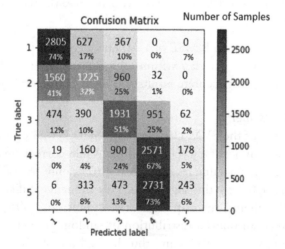

Overall Classification Accuracy: 46%

Fig. 2.5 Confusion matrix measuring performance of the SVM with SGD classifier algorithm.

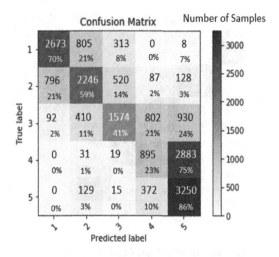

Fig. 2.6 Confusion matrix measuring performance of the decision tree classifier algorithm.

having the lowest accuracy (23%). Additionally, one of the highest misclassification rates (75%) is of the 4th quintile misclassified as the 5th suggesting the model is systematically misclassifying the mid-range production classes. Performance metrics in Table 2.4 support this, showing classes 3 and 4 to have the lowest F1 scores, with the recall of class 4 low (23%) due to misclassification as Class 5.

2.3.1.3 Random forest classifier

The RF classifier algorithm produces the highest accuracy outputs of all models, surpassing the best accuracy scores of SVM-SGD and DT models by 44.3% and 35.1%, respectively. At 91% accuracy, this model shows little discrepancy between precision and recall metrics, implying this model is highly predictive. Table 2.4 shows lower performance within the mid-production (3rd and 4th) classes, reflecting the systematic error seen in the confusion matrix (Fig. 2.7). This again illustrates the apparent systematic misclassification of mid-production classes, implying this error is due to the nature of the data and class scheme used for prediction, rather than algorithm error or selection.

The RF algorithm was selected as the best fit for our use case. RF is largely unaffected by noisy data input when using a large number of input variables, enabling greater accuracy of prediction.

2.3.1.4 Error analysis

Systematic misclassification within the mid-range classes of production success is likely due to the narrow ranges between the definition of classes. To address this error, the number of classes of production success was reduced from quintiles (5 classes) to

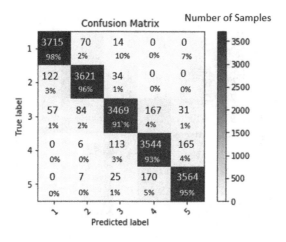

Overall Classification Accuracy: 91%

Fig. 2.7 Confusion matrix measuring performance of the random forest classifier algorithm.

quartiles (4 classes) of production to increase the range of data assigned within each prediction label and to reduce the number of mid-range classes that could be misclassified. Using the same input variables and hyperparameterization, SVM-SGD classifier accuracy increased by 8.7% to 54.9%, while decision tree classifier model accuracy increases by 17.8% to 72.7% using fewer classes of normalised production success. RF classifier remains the most accurate algorithm tested at 79.7% when using a system of quartile classes (Fig. 2.8). While a decrease in RF model accuracy and overall performance (Table 2.5) is observed from 91% accuracy using quintile classes (Table 2.4), the error accounting for this can be mitigated through data analysis and subsequent input feature optimization.

Use of the RF-embedded method of feature selection enables understanding of how each feature input into the model will decrease the Gini impurity of the final result within each decision tree, with the final importance of each variable determined as an average across the random forest as a whole[21], therefore explaining the variability of our target normalised initial production success. Table 2.6 examines all features available from the well log data after the data cleaning stage of the workflow, giving an overview of the distribution of importance between the features available for input into the model during fine-tuning where only features of greatest importance are used. The geological inputs used are those that can be modeled at the scale of the play from a regional geological model, with the method of calculation listed in Table 2.6. These are geological features that are either commonly linked to production performance, or could perceivably be linked to production performance in shale plays.

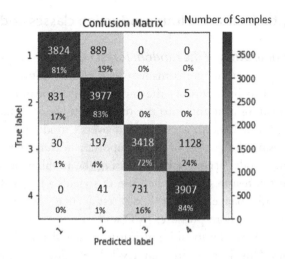

Overall Classified Accuracy: 79.7%

Fig. 2.8 Confusion matrix measuring performance of the random forest classifier algorithm predicting four classes of well production success.

Table 2.5 Performance metrics per quartile class calculated for the random forest model shown in Fig. 2.8.

Class	Precision	Recall	F1	Support
1	0.82	0.81	0.81	4713
2	0.78	0.83	0.8	4813
3	0.82	0.72	0.77	4773
4	0.78	0.84	0.8	4679
Avg/total	0.8	0.8	0.795	18978

Table 2.6 Feature importance weightings of the most influential input features derived from the random forest model in Fig. 2.8.

Feature label	Feature importance
Average thickness	0.543
Pore pressure	0.185
TVD	0.156
Resource concentration	0.0487
Geothermal gradient	0.0303
Reservoir pressure	0.0204
Maximum burial temperature	0.0152
GOR	0.00127
Porosity	0.000807

2.3.2 Further optimization for four production classes and application to shale plays

2.3.2.1 Further optimization of the random forest classifier

To optimize our RF algorithm performance on 4 classes, feature selection was conducted in each model to include only the most significant features driving play success within the formations as inputs into the model, and hence reduce the risk of data overfitting to noise. An increase in accuracy and overall model performance (Fig. 2.9) resulted from decreasing the cut off of most important features input into the model. Therefore, improving model accuracy by increasing the number of well log input features up to the four top features, in part because this yields a larger dataset size to train the model. Model performance peaks at 96.8% test accuracy when using the four top features (average thickness, pore pressure, TVD and resource concentration) but accuracy decreases by 1% when additional input features are added and plateaus at approximately 95% accuracy. This decrease in accuracy is likely due to overfitting through an increase in features with a low correlation to well production success, thus increasing the Gini impurity score of the overall Random Forest result.

The confusion matrix for the optimized RF model using the top 4 features extracted (Fig. 2.10) shows that the highest number of misclassifications is between class 3 and class 4. However, this optimized feature input yields the highest model accuracy with the highest precision and lowest random error spread through optimization of the

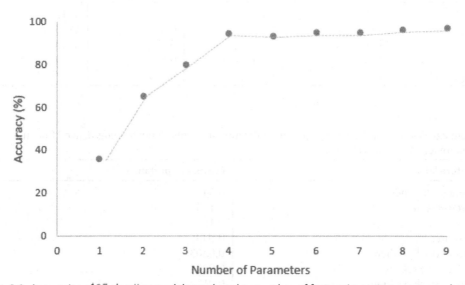

Fig. 2.9 Accuracies of RF classifier models produced as number of feature importance extracted well log attributes, and therefore dataset size, is increased.

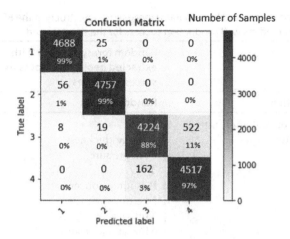

Overall Classified Accuracy: 96.7%

Fig. 2.10 Confusion matrix measuring performance of the most successful random forest classifier algorithm predicting four classes of well production success with four input variables (performance defined in Table 2.7).

number of output classes. The performance metrics and model hyperparameters used for this model are found in Tables 2.7 and 2.8, respectively.

2.3.2.2 Random forest prediction result for different North American shale plays

When applied to unseen test data, this highest accuracy RF model most accurately predicts the well success of the Wolfcamp play, specifically units B and C where accuracies greater than 90% are achieved (Table 2.9). All plays achieve accuracies of greater than 50%, however the range of accuracy scores differs by 38% between Wolfcamp B and Niobrara plays. Relative weighting of plays with different sample sizes does not appear to skew the model towards more accurate predictions in plays in which a greater proportion of data has been used to train the model.

Table 2.7 Performance metrics per quartile class calculated for the most successful random forest model produced shown in Fig. 2.10.

Class	Precision	Recall	F1	Support
1	0.99	1.00	0.99	4713
2	0.99	0.99	0.99	4813
3	0.98	0.97	0.98	4773
4	0.98	0.98	0.98	4679
Avg/total	0.98	0.98	0.98	18,978

Table 2.8 Model specifications and hyperparameters used in constructing the RF model in Fig. 2.10 which most accurately predicts well production success for the dataset used.

Model summary		Random forest classifier with top four features extracted geological inputs using production success quartiles
Classification algorithm		**Random forest**
Dataset	Number of geological inputs	4
	Input feature types	Average thickness
		Pore pressure
		TVD
		Resource concentration
Hyperparameters		N estimators: 1,400
		Min sample split: 5
		Min samples leaf: 1
		Max features: square root
		Max depth: 30
		Bagging: mean square error
Cross validation folds		10
R2 score		0.823
Test accuracy		0.968

Table 2.9 Classification accuracy testing results of each individual play using the Fig. 2.10 model, ordered by test accuracy result.

Play	Test accuracy	Precision	Recall	F1 Score	Total wells	% of data
Wolfcamp B	0.967	0.97	0.97	0.97	2414	3
Wolfcamp C	0.944	0.95	0.94	0.94	539	1
Bakken	0.885	0.89	0.89	0.88	16160	17
Wolfcamp D	0.718	0.72	0.72	0.72	625	1
Marcellus	0.68	0.68	0.68	0.68	9317	9
Wolfcamp A	0.667	0.67	0.67	0.66	3143	3
Woodford	0.651	0.65	0.65	0.65	4068	4
Eagle Ford	0.631	0.63	0.63	0.63	12072	19
Utica	0.602	0.6	0.6	0.6	1602	2
Niobrara	0.587	0.62	0.59	0.58	6294	7

2.4 Discussion

2.4.1 Algorithm overview and comparison

It is clear that RF is the most predictive algorithm out of all other algorithms we used for this dataset, and that selecting model features based on their relative importance produced the most accurate test result (97% accuracy). All algorithms using quintiles of success perform most successfully using the top four features of average thickness, pore pressure, TVD, and resource concentration. Low ranking feature extracted inputs such as porosity show no correlation to production in the shale plays included in this study, rather including these increases the risk of model overfitting to noise. Part of the reason RF performs best of the classification algorithms tested is that it is the most flexible when it comes to fine-tuning, including the ability to adjust the hyper-parameter of bootstrap samples applied to the Random Forest ensemble method of bagging used to improve variable selection. This enables advanced models to be built specific to the correlations between geological inputs and production metrics given as outputs during model construction. This flexibility enables dimensionality reduction techniques like PCA to be included in the machine learning coded process. The "white box" nature of RF classifier enables an introspective approach to model construction, emphasizing interpretability at all stages of model development to understand the nature of the data. We exploited this to extract information on the input features most impacting shale plays in general, and each play individually. This enables us to assess from a geological point of view if the machine learning results are meaningful.

Analysis of the confusion matrices for all algorithms produced using quintiles shows errors are most commonly concentrated within the 3rd and 4th of the quintile classes. Overall test accuracy is high in the initial RF model (Fig. 2.7), showing that RF classifier is excellent at predicting the highest and lowest classes of well successes, distinguishing between the extremes of production based on the input geological features. However, within classes where the narrow ranges of data in the input features exist, error is common. From a geological perspective this is likely a result of narrow ranges between "successful" and "unsuccessful" production as the data provided is sourced from only hydrocarbon-bearing wells, with no dry well data included.

As the number of output classes is reduced from quintiles to quartiles to decrease model error, a greater number of input features is needed to increase the accuracy of prediction. However, accuracy only increases to a defined value after which adding input features increases noise and decreases model accuracy. This is known as the 'Curse of Dimensionality' which states that as the number of variables and therefore dimensions increase, the amount of data needed to determine general trends also increases[22]. Increasing the number of variables proportionally increases the space between each data point. With an increased sparsity of training data, the ability of a hyperplane to separate points into classes improves as the divisions between clusters of data become clearer. However, this also means that features with a high relative separation risk being grouped in the same class. When applying a high dimensionality model trained on sparse data, the unknown instances could fall between data points used to train the model that are separated by large distances, increasing risk of misclassification.

2.4.2 Geological features of importance for each North American shale play

Distribution of each quintile class of success (Fig. 2.11) provides a relative scale of production success of each of the US shale plays, with the Marcellus and Bakken plays being the most successful (based on the available dataset). Fig. 2.12 shows clear clusters of data exist for the four geological features most related to initial production, demonstrating the narrow ranges of a highly bimodal spread in data and the clear separation of the Marcellus and Bakken play data points consistently plotting together. The spread of

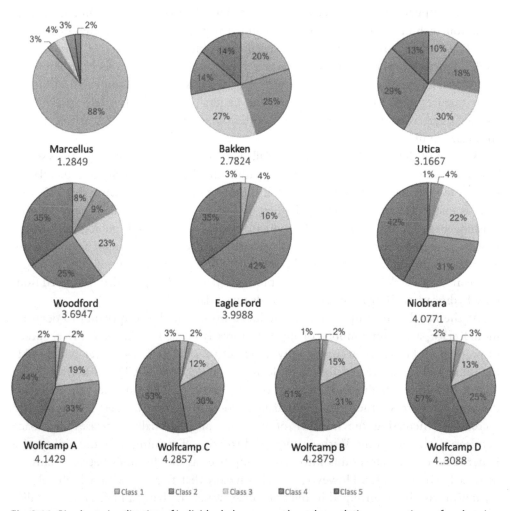

Fig. 2.11 Pie chart visualization of individual play success based on relative proportions of each quintile class of normalized initial production within each producing member of the shale plays studied. Ranking of the normalized production success of individual plays based on their average class of production (based on quintile distribution) shown in red beneath play names.

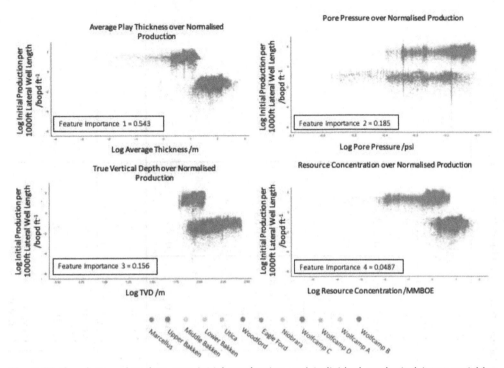

Fig. 2.12 Correlation plots between initial production and individual geological input variables, ranked based on their feature importance values. These features are used to produce the most successful (97% accuracy) RF model in Fig. 2.10.

data per play correlates with these four features being most deterministic in the success of production within the plays studied.

Feature Extraction for each play (Table 2.10) shows the relevance of each geological feature to the production from each shale play, enabling comparison between different plays and assessment as to whether certain attributes impact relative production success. The most successful shale plays (Fig. 2.11) are the Marcellus and Bakken plays. Pore pressure ranks consistently highly, averaging 0.261 across the Marcellus and Bakken. Pore pressure ranks low in carbonate plays (including the Wolfcamp, Niobrara, Woodford, and Eagle Ford) with the exception of the Utica play where this is the most important feature (score of 0.383). Within the Wolfcamp Units, porosity ranks highly averaging 0.388, however the spread of porosity importance is disparate across the other plays. Generally, all plays show a low importance of the geothermal gradient whereas reservoir and pore pressure have consistently high feature importance values. High pressure improves the mobility of hydrocarbon fluids and therefore has a direct impact on production, or well flow rate, as defined by Darcy's Law. High pressure will improve the mobility of viscous liquid phases in oil-producing plays such as the Middle Bakken where reservoir and

Table 2.10 Feature importance rankings of each geological input feature for each play derived through use of the 97% accurate Random Forest model (Fig. 2.10) constructed in this study.

Marcellus		Middle Bakken		Utica	
Feature label	*Feature importance*	*Feature label*	*Feature importance*	*Feature label*	*Feature importance*
Average Thickness		Pore Pressure	0.383	TVD	0.599
Porosity		GOR	0.172	Pore Pressure	0.242
Resource Concentration		Max. Burial Temperature	0.138	Reservoir Pressure	0.0511
TVD		TVD	0.109	Resource Concentration	0.0429
Pore Pressure		Average Thickness	0.0716	Max. Burial Temperature	0.019
Max. Burial Temperature		Porosity	0.0635	Geothermal Gradient	0.0163
Reservoir Pressure		Resource Concentration	0.0623	Average Thickness	0.0156
GOR		Reservoir Pressure	0	GOR	0.0145
Geothermal Gradient		Geothermal Gradient	0	Porosity	0

Woodford		*Eagle Ford*		*Niobrara*	
Feature label	*Feature importance*	*Feature label*	*Feature importance*	*Feature label*	*Feature importance*
Reservoir Pressure	0.197	Reservoir Pressure	0.299	Reservoir Pressure	0.381
TVD	0.177	Max. Burial Temperature	0.211	GOR	0.192
Max. Burial Temperature	0.257	Porosity	0.132	Resource Concentration	0.146

Feature label	Feature importance
GOR	0.128
Porosity	0.122
Average Thickness	0.12
Resource Concentration	0.0993
Pore Pressure	0
Geothermal Gradient	0

Feature label	Feature importance
Average Thickness	0.108
GOR	0.098
Resource Concentration	0.0859
TVD	0.0665
Pore Pressure	0
Geothermal Gradient	0

Feature label	Feature importance
TVD	0.139
Average Thickness	0.072
Max. Burial Temperature	0.0705
Porosity	0
Pore Pressure	0
Geothermal Gradient	0

Wolfcamp A

Feature label	Feature importance
Porosity	0.229
Reservoir Pressure	0.181
Average Thickness	0.147
GOR	0.12
Resource Concentration	0.115
TVD	0.108
Max. Burial Temperature	0.1
Pore Pressure	0
Geothermal Gradient	0

Wolfcamp C

Feature label	Feature importance
Porosity	0.561
Reservoir Pressure	0.267
Average Thickness	0.0909
Max. Burial Temperature	0.0474
GOR	0.0198
TVD	0.0142
Resource Concentration	0
Pore Pressure	0
Geothermal Gradient	0

Wolfcamp B

Feature label	Feature importance
Porosity	0.549
Reservoir Pressure	0.24
Resource Concentration	0.113
Average Thickness	0.0602
Max. Burial Temperature	0.03
TVD	0.00456
GOR	0.0035
Pore Pressure	0
Geothermal Gradient	0

Wolfcamp D

Feature label	Feature importance
Average Thickness	0.217
Porosity	0.212
Reservoir Pressure	0.148
GOR	0.126
TVD	0.11
Resource Concentration	0.0955
Max. Burial Temperature	0.0924
Pore Pressure	0
Geothermal Gradient	0

pore pressure have a high combined feature importance score of 0.812. Pore pressure results in increased overpressure produced through petroleum generation and compaction disequilibrium. This is intrinsically linked to basin-wide evolution and preservation during periods of uplift. There is a large disparity in the importance weightings observed (Table 2.10) relating to the relative success of the Middle Bakken member, in which the majority of completion landing zones in the Bakken play exist[23]. The Bakken play success is largely defined by production from mappable sweet spots, for example, the Parshall Field producing from areas of abnormally high pressures. These are a cumulative result of structural evolution producing a thermally mature hydrocarbon kitchen and migration of hydrocarbons into sealed pressure cells defined by heterogeneous lithologies[24]. Additionally, gas sorption in shales decreases with decreased temperature and pressure during uplift, with gas loss potentially causing undersaturation[25]. This can explain the relative feature importance of pore pressure as this reduces effective stress making the shale easier to fracture during completion, improving production[26].

The Marcellus Shale is the most successful gas play within this study. Geological production drivers are relatively equal across the play, demonstrated by an even distribution of feature importance values derived from the RF algorithm, reflecting the lateral continuity and overall homogenous nature of the Marcellus Shale. Use of machine learning to predict indicators of production success in the Marcellus has been previously demonstrated by Wang and Carr[27] using a SVM model to predict lithofacies. Biogenic silica is interpreted to increase in concentration in the SW and NE of the basin, making these areas sweet spots for hydraulic fracture stimulation through high brittle quartz content enabling effective fracture propagation through the shale[27]. Average thickness ranks as the factor most influencing production in the Marcellus as this becomes exceptionally thin, reaching 15-30 m in the SW area of the basin. Vertical extent of the shale limits production as the induced fractures can extend beyond the thickness of the formation in the SW. The influence of thickness on the density of high production wells is seen in the NE area (Fig. 2.13). In the SW, production within the thin condensed section is a factor of higher TOC and NTG ratios than in the NE core development area. The SW core is more calcite-rich than the NE, with a lower volume of clay minerals clogging pore throats[28]. Therefore, porosity is highly influential in production from the thinner shale, as identified using feature extraction. These previously defined relationships are re-enforced through use of our machine learning approach. This improves our understanding of the geological features influencing specific shale plays, while the ability of the algorithm to predict previously independently established relationships demonstrates the accuracy and applicability of the model in unconventional production.

Gas sweet spots are mapped within the Eagle Ford shale in the North West Burgos Basin, where reservoir pressure remains too low for high oil production rates[29]. Reservoir pressure shows the highest importance score of 0.299 for the Eagle Ford,

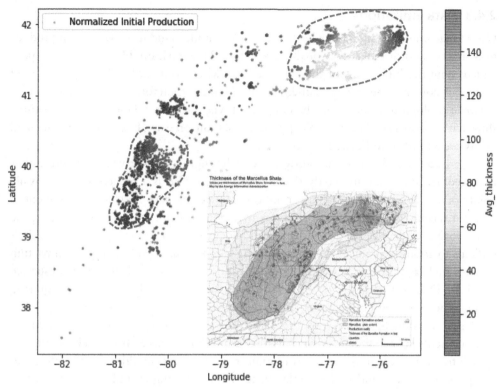

Fig. 2.13 Heat map showing the relationship between initial production and average play thickness as a function of well location within the Marcellus Shale. Red outlines mark the extent of the two core areas of production. Inset map showing aerial extent of the shale (from [31]).

reflecting the limiting nature of hydrocarbon maturity on production. Features relating to maturity such as maximum burial temperature and GOR also show importance to production success. As gas phase hydrocarbons are generated, the 'gas-push' effect of buoyant hydrocarbons migrating to the surface helps reduce overpressure and drive production. GOR is a significant feature driving production in the Utica play, suggesting success in gas plays such as this and the highly successful Marcellus play is linked to rapid initial rates of production from mature gas phase hydrocarbons. This has implications for production over longer timescales as this success may be short-lived with EUR (Expected Ultimate Recovery) of wells not reflecting the initial success found in this study.

Previous understanding of these plays from successful and extensive production helps to validate play and feature assessment through the machine learning model constructed, as this corroborates that the most impacting extracted features are the limiting factors in production historically in these shale plays.

2.4.3 Data limitations

Use of machine learning has enabled us to further understand unconventional production success, and to build a reasonable predictive RF algorithm. However, it is important to note the limitations of the data available within this study. Many features used in model construction are fundamentally interlinked, therefore existing relationships between geological features must be considered when assessing these apparent relationships with production. For example, porosity, reservoir pressure, burial temperature, and GOR are all linked to burial depth.

The features assessed in our study are largely engineering-driven in nature rather than geological and are implicitly tied to production. This is a reflection of the availability of data on shale plays, where common engineering data exists for over the entire play but specific geological feature vectors may not exist in each play. For the purpose of exploring undeveloped shale formations where little engineering data exists, it is difficult to robustly predict well success using our RF algorithm as highly impacting features such as pore and reservoir pressure cannot be measured until the subsurface is drilled. Due to the nature of tight shale plays and associated difficulty in acquiring accurate porosity and permeability data, pressure data is often calculated through basin modeling, introducing error into the algorithm which predicts outcomes based on measurements with inherent error. It is also important to note that a lack of singular overall trend in geological features important to play success could indicate differences in operator methods and well completion techniques having an influence on production in addition to the heterogeneous nature of the subsurface geology.

2.5 Conclusion

To conclude, this study has demonstrated the use of machine learning as part of a multidisciplinary approach to understanding the intrinsic complexities of shale plays. Based on the results, the most accurate model constructed provides a high accuracy (97%) method of prediction of new instance well successes for shale plays where well data has already been collected and used as part of the model training process. This model also provides a method of understanding which measurements are most important to collect for each play, and which features do not contribute to the success of a play. This has application within the petroleum industry as a means to streamline production and eliminate resource expenditure on unnecessary data collection. However, due to the lack of general trends observed spanning all the shale plays for which data was collected, re-enforcing the complex nature of the geology, this model may only be a robust predictive tool when applied to the specific plays from which data was used to train the algorithm and for which the algorithm has been specifically hyperparameterized. Rather than providing a model applicable to all shale plays, we propose using the Random Forest classifier to produce a predictive tool for each play using large unconventional datasets. The use and application of the models presented within this dataset show value as a means of determining spatial distribution of success. This has use as an interpretive tool to analyse the success of proposed wells in exploration of new plays through processing large and complex datasets to extract valuable information to optimise measurement collection.

Feature importance data also suggests classic unconventional shale gas plays such as the Marcellus Shale are likely to be equally influenced by a wide range of factors, whereas understanding of play-specific geology and structural evolution in tight oil plays like the Bakken play may efficiently explain the distribution of importance ascribed to certain factors. Although a large influence on production success is linked to geological factors such as depositional and tectonic history dictating play properties like mineralogy and pressure, the overall success of a hydraulically fractured well in these plays is also susceptible to the influence of production methods. We find that the classification of shale plays is complex and based on a wide range of factors. The Random Forest classifier model constructed provides a means to contribute to unraveling this complexity as part of a holistic approach including analysis of geology and production metrics in play assessment.

2.6 Acknowledgments

This study was conducted in the framework of a Halliburton Science and Technology for Exploration and Production Solutions (STEPS) initiative project awarded to CJ. STEPS projects offer students an opportunity to conduct a research project focused on real-world data, while receiving industry-relevant training and mentorship from the dedicated STEPS team. Each year, research projects are designed around a theme determined by the STEPS Scientific Advisory Board. Projects are designed to follow the 'learning while-doing' ethos, enabling deep engagement with Landmark software, product suites and knowledge base.

We are grateful to Rystad Energy for permission to use production data from ShaleWellCube within this project, and to Halliburton for the geological data on shale plays, their mentorship and also Xinli Jia for her constant IT support during the project.

We thank Shuvajit Bhattacharya (guest editor) and Haibin Di (guest editor) for their insightful comments.

References

1. McGlade C, Speirs J, Sorrell S. Methods of estimating shale gas resources - comparison, evaluation and implications. *Energy* 2013;**59**:116–25. http://dx.doi.org/10.1016/j.energy.2013.05.031.
2. Hefner R. The United States of gas: why the shale revolution could have happened only in America. *Foreign Affairs* 2014;**93**(3):9–14.
3. Middleton RS, Gupta R, Hyman JD, Viswanathan HS. The shale gas revolution: barriers, sustainability, and emerging opportunities. *Appl Energy* 2017;**199**:88–95. http://dx.doi.org/10.1016/j.apenergy.2017.04.034.
4. Weijermars R. Economic appraisal of shale gas plays in Continental Europe. *Appl Energy* 2013;**106**:100–15. http://dx.doi.org/10.1016/j.apenergy.2013.01.025.
5. Panja P, Velasco R, Pathak M, Deo M. Application of artificial intelligence to forecast hydrocarbon production from shales. *Petroleum* 2018;**4**(1):75–89. https://doi.org/10.1016/j.petlm.2017.11.003.

6. Flender S (2019). Data is not the new oil. 2019. Available from: https://towardsdatascience.com/data-is-not-the-new-oil-bdb31f61bc2d.

7. Cracknell M, Reading A. Geological mapping using remote sensing data: a comparison of five machine learning algorithms, their response to variations in the spatial distribution of training data and the use of explicit spatial information. *Comput Geosci* 2014;**63**:22–33.

8. Géron A (2017). Hands-On Machine Learning with Scikit-Learn & TensorFlow. Sebastopol, CA: O'Reilly Media, Inc.

9. Schafer JL, Graham JW. Missing data: our view of the state of the art. *Psychol Methods* 2002;**7**(2):147–77.

10. Ringnér M. What is principal component analysis? *Nat Biotechnol* 2008;**26**(3):303–4.

11. Wold S, Esbensen K, Geladi P (1987). Principal Component Analysis. Elsevier B.V., 2, pp. 37-52.

12. Cadima J, Jolliffe I. Variable selection and the interpretation of principal subspaces. *J Agric, Biol Environ Stat* 2001;**6**(1):62–79.

13. Cortes C, Vapnik V. Support-vector networks. *Machine Learning* 1995;**20**(3):273–97.

14. Karatzoglou A, Meyer D, Hornik K. Support vector machines in R. *J Stat Softw* 2006;**15**(9):1–28. https://doi.org/10.18637/jss.v015.i09.

15. Du W, Du W, Zhan Z, Zhan Z. Building decision tree classifier on private data. *Proceedings of the IEEE international conference on Privacy, security and data mining-Volume 14* 2002:1–8. http://portal.acm.org/citation.cfm?id=850784.

16. Pradhan B. A comparative study on the predictive ability of the decision tree, support vector machine and neuro-fuzzy models in landslide susceptibility mapping using GIS. *Comput Geosci* 2013;**51**:350–65.

17. Ceballos F (2019). Scikit-Learn Decision Trees Explained. 2019. Available from: https://towardsdatascience.com/scikit-learn-decision-trees-explained-803f3812290d.

18. Breiman L. Statistical modeling: the two cultures (with comments and a rejoinder by the author). *Stat Sci* 2001;**16**(3):199–231.

19. Menze B, Kelm B, Masuch R, Himmelreich U, Bachert P, Petrich W et al. A comparison of random forest and its Gini importance with standard chemometric methods for the feature selection and classification of spectral data. *BMC Bioinf* 2009;**10**(1):213.

20. Rodriguez-Galiano V, Ghimire B, Rogan J, Chica-Olmo M, Rigol-Sanchez J. An assessment of the effectiveness of a random forest classifier for land-cover classification. *ISPRS J Photogramm Remote Sens* 2012;**67**:93–104.

21. Saeys Y, Abeel T, Van de Peer Y. Robust feature selection using ensemble feature selection techniques. *Machine Learning and Knowledge Discovery in Databases* 2008;**5212**:313–25.

22. Hughes G. On the mean accuracy of statistical pattern recognizers. *IEEE Trans Inf Theory* 1968;**14**(1):55–63.

23. Kumar S, Hoffman T, Prasad M. Upper and lower Bakken shale production contribution to the middle Bakken reservoir. Unconventional Resources Technology Conference, Denver, Colorado, 12-14 August 2013, 1581459, 2013:1–1.

24. Jarvie DM. Shale resource systems for oil resource systems: Part 2 - shale-oil resource systems. *AAPG Memoir.* 2012;**97**:89–119.

25. Hao F, Zou H, Lu Y. Mechanisms of shale gas storage: implications for Shale gas exploration in China. *AAPG Bull* 2013;**97**(8):1325–46.

26. Gong C, Rodriguez L. Challenges in pore pressure prediction for unconventional petroleum systems AAPG Hedberg Conference, The Future of Basin and Petroleum Systems Modelling 2016:42018.

27. Wang G, Carr T. Methodology of organic-rich shale lithofacies identification and prediction: a case study from Marcellus Shale in the Appalachian basin. *Comput Geosci* 2012;**49**:151–63.

28. Zagorski WA, Bowman DC, Emery M & Wrightstone GR (2011). An overview of some key factors controlling well productivity in core areas of the Appalachian Basin Marcellus Shale Play *. 110147. 90122.

29. Yallup C, Bromhead A. Emerging unconventional resource plays in the onshore Gulf of Mexico: Assessing Agua Nueva and Tuscaloosa play potential. *Aapg Ice* 2018;**2018**:11166.

30. EIA.gov. (2011). Review Of emerging resources: U.S. Shale gas And Shale oil plays. Available at: https://www.eia.gov/analysis/studies/usshalegas/. Accessed on July 1st 2018.

31. US Energy Information Administration (2017). Marcellus Shale play. (January). pp. 1–14. Access date: July 1st 2018. URL: chrome- extension://efaidnbmnnnibpcajpcglclefindmkaj/viewer.html?pdfurl=https%3A%2F%2Fwww.eia.gov%2Fmaps%2Fpdf%2FMarcellusPlayUpdate_Jan2017.pdf&clen=1430404&chunk=true.

32. Engle MA, Reyes FR, Varonka MS, Orem WH, Ma L, Ianno AJ, Schell TM, Xu P, Carroll KC. Geochemistry of formation waters from the Wolfcamp and "Cline" shales: insights into brine origin, reservoir connectivity, and fluid flow in the Permian Basin, USA. *Chem Geol* 2016;**425**:76–92.

33. Zakhour N, Shoemaker M, Lee D. Integrated workflow using 3D seismic and geomechanical properties with microseismic and stimulation data to optimize completion methodologies: Wolfcamp shale-oil play case study in the Midland BasinSPE Eastern Regional Meeting 2015;2015.

34. Romero AM, Philp RP. Organic geochemistry of the Woodford Shale, southeastern Oklahoma: How variable can shales be? *AAPG Bull* 2012;**96**(3):493–517.

35. Kennedy MJ, Pevear DR, Hill RJ. Mineral surface control in organic black shale. *Science* 2002;**295**(5555):657–60.

36. Sonnenberg SA (2012). The Niobrara Petroleum System, Rocky Mountain Region. In: Tulsa Geological Society, AAPG Publishing, Tulsa. 2012.

37. ter Heege J, Zijp M, Nelskamp S, Douma L, Verreussel R, ten Veen J, de Bruin G, Peters R. Sweet spot identification in underexplored shales using multidisciplinary reservoir characterization and key performance indicators: Example of the Posidonia Shale Formation in the Netherlands. *J Nat Gas Sci Eng* 2015;**27**:558–77. http://dx.doi.org/10.1016/j.jngse.2015.08.032.

38. Lehmann D, Brett CE, Cole R, Baird G. Distal sedimentation in a peripheral foreland basin: Ordovician black shales and associated flysch of the western Taconic foreland, New York State and Ontario. *Geol Soc Am Bull* 1995;**107**(6):708–24.

39. Pitman JK, Price LC, Lefever JA (2001). Diagenesis and Fracture Development in the Bakken Formation, Williston Basin. US Geological Survey Professional Paper. 1653.

PART 2

Deep learning approaches

CHAPTER 3

Recurrent neural network: application in facies classification

Miao Tian, Sumit Verma
Department of Geosciences, The University of Texas Permian Basin, Odessa, TX, USA

Abstract

Most of the geological processes are gradual, which results in gradual variation in the lithofacies, both spatially and temporally. In machine learning framework, unlike the common deep neural network (DNN), the recurrent neural network (RNN) honors the spatio-temporal relationships. There are different RNN models, some of which are more useful for geoscience problems than others. The simple RNN model remembers the output of previous steps and feeds it into the computation of the temporally or spatially next step as an input. However, in the simple RNN model, the effect of the farther steps are reduced exponentially, which does not allow to capture long-term dependencies. RNN models, such as long short-term memory (LSTM) and bidirectional long short-term memory (Bi-LSTM) can provide better results in many time series-related problems in geosciences, by employing the forget, update and output gates inside its special memory cell. Although the LSTM and Bi-LSTM models work well with the large data sets, they present a challenge of complex parameterization. Simple RNN, LSTM, and GRU work for data with 1D dependency, for example, well logs. The convolutional recurrent neural network (ConvRNN) model, effectively combines the CNN and RNN for the data with multidimensional dependencies, for example, 3D seismic data. We performed a facies classification case study based on a Permian Basin (United States) well log data set and compared the results between traditional DNN and Bi-LSTM. The Bi-LSTM model provides a better facies classification results compared to DNN.

Keywords

Deep learning; Facies classifiction; Long short-term memory; Spatio-temporal data; Recurrent neural network

3.1 Introduction

Most of the machine learning methods have been proven to be effective in specific geoscience areas. For example, support vector machine (SVM) and random forest (RF) provide high classification accuracies ([4,5,10,29,55]); deep neural network (DNN) provides a more reasonable reconstruction of the missing well logs ([22,38]). However, there are still some inherent challenges arising from the nature of geological process. Geological objects, such as facies, generally have spatio–temporal dependency[17,33,44]. In addition to this, the current data collection procedure introduces the convolutional effect between the observations of adjacent measuring positions.

Advances in Subsurface Data Analytics
DOI: https://doi.org/10.1016/B978-0-12-822295-9.00013-3

In many geosciences studies, the classic machine learning methods which take the samples as element-independent events provide the prediction for each sample just based on its own measurement set[45]. On one hand, this data processing is inconsistent with traditional manual methods of geologic analysis and interpretation. For example, geologists prefer to determine the channel deposits from seismic time and stratal slices through the curved shapes formed by the variation of the local attribute values but not just rely on the attribute value at certain locations[34,37]. On the other hand, the obtained measurements are influenced by the convolutional effect due to limits of the measuring techniques and instruments. For instance, affected by surrounding sediments, each recorded well logging curve is a weighted summation of the corresponding logging responses of adjacent sediments[33,48]. Hence, we will get the predictions with lower global accuracy and less physical significance if the models take the samples as the element-wise independent events.

In the example shown in Fig. 3.1, we assume that there are three lithologies in a section of a vertical borehole: type A, type B, and type C. The true 1D profile is displayed in Fig. 3.1i, two different predictions are displayed in Fig. 3.1ii and Fig. 3.1iii. According to the comparisons between them, the two predictions have the same local accuracy (12 out of 15 points). We have the correct lithology order and rational thickness for each layer in prediction ii, whereas the disturbed lithology order and incorrectly predicted thin layers in prediction iii. Without any doubt, the prediction ii provides a higher global accuracy over this section and conforms to the reality better. Therefore, we need to comprehend the data structure of geoscience observations and take advantage of the models, which consider the additional information hidden in the structure to improve the model performance.

In this chapter, we discuss different data types (chronological and spatial) and recurrent neural network (RNN), specifically, bidirectional long short-term memory (Bi-LSTM). RNN is one of the advanced neural network methods with improved architecture. Instead of the neurons in traditional DNN layers, a kind of sceptical unit named memory cell is employed in RNN layers to remember the hidden information

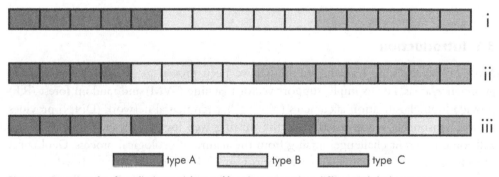

Fig. 3.1 *An example of predictions with equal local accuracy but different global accuracy.*

extracted from adjoining measurement locations. The output of each memory cell is a integrated result of that former memories and the current input. The RNN model is specially designed for sequential data processing. Hence, RNN provides prediction with a great local and global accuracy by taking the data coupling into account. We demonstrate the applicability of traditional DNN and Bi-LSTM (a type of RNN) for well-log-based facies classification in the Wolfcamp Formation in the Permian Basin, United States. Our results show that the Bi-LSTM model has higher F-measure scores, global accuracy, and less loss on the test dataset, comapred to traditional DNN.

3.2 Data types

In many geological studies such as sedimentology, basin analysis, plate tectonics, hydrocarbon migration, and reservoir simulation, the subsurface of the research area is investigated as a whole system, which has spatio-temporal dependencies. Hence, we prefer to study the subsurface system in the three-dimensional (3D) space. In 3D geologic models, we separate the continuous local subsurface space into a discrete regularly-spaced 3D corner-point grids. Our target is to obtain interesting geological attributes for all grids by inversion methods based on limited observations and the full covered indirect measurements.

In different geologic times, different sediments deposit as well as different geologic structures form. It results in the formation of different geological features. Moreover, most of the geological features show lateral continuity as well as vertical continuity[52]. For example, the flood plain clays of meandering river systems generally have a wide extent along the plain[2,49]. In simple terms, there are structural and stratigraphic similarities amongst the adjacent rocks deposits, both in lateral and vertical directions. So, to mimic the geological reality, machine learning algorithms must honor the spatio-temporal correlation. Further, it is important to identify the type and the corresponding characteristics of the measurement data to make utmost use of relevant deep learning methodologies[32].

Based on the data structure and dependency types, two kinds of data sets are widely used in hydrocarbon exploration and production[28]. They are chronological and spatial data sets. We introduce both of these data sets in this sections to improve the readability of the following methodology part.

3.2.1 Chronological data sets

A time series is a sequence of observation sets indexed in time order. In general, the measurements in a series are taken at equally spaced time intervals[19]. For example, a seismic trace is a time series indexed by travel time, with a regular sampling interval. Hence a 3D seismic volume can be regarded as a set of 1D time series arranged in spatial order. Well logging data is a special case of time series or depth series. The logging curves are collected at regular sampling rate but indexed by depth rather than time[28]. Although both the seismic traces and the well logs must contain information about their fixed surface locations (X and Y, or inline and crossline), they are indexed by the third domain (Z), namely time or depth. Such time and depth series data are commonly used in subsurface studies. However, the time series actually can be indexed by any domain. For example, a sequence of a certain seismic attribute with the same travel time and crossline index can be indexed by inlines. As we mentioned before, the subsurface sediments exert a distinct sequential pattern in certain directions. Hence there are natural order of dependencies among the steps of observations.

3.2.2 Spatial data sets

Different from the chronological data that represent 1D profiles like boreholes and seismic traces, the measurements of geological and geophysical attributes obtained over the 2D and 3D spaces are characterized as spatial data[15,28,46]. In geologic studies, the most common spatial data is the seismic volumes, satellite data, and fiber-optic data along horizontal wells. Because the geological events exist at a continuous spatio-temporal field, the observations collected from the discrete grids exert stronger spatial connectivity.

3.3 Recurrent neural network (RNN) methods

There are two ways to guide the model, so it takes the spatio-temporal dependencies into account. One is taking the order index as a input measurement of the data set[10]. Another one is representing the sequential information implicitly by using special model architectures[16]. Neural networks with memory are framed in the latter method. The spatio-temporal correlations impact on data processing by considering the filtered information from the previous measurements[9,41]. The data processing system is then equipped with the dynamic properties to reflect the sequential variations. The neural networks with sequential memory increases the efficiency of the method for temporal data analysis[30]. RNNs have a recurrent connection layer in the network to remember the hidden information extracted from the preceding sequential steps and deliver it to the current step. The nodes of this connection are memory cells which determine the memory capability and the memory content[20]. A neural network which contain one or more recurrent connection layers is known as a recurrent neural network[6]. Various RNN models have been developed in recent years with their unique memory cells inside to take the spatio-temporal dependencies into account during data processing[6,11,16,26,30].

In this section, we first describe the frequently-used data structures in geoscience studies and define the corresponding notations. Then we introduce different RNN models with their particular memory cells or network structures.

3.3.1 Notation definition

The special data structures of trainable input sets are required by most of the RNN models[1,7]. To exhibit the mathematical expressions more clearly and avoid the confusions caused by content description. We define the unified notations for study objects and required data in this section.

As we mentioned in Section 3.2, the 3D subsurface space is separated into grids. Hence, the data set for each attribute is always displayed as a 3D matrix in traditional geoscience researches. For example, the numerical attribute volume like seismic amplitude and the categorical attribute volume like lithofacies. There are three kinds of sequence structures can be used to describe this 3D matrix:

- A set of 1D vectors which along a certain domain (Fig. 3.2). The value of each step in vectors is a constant for one attribute.
- A set of sequential 2D matrices. They are indexed along a certain domain (Fig. 3.3). Each step inside is a 2D matrix for one attribute.
- A set of sequential 3D matrices. They are indexed along a certain domain (Fig. 3.4). Each step inside is a little 3D matrix for one attribute.

In traditional convolutional neural network, researchers take the target value of the centre grid of each 2D and 3D matrix as the label for the whole matrix[51,53,54].

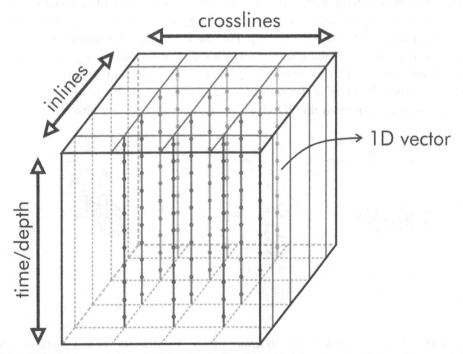

Fig. 3.2 *The 1ˢᵗ schematic diagram for a 3D seismic volume.* The data structure that can be regarded as a set of 1D vectors (in z or time/depth domain) spread laterally (in x, y or latitude, longitude domain).

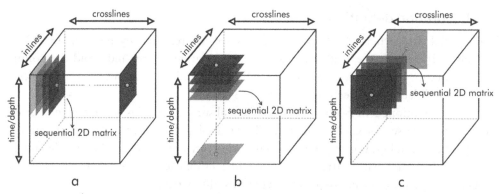

Fig. 3.3 *The 2ⁿᵈ schematic diagram for a 3D seismic volume.* The data structure that can be regarded as: (A) a set of sequential 2D matrices (or vertical crossline sections) with the two dimensions being inline and time; (B) a set of sequential 2D matrices (or time or depth slices) with the two dimensions being inline and crossline; (C) a set of sequential 2D matrices (or vertical inline sections) with the two dimensions being cross line and time.

Hence, independent of the data types, the labels and predictions are 1D vectors with the interest attribute value inside. The 1D profile along a certain domain is meshed into $\mathcal{T} = \{1, \ldots, T\}, t \in \mathcal{T}$ is the step index. At each t, we can get an observation set according to the available data set. For example, in the studies based on well log data, the observation vector $\mathbf{d}_t = (d_{t,1}, \ldots, d_{t,n})$ consists of n well logs at each step $t \in \mathcal{T}$. The complete observation set is $d = \{\mathbf{d}_t; t = 1, \ldots, T\}$. For the cases with 2D seismic profiles, the observation set of one attribute for each step $t \in \mathcal{T}$ is a 2D matrix $\mathbf{d}_{t;2d} = (d_{t;i,j}) \in \mathbb{R}^{k \times l}$ where k and l are the matrix ranges along the first and the second domain. Both k and l are smaller than the range of the whole seismic profile. For the cases with available 3D seismic volumes, apart from the 2D observation, there also is a 3D observation matrix

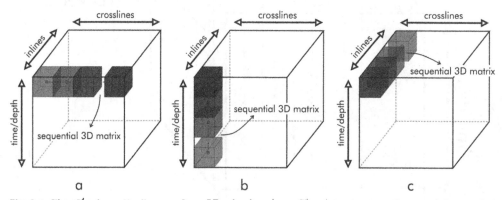

Fig. 3.4 *The 3ʳᵈ schematic diagram for a 3D seismic volume.* The data structure that can be regarded as: (A) a set of sequential 3D matrices along the crossline domain; (B) a set of sequential 3D matrices along the time/depth domain; (C) a set of sequential 3D matrices along the inline domain.

for each attribute $\mathbf{d}_{t;3d} = (d_{t;i,j,z}) \in \mathbb{R}^{k \times l \times m}$ at $t \in \mathcal{T}$, where m is the matrix range along the third domain.

At each grid of the 1D profile described above, we assign the attribute of interest φ_t. Hence the target of the geoscience research, such as property estimation (e.g., porosity and permeability) and category classification (e.g., facies and fault) is to assess the complete attribute profile denoted by $\boldsymbol{\varphi} : \{\varphi_t; t = 1,...,T\}$ given the observation set $\mathbf{d}$, i.e. $[\boldsymbol{\varphi} \mid \mathbf{d} \mid]$.

3.3.2 Simple recurrent neural network (simple RNN)

Simple RNN was the first proposed RNN to be implemented[16,30]. The simple RNN models employ a simple memory cell to remember the captured hidden state of the last step in the sequence. This hidden state represents the order dependencies in previous steps implicitly. Combined with the current input, the hidden state from last step participates in the calculation of the hidden state at the present step. Hence compared to the classic DNN, the sequential order is crucial for RNN models.

The detailed calculation of each recurrent layer $l = (1,...,L)$ of the simple RNN model is expressed as

$$\mathbf{h}_t^l = \sigma_h(\mathbf{W}_h^l \cdot \mathbf{y}_t^{l-1} + \mathbf{U}^l \cdot \mathbf{y}_{t-1}^l + \mathbf{b}_h^l)$$

$$\mathbf{y}_t^l = \sigma_y(\mathbf{W}_y^l \cdot \mathbf{h}_t^l + \mathbf{b}_y^l)$$

where $\mathbf{y}_t^l$ is a vector of n_l length, computed at step t for the l-th recurrent layer. n_l represents the output dimension of the recurrent layer l. As the initial conditions, $\mathbf{y}_t^0$ is the input observations, $\mathbf{d}_t$ and $\mathbf{y}_0^l$ are zero vectors with n_l length. $\mathbf{h}_t^l$ is an n_l-vector indicates the memory cell of layer l at step t. In layer l, the memory cell integrates both the input of step t (denoted as $\mathbf{y}_t^{l-1}$) and the captured hidden state of step $t - 1$ (denoted as $\mathbf{y}_{t-1}^l$) to get computed state at step t as $\mathbf{y}_t^l$. The unknown model parameters $\mathbf{W}_y^l$ and $\mathbf{U}^l$ are $(n_l \times n_l)$-matrices, $\mathbf{W}_h^l$ is a $(n_l \times n_{l-1})$-matrix while $\mathbf{b}_h^l$ and $\mathbf{b}_y^l$ are n_l-bias vectors. σ_h and σ_y are the activation functions to enhance the nonlinear representation of each layer (Fig. 3.5).

The output of the last recurrent layer L is an artificial state that implicitly represents the unidirectional sequential dependency in the observations. It is then processed by a

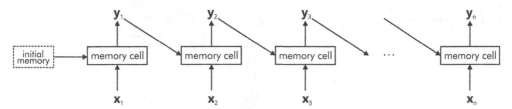

Fig. 3.5 *The schematic diagram for a simple RNN layer.*

following classifier or regressor to obtain the predictions. The DNN model is the most commonly used method[13,22,44,45]. The final state vector $\mathbf{y}^L$ is then processed through DNN layers $l_d = 1,\ldots,L_d$ of dimension n_{l_d} as

$$\mathbf{z}_t^{l_d} = \sigma_d(\mathbf{W}_d^{l_d} \cdot \mathbf{z}_t^{l_d-1} + \mathbf{b}_d^{l_d})$$

$\mathbf{z}_t^{l_d}$ is the n_{l_d}-middle state vector. Its initial condition $\mathbf{z}_t^0 = \mathbf{y}_t^{L_r}$. The unknown model parameters here are $\mathbf{W}_d^{l_d}$, a $(n_{l_d} \times n_{l_d-1})$-weight matrix and $\mathbf{b}_d^{l_d}$, an n_{l_d}-bias vector. σ_{l_d} is the activation function. Relu[35] is a general choice for the layer $l_d = \{1,\ldots,L_d-1\}$. For classification, the final prediction is

$$\hat{\varphi}_t = \underset{\mathbf{W},\mathbf{U},\mathbf{b}}{\mathrm{argmax}}\{\mathbf{z}_t^{L_d}\}$$

where σ_{L_d} is the softmax function. For regression, the final prediction is

$$\hat{\varphi}_t = \mathbf{z}_t^{L_d}$$

where σ_{L_d} is the linear function with unit weights.

Different from the traditional back-propagation algorithm employed by the DNN model, for the RNN model an optimization method known as the back-propagation through time (BPTT) is implemented to calculate the gradients.

Let us consider a simple RNN model with just one recurrent layer inside (Fig. 3.6). The loss function for such model based on just one training sample is defined as

$$E(\varphi,\hat{\varphi}) = \sum_t E_t(\varphi_t,\hat{\varphi}_t)$$

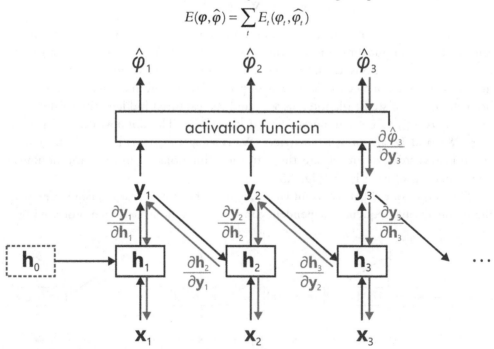

Fig. 3.6 *The schematic diagram for a brief simple RNN model with just one recurrent layer inside.*

This RNN optimization problem can be solved by the optimizer based on gradient descent methods through partial derivatives. Notice that, the computation of $\mathbf{W}$ estimation here is similar to that in the basic gradient descent method. Hence the evaluation of $\mathbf{U}$ is the key point of the BPTT method. The gradient of E is

$$\frac{\partial E}{\partial \mathbf{U}} = \sum_t \frac{\partial E_t}{\partial \mathbf{U}}$$

According to the chain rule, the gradient at step i is

$$\frac{\partial E_i}{\partial \mathbf{U}} = \frac{\partial E_i}{\partial \hat{\varphi}_i} \cdot \frac{\partial \hat{\varphi}_i}{\partial \mathbf{h}_i} \cdot \frac{\partial \mathbf{h}_i}{\partial \mathbf{U}}$$

$\mathbf{h}i$ depends on $\mathbf{y}_{i-1}$, whereas $\mathbf{y}_{i-1}$ depends on $\mathbf{y}_{i-2}$, and so on. Then we have

$$\frac{\partial E_i}{\partial \mathbf{U}} = \sum_{t=1}^{i} \frac{\partial E_i}{\partial \hat{\varphi}_i} \cdot \frac{\partial \hat{\varphi}_i}{\partial \mathbf{h}_i} \cdot \frac{\partial \mathbf{h}_i}{\partial \mathbf{y}_t} \cdot \frac{\partial \mathbf{y}_t}{\partial \mathbf{h}_t} \cdot \frac{\partial \mathbf{h}_t}{\partial \mathbf{U}}$$

We can expand $\dfrac{\partial \mathbf{h}_i}{\partial \mathbf{y}_t}$ at $t = 1,\ldots,i$ using chain rule also

$$\frac{\partial \mathbf{h}_i}{\partial \mathbf{y}_1} = \frac{\partial \mathbf{h}_i}{\partial \mathbf{y}_{i-1}} \cdot \frac{\partial \mathbf{y}_{i-1}}{\partial \mathbf{h}_{i-1}} \cdot \frac{\partial \mathbf{h}_{i-1}}{\partial \mathbf{y}_{i-2}} \cdots \frac{\partial \mathbf{h}_2}{\partial \mathbf{y}_1}$$

We can rewrite the gradient at step i as

$$\frac{\partial E_i}{\partial \mathbf{U}} = \sum_{t=1}^{i} \frac{\partial E_i}{\partial \hat{\varphi}_i} \cdot \frac{\partial \hat{\varphi}_i}{\partial \mathbf{h}_i} \cdot \left(\prod_{j=t+1}^{i} \frac{\partial \mathbf{h}_j}{\partial \mathbf{y}_{j-1}} \cdot \frac{\partial \mathbf{y}_{j-1}}{\partial \mathbf{h}_{j-1}} \right) \cdot \frac{\partial \mathbf{h}_1}{\partial \mathbf{U}}$$

The update of $\mathbf{U}$ is based on its gradient,

$$\mathbf{U} \leftarrow \mathbf{U} - \eta \frac{\partial E}{\partial \mathbf{U}}$$

where η is the learning rate.

Fig. 3.6 is the schematic diagram for the brief example above to help understand the BPTT method more clearly. The activation functions that frequently used in RNN model are tanh and sigmoid. Fig. 3.7 displays the curves of 1D tanh and sigmoid with their derivatives. The derivatives approach a flat line at both ends. When the gradient falls into this part, it will get a value close to zero and then drives the gradients in previous steps toward zero. As a consequence, a small gradient will lead a exponentially fast shrinking in previous gradients. Which in simple terms mean, the inputs at previously far away steps contributes significantly less to the output of the current step, as compared the just previous step. Such that, contribution from the steps much further before would be negligible. This suggests that, there is a vanishing gradient problem, which may restrict the long-range memory of the simple RNN model[24,25]. The model may ends up without learning long-term dependencies[3]. Therefore, we need more advanced versions of memory cell to deal with this problem and efficiently learning the dependencies in large span.

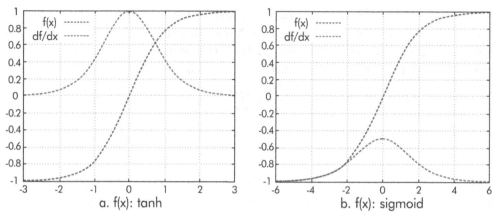

Fig. 3.7 *The schematic diagram for 1D tanh and sigmoid with their derivatives.*

3.3.3 Long short-term memory (LSTM)

We described the improvements of the RNN framework and also mentioned the disadvantages of the simple RNN model in the previous section. However, in geosciences, the long-term dependencies prevail in observation sequences[36]. For example, in meandering stream deposit, the oscillating channels generate a set of interbedding of coarse grained channel sands and fine-grained sediments. The well logs from the boreholes which drill through the interbedded deposits show obvious long-term dependencies because of the semblable characteristics of similar sediments (Fig. 3.8). The long-term dependencies assist models capture the special data features and sequential correlations hidden behind the observations of the similar sediments deposited at different locations and improve the model performance based on the captured patterns[44]. Hence we need RNN models equipped with the long-term memory and the capacity for regulating the memory impact.

The LSTM model proposed by[26] added a more sophisticated memory cell with several gate units to calculate the memory contents and control the corresponding impacts of those memory contents. The most common memory cell of each LSTM layer $l_m = \{1,\ldots,L_m\}$ is defined as

$$\tilde{\mathbf{c}}_t^{l_m} = \sigma_{\tanh}(\mathbf{W}_c^{l_m} \cdot [\mathbf{a}_{t-1}^{l_m}, \mathbf{a}_t^{l_m-1}] + \mathbf{b}_c^{l_m})$$

$$\Gamma_u^{l_m} = \sigma_{\text{sigmoid}}(\mathbf{W}_u^{l_m} \cdot [\mathbf{a}_{t-1}^{l_m}, \mathbf{a}_t^{l_m-1}] + \mathbf{b}_u^{l_m})$$

$$\Gamma_f^{l_m} = \sigma_{\text{sigmoid}}(\mathbf{W}_f^{l_m} \cdot [\mathbf{a}_{t-1}^{l_m}, \mathbf{a}_t^{l_m-1}] + \mathbf{b}_f^{l_m})$$

$$\Gamma_o^{l_m} = \sigma_{\text{sigmoid}}(\mathbf{W}_o^{l_m} \cdot [\mathbf{a}_{t-1}^{l_m}, \mathbf{a}_t^{l_m-1}] + \mathbf{b}_o^{l_m})$$

$$\mathbf{c}_t^{l_m} = \Gamma_u^{l_m} \otimes \tilde{\mathbf{c}}_t^{l_m} + \Gamma_f^{l_m} \otimes \mathbf{c}_{t-1}^{l_m}$$

$$\mathbf{a}_t^{l_m} = \Gamma_o^{l_m} \otimes \sigma_{\tanh}(\mathbf{c}_t^{l_m})$$

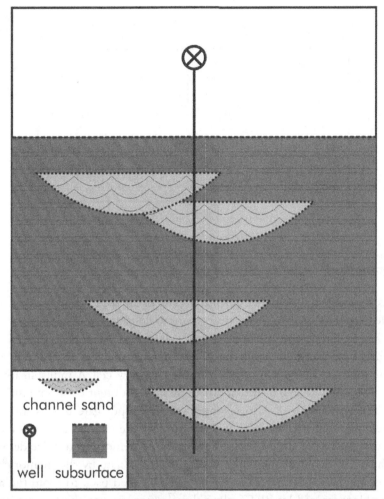

Fig. 3.8 *A well drilled through a fluvial deposit.* The well path meets several channel sands which have similar well logging responses. The observation characteristics of one channel sands summarized by the RNN model can be used in the classification for another channel sands far away.

where the operator $\otimes$ denotes the element-wise product. $\tilde{\mathbf{c}}_t^{l_m}$ is a n_{l_m}-vector indicates the candidate memory content at step t of the l_m-th LSTM layer. $\mathbf{c}_t^{l_m}$ is the n_{l_m}-current memory vector at step t of the l_m-th layer. Likewise $\mathbf{a}_t^{l_m}$ is a n_{l_m}-captured state vector at step t of the l_m-th layer. $\mathbf{a}_{t-1}^{l_m}$ is the n_{l_m}-captured state vector taken from the previous step $t - 1$. $\mathbf{a}_t^{l_m-1}$ is the n_{l_m-1}-input vector fed by last layer $l_m - 1$. The initial conditions are $\mathbf{a}_0^l = 0\mathbf{i}_{n_l}$ and $\mathbf{a}_t^0 = \mathbf{b}_t$. $\Gamma_u^{l_m}$, $\Gamma_f^{l_m}$ and $\Gamma_o^{l_m}$ are the nonlinear gate units respectively named update gate, forget gate and output gate. All of them are n_{l_m}-vector. Then unknown model parameters $\mathbf{W}_c^{l_m}$, $\mathbf{W}_u^{l_m}$, $\mathbf{W}_f^{l_m}$ and $\mathbf{W}_o^{l_m}$ are $(n_{l_m} \times n_{l_m-1})$-weight matrices, while $\mathbf{b}_c^{l_m}$, $\mathbf{b}_u^{l_m}$, $\mathbf{b}_f^{l_m}$ and $\mathbf{b}_o^{l_m}$ are all n_{l_m}-bias vectors.

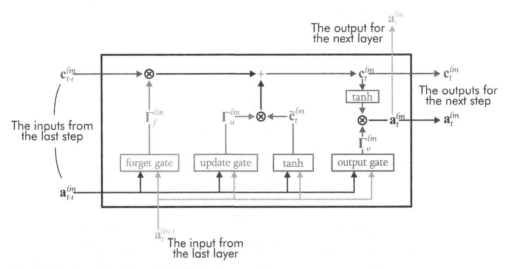

Fig. 3.9 *The data flow through the LSTM memory cell.*

Same as the simple RNN model, the final outputs of the last LSTM layer, denoted by $\mathbf{a}^{L_m}$, should be fed into a following classifier or regressor to do pattern identification or regression.

Fig. 3.9 is an illustration of the memory cell of LSTM models. There is a multiplicative forget gate unit to avoid the disturbances caused by the current input for the existing memory content $\mathbf{c}_{t-1}^{l_m}$ stored in the memory cell. A multiplicative update gate unit to decide how much the candidate memory content $\tilde{\mathbf{c}}_t^{l_m}$ should influence the current memory content $\mathbf{c}_t^{l_m}$. Also, a multiplicative output gate unit is introduced to protect the computed state $\mathbf{a}_t^{l_m}$ from perturbation by the current memory content $\mathbf{c}_t^{l_m}$.

In this way, the LSTM model keeps the long-term memory. Also, it decides when and how much automatic effect should be on the captured state by the memory[21]. Hence, the vanished memory from far away steps can be bridged back to avoid the abnormal gradient problems caused by BPTT algorithm[27]. According to the specific studies, the LSTM model can learn to bridge minimal step lags in excess of 1,000 steps[18,26]. That is a huge progress compare with the standard RNN model whose minimal step lags are just 3 to 5 steps[18].

3.3.4 Gated recurrent unit (GRU)

While the LSTM model improves learning in time series processing, it introduces an increase in parameterization through the gate units[11]. Consequently, there is an added restriction on the application of the LSTM model especially in the cases with small data set.

The GRU model is proposed to reduce the external gate units while holding the long-term memory as well[6]. The GRU memory cell of each recurrent layer $l_g = \{1,\ldots,L_g\}$ is defined as

$$\tilde{\mathbf{c}}_t^{l_g} = \sigma_{\text{tanh}} \left(\mathbf{W}_c^{l_g} \cdot \left[\Gamma_r^{l_g} \otimes \mathbf{c}_{t-1}^{l_g}, \mathbf{c}_t^{l_g-1} \right] + \mathbf{b}_c^{l_g} \right)$$

$$\Gamma_u^{l_g} = \sigma_{\text{sigmoid}}\left(\mathbf{W}_u^{l_g} \cdot \left[\mathbf{c}_{t-1}^{l_g}, \mathbf{c}_t^{l_g-1}\right] + \mathbf{b}_u^{l_g}\right)$$

$$\Gamma_r^{l_g} = \sigma_{\text{sigmoid}}\left(\mathbf{W}_r^{l_g} \cdot \left[\mathbf{c}_{t-1}^{\mathbf{l_g}}, \mathbf{c}_t^{l_g-1}\right] + \mathbf{b}_r^{l_g}\right)$$

$$\mathbf{c}_t^{l_g} = \Gamma_u^{l_g} \otimes \tilde{\mathbf{c}}_t^{l_g} + \left(\mathbf{i}_{n_{l_g}} - \Gamma_u^{l_g}\right) \otimes \mathbf{c}_{t-1}^{l_g}$$

where $\tilde{\mathbf{c}}_t^{l_g}$ is a n_{l_g}-candidate memory vector for step t of the l_g-th GRU layer. $\mathbf{c}_{t-1}^{l_g}$ is the n_{l_g}-existing memory vector delivered from step $t-1$. $\mathbf{c}_t^{l_g-1}$ is a n_{l_g-1}-input vector provided by layer $l_g - 1$. The corresponding initial conditions are $\mathbf{c}_0^{l_g} = 0\mathbf{i}_{n_{l_g}}$ and $\mathbf{c}_t^0 = \mathbf{d}_t$. $\Gamma_u^{l_g}$ and $\Gamma_r^{l_g}$ are the update gate unit and the reset gate unit. Both of them are n_{l_g}-vectors. The unknown model parameters $\mathbf{W}_c^{l_g}$, $\mathbf{W}_u^{l_g}$ and $\mathbf{W}_r^{l_g}$ are $(n_{l_g} \times n_{l_g-1})$-weight matrices, while $\mathbf{b}_c^{l_g}$, $\mathbf{b}_u^{l_g}$ and $\mathbf{b}_r^{l_g}$ are all n_{l_g}-bias vectors. Notice that, the memory contents $\mathbf{c}^{l_g}$, $l_g = \{1,...,L_g\}$ are taken directly as the computed states to transport the sequential information inside and between layers. Compared to the LSTM memory cell, the GRU model has just two gate units that process the data without a separate memory content or a computed state. No output gate to control how much the memory contents should be exposed[8,21]. Moreover, according to the mathematical expression of $\mathbf{c}_t^{l_g}$, the current memory content is a linear sum of the existing memory and the candidate memory (see Fig. 3.10). Then the identical algorithms for classification or regression are following to process the final output $\mathbf{c}^{L_g}$.

So far, many research publications verify that the GRU model exhibits comparable performance to the LSTM model[8,23]. The model structure is probably the more

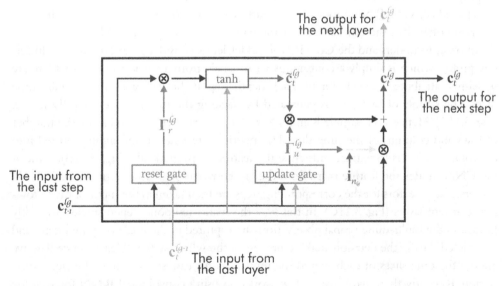

Fig. 3.10 *The data flow through the GRU memory cell.*

important factor for prediction than the architecture of memory cells. With certain smaller data sets, the GRU model has been proved to achieve better performance than the LSTM model because of the less parameter complexity[31].

3.3.5 Convolutional recurrent neural network (ConvRNN)

Up to now, the models introduced in this chapter have a focus on the cases with the observation set at each step as a sequential 1D vector like well logging measurements. In other words, the model just takes the 1D dependencies into account during the processing. However, for the cases with multi-dimensional observations, we need to incorporate parameters related to spatial connectivity in our model.

As the CNN section describes, the convolution operation is an effective way to extract the spatial local correlation from the matrix-like inputs by enforcing the sparse local connectivity patterns. Various grades of the local correlation can be captured by stacking multiple convolutional layers. The outputs from the last convolutional layer are regarded as the states that represent the global spatial dependencies of input matrices. Implementing 2D or 3D convolution operation is a common choice for some geoscience studies such as seismic facies classification with the scattered labels. However, there is also the ordered dependencies among the observation matrices of the adjacent time/depth steps. So, for a continuous set of labels, for example, 1D profiles along well paths, the advanced methods which employ the spatio-temporal dependencies are needed to improve the prediction.

One of the effective methods is combining the convolution operation with the RNN model (ConvRNN) to process the multidimensional inputs. The primary algorithms employed in the ConvRNN model are all described in the contents above. Hence to help understand the ConvRNN model more clearly, we prefer to introduce it with a schematic diagram rather than the burdensome mathematical expressions (Fig. 3.11).

In order to understand the ConvRNN model, let us consider an interpreted 1D lithology profile, which is simply a continuous set of labels, from a vertical well. Also, there are n_p seismic attributes available for us. Each depth step at the well get n_p rectangular time slices. The sample of each step is generated by stacking the n_p time slices as a 3D matrix (Fig. 3.11A). Hence, each sample is an $(n_r \times n_c \times n_p)$-matrix, while n_r and n_c are the number of rows and columns of the time slice. The matrices are taken as the inputs and fed into convolutional layers to extract and refine the feature maps gradually (Fig, 3.11B). Same as the CNN model, the feature maps from the last convolutional layer are flattened into 1D vectors. These vectors for the corresponding steps are then taken as the input of the following recurrent layers (Fig. 3.11C). In this way, the spatio-temporal dependencies implicitly hidden in the multi-dimensional observations are captured in the final computed states and considered during the decision-making process of the whole system. Please notice that, we choose the time slices of each step as the input data for case study here and easier visualization. Recently there have been a few works on using convolutioal RNN for well log

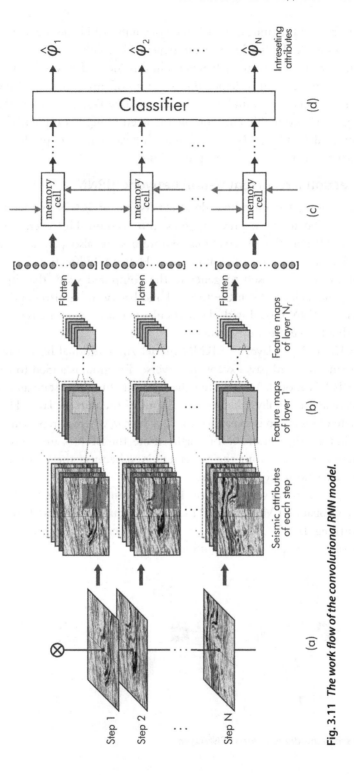

Fig. 3.11 *The work flow of the convolutional RNN model.*

interpretation from 3D seismic, in which the convolutional blocks are applied on seismic data as 3D cubes while the RNN for depth dependency, see[12].

However, there are still some deficiencies in this method. Several 3D matrices need to be processed simultaneously, hence the structure of this ConvRNN model is more complex with the very large number of parameters (see Fig. 3.11). ConvRNN's limitations include, costly and inefficient training, the need of high computational power and large amount of data. ConvRNN training with small data sets produce complicated models, which result in unsatisfactory predictions[56].

3.3.6 Bidirectional recurrent neural network (BRNN)

In previous sections, we introduce diverse versions of RNN model that take the sequential dependencies from previous steps into account. However, the dependency information hidden in the observations of future steps also play an equally important role in the data processing for time series. Hence a bidirectional recurrent neural network (BRNN) is proposed to generate the computed states that incorporate the sequential dependencies in both the forward and backward recurrences[40]. Notice that, the BRNN is an RNN model with the particular network structure but not a new type of RNN with advanced memory cells.

As Fig. 3.12 displays a layer of BRNN model. An additional backward pass, identical to the normal forward pass except the inverse direction, is added to each recurrent layer of the BRNN model. At each step, the forward computed state and the backward computed state are integrally taken as the output of the current layer. Hence, the final computed state of each step is expanded to a $2n_l$-vector, where n_l represents the memory cell number in the l-th layer of RNN. Furthermore, the BRNN model is trainable with the optimization method from the general RNN because there are no interactions between the states from these two passes.

Nowadays, with the development of the computing hardware, the BRNN models like the bidirectional LSTM model and the bidirectional GRU model are more popular in studies with big data[44,50]. The long-term dependencies obtained from two directions assist in the long range pattern recognition.

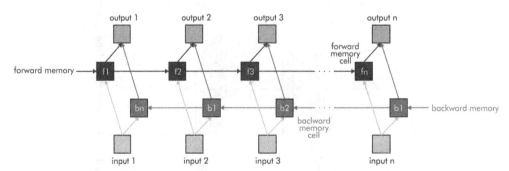

Fig. 3.12 *The schematic diagram for a BRNN layer.*

3.4 Case study: Bi-LSTM-assisted facies classification based on well logging data

In this section, we show a case study of the facies classification for 1D profile based on well logging data with the help of a bidirectional LSTM (Bi-LSTM) model. The facies prediction supplied by a DNN model with the same network architecture is taken as the benchmark in this case. According to the comparison, the Bi-LSTM model which takes the 1D spatial dependency into account provides a more accurate prediction.

3.4.1 Geological setting

The Midland Basin is a sub-basin of the well known Permian Basin in west Texas, United States. For our study, the target formation is Wolfcamp shale of Midland Basin (Fig. 3.13). The Wolfcamp shale was deposited from late Pennsylvanian to early Permian period known as Wolfcampian in a deep water basin surrounded by carbonate platform.

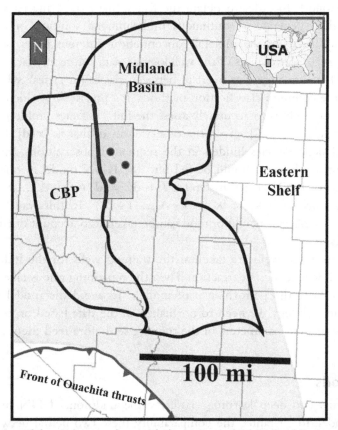

Fig. 3.13 *Map view of the study area along with the outlines of the Midland Basin and Central Basin Platform (CBP).* The study area is displayed with a rectangle filled with yellow. The location of the wells is shown as blue dots.[47]

The Wolfcamp shale consists of organic rich shale and argillaceous carbonates and has been producing hydrocarbons for many years[14,42,47].

3.4.2 Problem definition and notation

In this case study, we used gamma-ray (GR), bulk density (RHOB), neutron porosity (PHIN), and deep resistivity (RESD) logs from three vertical wells in the Wolfcamp Formation (Fig. 3.14). The sampling rate of the well logs was 0.5 foot (~0.154 m). These wells were approximately 1 km apart. The details about these curves are shown in Table 3.1. All the observations are measured along 1D well paths which are respectively discretized to $T = \{1,...,T\}$. Normally, we express the observation set of each $t \in T$ as a 1D vector $\mathbf{d}_t = \{d_{t,1},...,d_{t,4}\}$. We can represent the complete set of well logs as $\mathbf{d} = \{\mathbf{d}_t; t = 1,...,T\}$. At each depth step of the three wells, we assign one of the four facies $\kappa_t \in \Omega_\kappa$: {Class 1 (C1, Organic rich siliceous mudstone), Class 2 (C2, Organic rich mudstone), Class 3 (C3, Organic rich calaereous mudstone), Class 4 (C4, Calcareous siliceous mudstone)}. In order to obtain the facies, first a joint inversion of well logs was performed to generate a continuous multi-mineral solution of the Wolfcamp section. Second, based on the mineral concentration different facies were identified (Table 3.2). Pairwise scatter plots of the well logging data colored by facies are displayed in Fig. 3.15. There are large data overlaps between classes (or facies), which presents a challenge for the common classification methods in a precise facies identification. The target of this case study is to accurately assess the full 1D facies profile represented by the vector $\kappa : \{\kappa_t; t = 1,...,T\}$ given the observations $\mathbf{d}$, that is, $[\kappa \mid \mathbf{d}]$. To capture the implicit spatial dependencies hidden in the sequential observations, the observation sets of adjacent depth steps should be fed into the RNN model as one sample. The observation set for RNN at depth step $t \in T$ is $\mathbf{d}_t^{\Delta_t} = \{\mathbf{d}_t; t = t - \Delta_t,...,t + \Delta_t\}$ and the corresponding label set is $\kappa_t^{\Delta_t} = \{\kappa_t; t = t - \Delta_t,...,t + \Delta_t\}$. In this way, the 1D spatial dependencies in a $2\Delta_t + 1$ neighborhood can be taken into account during the model learning process.

Two of these three wells are taken as the training wells for the hidden information and spatial dependencies extraction. Then the remaining one is taken as the blind test well to evaluate the performance of models. To avoid the model over-reliance on some of the well logs, we need to normalize all the data based on the scaling and shifting transformation parameters of the training set before feed them into classifiers (Fig. 3.16).

3.4.3 Methods

Here we discuss two deep learning classifiers: the traditional DNN model and the Bi-LSTM model. To guarantee the comparability, these two neural network classifiers employ the same model architecture. The DNN model is based on a 64-128-32-4 general DNN layer frame. Whereas the Bi-LSTM model consists of 32-64 bidirectional

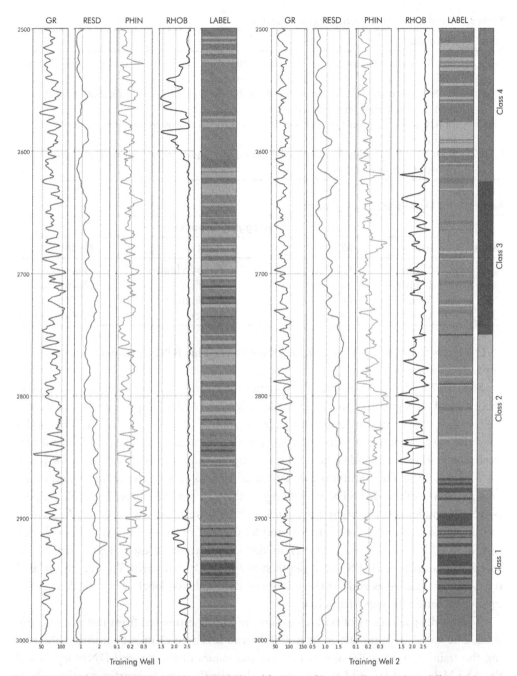

Fig. 3.14 *Well logs (GR, RESD, PHIN, and RHOB) and facies profiles in wells.* Note that, different classes are different facies in Wolfcamp Formation. The wells are hung from an unknown datum, due to confidentiality reasons.

Table 3.1 The details about the well logs in this study.

Logs	Definition
GR	GR log indicates indicator of the clay content. The facies generated in different depositional conditions exhibit various clay content.
RESD	RESD logs record the resistivity of the subsurface rocks. It is a indicator of fluid type and hydrocarbon-saturated rocks.
PHIN	PHIN is the neutron porosity log. It is sensitive to the hydrogen content in formations.
RHOB	RHOB is the density log. It measures the bulk density of the formation. Porosity can be estimated from this log.

Table 3.2 The logging responses of facies.

Facies	Well logging responses			
	GR	**RESD**	**PHIN**	**RHOB**
Class 1	Varies	Slightly higher	Slightly higher	Lower
Class 2	Higher	Medium	Medium	Slightly lower
Class 3	Varies	High	Medium	Higher
Class 4	Lower	Lower	Lower	Slightly higher

LSTM layers followed by 32-4 general DNN layers. This RNN model takes the spatial dependencies into account.

Figs. 3.17 and 3.18 show the dependence structures of the DNN and Bi-LSTM model. The DNN classifier provides the identification result of depth step t based on the corresponding observation set $\mathbf{d}_t$, that is, $[\hat{\kappa}_t | \mathbf{d}_t]$, $t \in \mathcal{T}$ (Fig. 3.17B). The RNN model takes the observation set $d_t^{\varnothing}$ in to consideration to generate the classification result $\hat{\kappa}_t$. Such observation set includes the corresponding and adjacent observations, as $[\hat{\kappa}_t | \mathbf{d}_t^{\Delta_t}]$, $t \in \mathcal{T}$ (Fig. 3.18B). The optimum Δ_t is estimated by maximizing the accuracy and minimizing the loss of the Bi-LSTM pretraining in this case study. As shown in Table 3.3, the Bi-LSTM model gets the highest accuracy and the lowest loss when $\Delta_t = 2$. According to this pretraining result, the Bi-LSTM model is based on $\Delta_t = 2$ neighborhood hence including $\mathbf{d}_t^{\Delta_t=2} : \{\mathbf{d}_t; t = t-2, \ldots, t+2\}$ in the classification at each depth step t. The acceptable sample for Bi-LSTM at depth step t is a (5×4)-matrix. The matching label is a 5-vector. The facies code at the middle of the predicted label vector is taken as the prediction for depth step t namely $\hat{\kappa}_t$.

The model parameters of both classifiers are jointly estimated from the two training wells. The initial learning rate is set to 0.001. And it can be changed automatically during the training process. We also assign dropout values of the general DNN layers and the LSTM layers as 0.2 and 0.1 to protect the classifiers from overfitting. The whole training process gets to stop once the total loss remains unchanged. Then the prediction result based on the data from blind test well is applied to evaluate the classification accuracy of models.

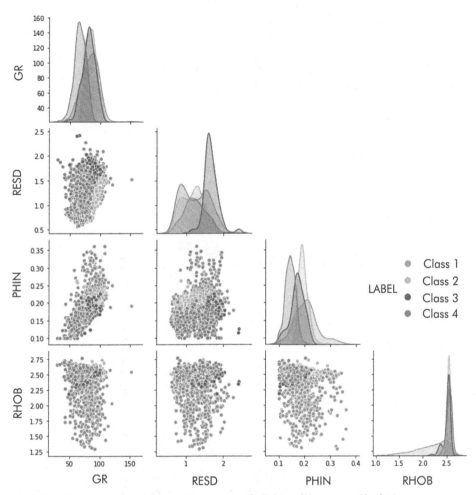

Fig. 3.15 *Pairwise scatter plots and histograms of all the well logs sorted by facies.*

3.4.4 Evaluation of the model performance

As we can see in Fig. 3.14, the training wells provide us with a category imbalanced data set. 46.2% and 29.6% of the logging interval are identified as C1 and C4. Just 16.3% and 7.9% are C2 and C3. To avoid the evaluation from over-reliance on the majority, the F-measure is employed to assess the model performance here[39,43]. For each class, we take the depth points with the corresponding class label as the positive set and others as negative set. Then we define the precision as

$$precision = \frac{\text{The number of the correctly predicted positive samples}}{\text{The number of all the predicted positive samples}}$$

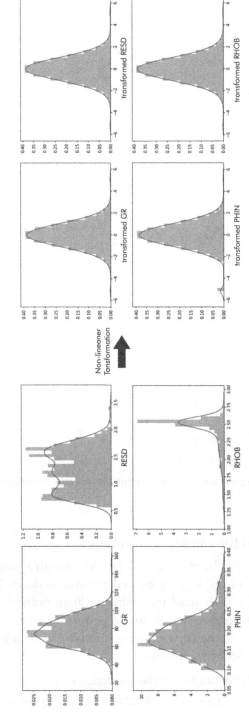

Fig. 3.16 *The test data are transformed based on the transformation parameters of the training data.*

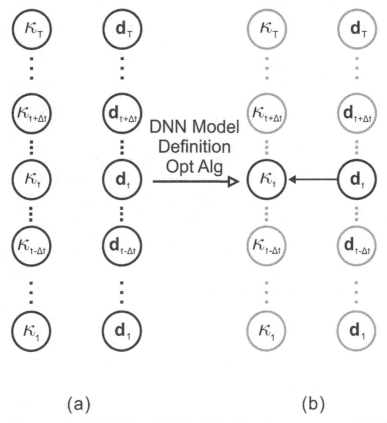

(a)   (b)

Fig. 3.17 *Dependence structure of DNN*; (A) the 1D facies profile with the corresponding well logs; (B) the prediction process of DNN.

and the recall as

$$recall = \frac{\text{The number of the correctly predicted positive samples}}{\text{The number of true positive samples}}$$

The F-measure is the harmonic mean of precision and recall:

$$F - measure = 2 \times \frac{\text{precision} \times \text{recall}}{\text{precision} + \text{recall}}$$

where F-measure $\in [0,1]$. The model with a higher F-measure is more robust for the classification problem here.

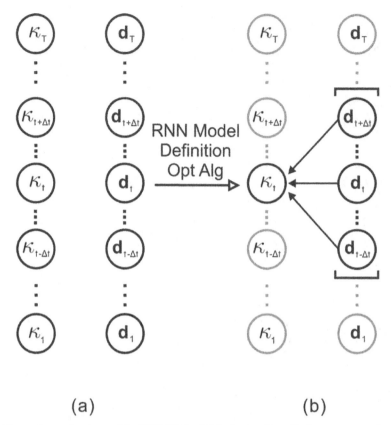

(a)   (b)

Fig. 3.18 *Dependence structure of the RNN*; (A) the 1D facies profile with the corresponding well logs; (B) the prediction process of RNN.

Table 3.3 The optimization result of Δ_t by pretraining.

	Model performance	
Δ_t	Accuracy	Loss
1	0.887	0.358
2	0.896	0.387
3	0.864	0.436
4	0.852	0.490

3.5 Results and discussion

Table 3.4 shows the model structure of the chosen Bi–LSTM and DNN classifier. The merge mode of the bidirectional LSTM layer is concat in this case. The 128–output vector of each step is constituted by 64 forward states and 64 backward states as mentioned in Section 3.3.6 to guarantee the identical model architecture of the Bi–LSTM and DNN classifier.

Table 3.4 The model structure of the Bi-LSTM and DNN classifier.

Model	Layer description	Type	Parameters	Output
Bi-LSTM	(5 × 4)-input matrix	Input layer	–	5 × 4
	32 memory cells	Bidirectional LSTM layer	Merge mode: concat	5 × 64
	64 memory cells	Bidirectional LSTM layer	Merge mode: concat	5 × 128
	32 neurons	DNN layer	Dropout=0.2	5 × 32
	4 neurons	Output layer	–	5 × 4
DNN	(1 × 4)-input vector	Input layer	–	1 × 4
	64 neurons	DNN layer	Dropout = 0.2	1 × 64
	128 neurons	DNN layer	Dropout = 0.2	1 × 128
	32 neurons	DNN layer	Dropout = 0.2	1 × 32
	4 neurons	Output layer	–	1 × 4

Fig. 3.19 display the blind test result. It contains from the left to right, the four well logs, the true 1D facies profile, the Bi-LSTM prediction and the DNN prediction. According to this result exhibition, both the classifiers provide satisfactory prediction for C1. However, the DNN model over-predicts the chaotic C2 layers inserted in C4 intervals such as 2,520–2,590 m, 2,895 m, and 2,950 m. Furthermore, the Bi-LSTM model has a better performance on the prediction of C3. Especially, the delicate facies transformations between C1 and C3 are only captured by the Bi-LSTM classifier (see 2590 m and 2930 m). Both the two models pick up the C4 very well.

Table 3.5 shows the F-measures and the global measures of the two models separately. It provides us with quantitative information to compare the model performances. According to the F-measures, the DNN model did a better job with the training set, especially in the identification of C1, C2, and C4. Combined with the lower training C3 F-measure score, we can infer that some C3 samples are identified as other classes by the DNN model. It also happened in the prediction of the test data. DNN model fails to recognize the C2 samples from the test data accurately. Although the training C1, C2 and C4 F-measures of the Bi-LSTM model are slightly lower than the DNN model, the Bi-LSTM model takes a balance between the four classes. That means the generalized patterns hidden in the training set are well captured by the Bi-LSTM model. Moreover, the test F-measures of C1 and C2 of the Bi-LSTM model are extremely close to the corresponding scores with training set. Hence, the Bi-LSTM model avoids the overfitting successfully at least for these classes. The test C4 F-measure score is anomalously higher than the training score. According to the visualized predictions (Fig. 3.19), we can attribute this to the C3 under-prediction and the corresponding C4 over-prediction of the Bi-LSTM model. In other words, the RNN model regards several C3 layers as C4 layers improperly (see 2,650 m, 2,930 m, and 2,980 m). The low test C3 F-measure score also coincides with this phenomenon. We also calculate the global accuracy and loss, which are very important coefficients to evaluate the model performance, for both models with training set and test set. The Bi-LSTM model is superior to both the

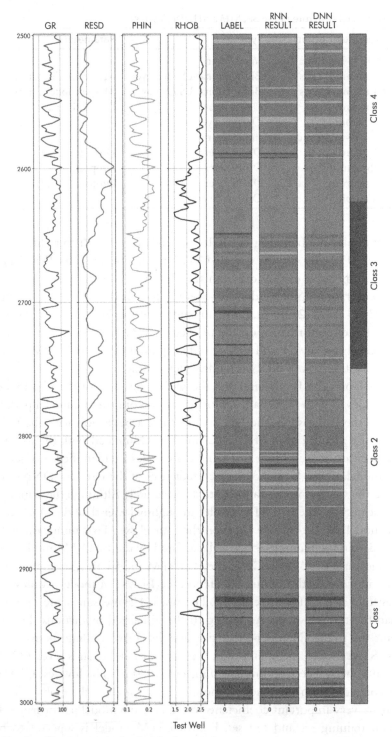

Fig. 3.19 *The classification results for the blind test well.*

Table 3.5 Model comparison by evaluation coefficients.

Model	Data set	F-measures				Global measures	
		C1	C2	C3	C4	Accuracy	Loss
DNN	Training set	0.974	0.900	0.760	0.966	0.944	0.204
	Test set	0.949	0.731	0.577	0.934	0.900	0.556
Bi-LSTM	Training set	0.969	0.884	0.907	0.959	0.948	0.202
	Test set	0.956	0.874	0.64	0.963	0.936	0.387

accuracy scores with training set and test set. In addition, it also produces lower training and test loss scores. All the presentations above indicate that the data sequence and the corresponding spatial coupling are the crucial factors for pattern recognition of time series data. The classifiers such as the Bi-LSTM model which capture and employ these two factors during the decision process are more general and robust in this study.

3.6 Conclusion

In this study, we implement two deep learning classifiers (RNN and Bi-LSTM) on a well logging data set from three real wells in the Permian Basin to classify the facies along well paths. Traditional DNN model takes each sample as an independent event. Bi-LSTM model takes the vertical spatial dependencies between adjacent samples into account. The predictions comparison of the blind well test verifies that the Bi-LSTM classifier is equipped with higher test F-measure scores, global accuracy and less loss. The visualized predicted results also provide an intuitive reference. Hence the adjoining data sequence and the corresponding spatial dependencies are apparently helpful in pattern recognition and the relevant classification. Limited by the small data set from just three wells, the results and discussion may not be generalized enough. More experiments based on big real data sets are needed for future research.

References

1. Abadi M., Agarwal A., Barham P., Brevdo E., Chen Z., Citro C., et al. TensorFlow: large-scale machine learning on heterogeneous systems. 2015. Software available from tensorflow.org. https://www.tensorflow.org/.
2. Aslan A., Autin W.J. Evolution of the holocene mississippi river floodplain, ferriday, louisiana; insights on the origin of fine-grained floodplains. *J Sediment Res* 1999;**69**(4):800–15.
3. Bengio Y., Simard P., Frasconi P. Learning long-term dependencies with gradient descent is difficult. *IEEE Trans Neural Netw* 1994;**5**(2):157–66.
4. Bhattacharya S., Carr T.R., Pal M. Comparison of supervised and unsupervised approaches for mudstone lithofacies classification: Case studies from the bakken and mahantango-marcellus shale, usa. *J Nat Gas Sci Eng* 2016;**33**:1119–33.
5. Bressan T.S., de Souza M.K., Girelli T.J., Junior F.C. Evaluation of machine learning methods for lithology classification using geophysical data. *Comput Geosci* 2020:104475.

6. Cho K., Van Merriënboer B., Gulcehre C., Bahdanau D., Bougares F., Schwenk H. Learning phrase representations using RNN encoder-decoder for statistical machine translation. *arXiv preprint arXiv:1406.1078* 2014.

7. Chollet F., et al., 2015. Keras. https://github.com/fchollet/keras.

8. Chung J., Gulcehre C., Cho K., Bengio Y. Empirical evaluation of gated recurrent neural networks on sequence modeling. *arXiv preprint arXiv:1412.3555* 2014.

9. Cleeremans A., Servan-Schreiber D., McClelland J.L. Finite state automata and simple recurrent networks. *Neural Comput* 1989;**1**(3):372–81.

10. Cracknell M.J., Reading A.M. Geological mapping using remote sensing data: A comparison of five machine learning algorithms, their response to variations in the spatial distribution of training data and the use of explicit spatial information. *Comput Geosci* 2014;**63**:22–33.

11. Dey R., Salemt F. M., 2017. Gate-variants of gated recurrent unit (GRU) neural networks. In: 2017 IEEE 60th international midwest symposium on circuits and systems (MWSCAS). IEEE, p. 1597–1600.

12. Di H., Chen X., Maniar H., Abubakar A. Semi-supervised seismic and well log integration for reservoir property estimationSEG Technical Program Expanded Abstracts 20202020Society of Exploration Geophysicists:2166–70.

13. Dorrington K.P., Link C.A. Genetic-algorithm/neural-network approach to seismic attribute selection for well-log prediction. *Geophysics* 2004;**69**(1):212–21.

14. EIA, 2020. Permian basin part 2: Wolfcamp shale play of the midland basin geology reviewhttps://www.eia.gov/maps/pdf/Permian_Wolfcamp_Midland_EIA_reportII_09092020.pdf.

15. Eidsvik J., Avseth P., Omre H., Mukerji T., Mavko G. Stochastic reservoir characterization using prestack seismic data. *Geophysics.* 2004;**69**(4):978–93.

16. Elman J.L. Finding structure in time. *Cognit Sci* 1990;**14**(2):179–211.

17. Fjeldstad T., Omre H. Bayesian inversion of convolved hidden Markov models with applications in reservoir prediction. *IEEE Trans Geosci Remote Sens* 2019;**58**(3):1957–68.

18. Gers F.A., Schmidhuber J., Cummins F., 1999. Learning to forget: continual prediction with lstm .

19. Gossel W., Laehne R. Applications of time series analysis in geosciences: an overview of methods and sample applications. *Hydrol Earth Syst Sci Discuss.* 2013;**10**(10).

20. Graves A., Liwicki M., Fernández S., Bertolami R., Bunke H., Schmidhuber J. A novel connectionist system for unconstrained handwriting recognition. *IEEE Trans Pattern Anal Mach Intell.* 2008;**31**(5):855–68.

21. Greff K., Srivastava R.K., Koutník J., Steunebrink B.R., Schmidhuber J. LSTM: a search space odyssey. *IEEE Trans Neural Netw Learn. Syst* 2016;**28**(10):2222–32.

22. Hampson D.P., Schuelke J.S., Quirein J.A. Use of multiattribute transforms to predict log properties from seismic data. *Geophysics* 2001;**66**(1):220–36.

23. Heck J.C., Salem F.M., 2017. Simplified minimal gated unit variations for recurrent neural networks. In: 2017 IEEE 60th International Midwest Symposium on Circuits and Systems (MWSCAS). IEEE, p. 1593–1596.

24. Hochreiter S. Recurrent neural net learning and vanishing gradient. *Int J Uncertainity, Fuzziness Knowledge-Based Syst* 1998;**6**(2):107–16.

25. Hochreiter S. The vanishing gradient problem during learning recurrent neural nets and problem solutions. *Int J Uncertainty, Fuzziness Knowledge-Based Syst* 1998;**6**(02):107–16.

26. Hochreiter S., Schmidhuber J. Long short-term memory. *Neural Comput* 1997;**9**(8):1735–80.

27. Hochreiter, S., Schmidhuber, J. LSTM can solve hard long time lag problems. In: Proceedings of the 9th International Conference on Neural Information Processing Systems. MIT, 1996:473–479.

28. Holdaway K.R., Irving D.H. Enhance Oil and Gas Exploration with Data-Driven Geophysical and Petrophysical Models2017John Wiley & Sons.

29. Horrocks T., Holden E.-J., Wedge D. Evaluation of automated lithology classification architectures using highly-sampled wireline logs for coal exploration. *Comput Geosci* 2015;**83**:209–18.

30. Jordan M.I. Serial order: a parallel distributed processing approach, Advances in Psychology, Vol. 121. Elsevier; 1997. p. 471–95.

31. Jozefowicz R., Zaremba W., Sutskever I. An empirical exploration of recurrent network architectures. In: International conference on machine learning; 2015. p. 2342–2350.

32. Karpatne A., Ebert-Uphoff I., Ravela S., Babaie H.A., Kumar V. Machine learning for the geosciences: Challenges and opportunities. *IEEE Trans Knowl Data Eng* 2018;**31**(8):1544–54.
33. Lindberg D.V., Rimstad E., Omre H. Inversion of well logs into facies accounting for spatial dependencies and convolution effects. *J Pet Sci Eng* 2015;**134**:237–46.
34. Lyu B., Qi J., Li F., Hu Y., Zhao T., Verma S. Multispectral coherence: which decomposition should we use? *Interpretation* 2020;**8**(1):T115–29.
35. Nair V., Hinton G. E., 2010. Rectified linear units improve restricted Boltzmann machines. In: ICML.
36. Nichols G. Sedimentology and Stratigraphy. John Wiley & Sons; 2009.
37. Raef A., Meek T., Totten M. Applications of 3d seismic attribute analysis in hydrocarbon prospect identification and evaluation: verification and validation based on fluvial palaeochannel cross-sectional geometry and sinuosity, ness county, kansas, usa. *Marine Pet Geol* 2016;**73**:21–35.
38. Saggaf M., Nebrija E.L. Estimation of missing logs by regularized neural networks. *AAPG Bull* 2003;**87**(8):1377–89.
39. Sasaki Y. The truth of the f-measure. *Teach Tutor Mater* 2007:2–3.
40. Schuster M., Paliwal K.K. Bidirectional recurrent neural networks. *IEEE Trans Signal Process* 1997;**45**(11):2673–81.
41. Servan-Schreiber D., Cleeremans A., McClelland J.L. Learning sequential structure in simple recurrent networks. Advances in Neural Information Processing Systems 1. Morgan Kaufmann Publishers Inc; 1989. p. 643–652.
42. ShaleEeperts, 2020. Wolfcamp shale overview. https://www.shaleexperts.com/plays/wolfcamp-shale/Overview.
43. Tian M., Li B., Xu H., Yan D., Gao Y., Lang X. Deep learning assisted well log inversion for fracture identification. *Geophys Prospect* 2021;**69**(2):419–33.
44. Tian M., Omre H., Xu H. Inversion of well logs into lithology classes accounting for spatial dependencies by using hidden Markov models and recurrent neural networks. *J Pet Sci Eng* 2020;**196**:107598.
45. Tian M., Xu H., Cai J., Wang J., Wang Z. Artificial neural network assisted prediction of dissolution spatial distribution in the volcanic weathered crust: A case study from Chepaizi bulge of Junggar basin, northwestern china. *Marine Pet Geol* 2019;**110**:928–40.
46. Tjelmeland H., Luo X., Fjeldstad T. A Bayesian model for lithology/fluid class prediction using a Markov mesh prior fitted from a training image. *Geophys Prospect* 2019;**67**(3):609–23.
47. Verma S., Scipione M. The early paleozoic structures and its influence on the permian strata, midland basin: Insights from multi-attribute seismic analysis. *J Nat Gas Sci Eng* 2020;**82**:103521.
48. Verma S., Zhao T., Marfurt K.J., Devegowda D. Estimation of total organic carbon and brittleness volume. *Interpretation* 2016;**4**(3):T373–8510.1190/INT-2015-0166.1.
49. Woodroffe C., Chappell J., Thom B., Wallensky E. Depositional model of a macrotidal estuary and floodplain, South Alligator River, Northern Australia. *Sedimentology* 1989;**36**(5):737–56.
50. Zeng L., Ren W., Shan L. Attention-based bidirectional gated recurrent unit neural networks for well logs prediction and lithology identification. *Neurocomputing* 2020;**414**:153–71.
51. Zhang G., Wang Z., Chen Y. Deep learning for seismic lithology prediction. *Geophys J Int* 2018;**215**(2):1368–87.
52. Zhang Z., Zhu X., Zhang R., Li Q., Shen M., Zhang J. To establish a sequence stratigraphy in lacustrine rift basin: a 3d seismic case study from paleogene baxian sag in bohai bay basin, china. *Marine Pet Geol* 2020;**120**:104505.
53. Zhao T. Seismic facies classification using different deep convolutional neural networksSEG Technical Program Expanded Abstracts 2018. Society of Exploration Geophysicists; 2018. p. 2046–50.
54. Zhao T. 3D convolutional neural networks for efficient fault detection and orientation estimationSEG Technical Program Expanded Abstracts 2019. Society of Exploration Geophysicists; 2019. p. 2418–22.
55. Zhao T., Verma S., Qi J., Marfurt K.J. Supervised and unsupervised learning: how machines can assist quantitative seismic interpretation2015:1734–8.
56. Zihlmann M., Perekrestenko D., Tschannen M., 2017. Convolutional recurrent neural networks for electrocardiogram classification. In: 2017 Computing in Cardiology (CinC). IEEE, p. 1–4.

CHAPTER 4

Recurrent neural network for seismic reservoir characterization

Mingliang Liu[a], Philippe Nivlet[b], Robert Smith[b], Nasher BenHasan[b], Dario Grana[a]
[a]Department of Geology and Geophysics, School of Energy Resources, University of Wyoming, Wyoming, United States
[b]EXPEC ARC, Saudi Aramco, Dhahran, Saudi Arabia

Abstract

Seismic reservoir characterization aims to estimate the rock and fluid properties of the subsurface, such as velocities, density, porosity, mineral fractions and fluid saturations, from seismic and well log data. Mathematically, this task is an inverse problem which attempts to find the most likely reservoir models by minimizing the difference between the predicted seismic response and the measured data using optimization algorithms. The conventional model-driven methods based on physical relations usually require pre and postprocessing of data and parameter calibrations. With the advance of artificial intelligence and high-performance computing, many data-driven methods based on deep learning have been investigated and developed to make some procedures in data processing automatic. In this chapter, we review recurrent neural networks, a powerful and efficient algorithm to process time series signals, like seismic data, and present their applications in seismic reservoir characterization.

Keywords

Deep learning; Long short-term memory; Recurrent neural network; Reservoir characterization; Seismic inversion

4.1 Introduction

Seismic reservoir characterization is a critical modeling step in several subsurface problems, such as hydrocarbon exploration and production, carbon dioxide sequestration and geothermal resources[1,2]. It aims to accurately estimate rock and fluid properties from seismic data and borehole well logs. The procedure generally includes two steps. The first step is seismic inversion to predict elastic properties, including P- and S-wave velocities and density from seismic data. The second step is petrophysical or rock physics inversion to transform the elastic properties into petrophysical parameters, such as porosity, clay volume, fluid saturations, and lithological facies.

The majority of the modeling methods currently applied for seismic inversion are model-based approaches where the prediction of the properties of interest is formulated as an inverse problem. These methods assume that the physical relations (e.g., seismic and rock physics models) that link the subsurface model to the geophysical observations are known. The solution of the inverse problem can then be obtained by iteratively

Advances in Subsurface Data Analytics
DOI: https://doi.org/10.1016/B978-0-12-822295-9.00010-8
95

updating the model variables according to the residuals, i.e., the difference between model predictions and measured data, until the predefined convergence criteria are met. In general, the approaches can be categorized into two main groups: the deterministic and probabilistic approaches[3,4]. In deterministic methods, a single solution is obtained by minimizing a pre-defined objective function using gradient-based optimization algorithms, such as the steepest descent method, Newton method and quasi-Newton methods. However, deterministic approaches often depend on the initial model and might lead to local minima of the optimization problem. Moreover, deterministic methods do not quantify the uncertainty of the predicted model. In probabilistic methods, the solution is given by a probability density function (PDF) of the variables of interest conditioned on the measured data. In geoscience applications, the posterior distribution is often obtained in a Bayesian framework, which integrates the prior knowledge of the model (e.g., geological information) with the available geophysical measurements (e.g., seismic data and well logs). The use of probability distributions allows the quantification of the uncertainty associated with the model predictions. For linear relations and Gaussian prior distributions, analytical solutions of the Bayesian inverse problems can be obtained and multiple model realizations can be efficiently sampled from the posterior PDF[5]. For non-linear and non-Gaussian cases, it is necessary to adopt numerical approaches to approximately infer the posterior PDFs, such as variational inference[6], Markov chain Monte Carlo (McMC)[7,8] and ensemble-based methods[9,10].

Model-driven inverse methods provide robust solutions for many practical applications in seismic reservoir characterization, but they usually require data processing, such as pre-conditioning of pre-stack seismic data, seismic-well-tie, wavelet estimation and rock physics modeling. Furthermore, the inversion of large seismic volumes is computationally expensive. With the advance of deep learning in recent years, many efforts have been made in automatic quantitative interpretation of seismic data. A comprehensive end-to-end solution of the modeling problem to directly transform seismic data into reservoir properties using deep learning is still missing; however, progress has been made to accomplish some of the tasks using deep learning instead of manual processing. For instance, deep auto-encoders have been applied for unsupervised seismic facies classification[11] and noise attenuation[12]. Similarly, convolutional neural networks (CNNs) have been used for seismic facies classification[13,14], fault detection[15-19], and salt segmentation[20,21]. Generative adversarial networks (GANs) have been adopted for seismic data interpolation[22] and interpretation[14]. Several papers also included packages and code available in the public domain[23-25].

Deep learning has also been successfully applied and shown significant potential in seismic interpretation and seismic noise attenuation. In recent years, research efforts have been made to apply these methods to more challenging problems, such as full waveform inversion for the estimation of reservoir properties from seismic data. Alfarraj and AlRegib[26] proposed a semi-supervised method with seismic forward modeling as a physical constraint based on recurrent neural networks (RNNs) to estimate elastic impedance and presented its application to the Marmousi2 model[27]. Mustafa et al.[28,29] introduced a sequential modeling method based on temporal convolutional network to predict acoustic impedance. Das et al.[30] applied CNNs to estimate reservoir properties from post-stack seismic data and achieved robust results on the real data set of the Volve field, offshore Norway. Di et al.[31] proposed a semi-supervised deep neural network for the estimation of reservoir properties. Recent works extended the application of deep learning to full waveform inversion[32,33].

In the following, we investigate the applicability and performance of RNNs to reservoir characterization, for the estimation of petrophysical and elastic properties as well as lithological facies from seismic and well log data.

4.2 Methodolgy

The seismic reservoir characterization problem can be formulated as a mathematical model of the form:

$$d = F(m) + e,$$

(4.1)

Where, d represents the observations; m represents the model parameters; F represents the forward model mapping m to d; and e is the observation error term. The goal is to recover the unknown model variables from the data assuming that the forward model provides an accurate approximation.

From a deep learning perspective, the goal is to find an accurate approximation $G \cong F^{-1}$ to the inverse function of the forward model F. This assumes that a training dataset is available, i.e., a set of observations and model variables, is known at some locations. In seismic reservoir characterization, the input sample is generally a one-dimensional seismic trace, which is a time series signal, hence we can adopt the RNN approach[34]. Unlike feed-forward neural networks, such as multiple layer perceptron (MLP) and CNNs, RNNs are a specialized class of deep neural networks which allow information cycles through a feedback loop and thus exhibit temporal dynamic behavior. With such an architecture, RNNs have the ability to account for the input at the current time step and the information learned from previous time steps. Therefore, they can capture temporal dependencies in sequential data.

To allow information to pass through time steps, RNNs include hidden layers serving as state memory to store the information learned from previous time steps. Fig. 4.1 illustrates the basic unit used in RNNs. The variables x_t and y_t with the time step index t ranging from 1 to T represent the input and output sequences, respectively, and the variable s_t denotes the hidden state. At a specific time step t, the hidden state and output are computed as:

$$s_t = g(W_{ss}s_{t-1} + W_{xs}x_t + b_s),$$

(4.2)

$$y_t = h(W_{sy}s_t + b_y),$$

(4.3)

where W_{xs}, W_{ss} and W_{sy} are the trainable weight matrices: W_{xs} connects the input to the state, W_{ss} connects the state at the previous time step to the state in the following time step, and W_{sy} connects the state to the output; b_s and b_y are bias vectors for the hidden state and output, respectively; and g and h is activation functions. Examples of activation functions include the hyperbolic tangent (tanh) or the normalized exponential softmax.

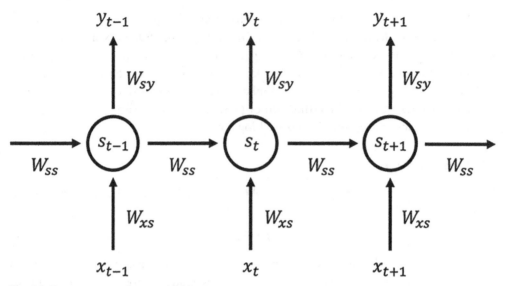

Fig. 4.1 Recurrent neural network unit.

The trainable weight matrices and bias vectors share parameters across different time steps, which is helpful to increase the generation capability of neural networks and allows applying the trained networks to non–observed inputs. s_t acts as a lossy summary of the past sequence of the input up to time step t by selectively weighting features relevant to the prediction. Eq. 4.2 is the classical form of a dynamic system driven by an external signal x. It is recurrent since the state at the time step t refers back to the state at the previous time step $t − 1$. The state variable plays a key role in RNNs with the ability to make predictions of future states from past information.

To account for both forward and backward information in the input sequence, we can combine two RNNs together: one from past to future and one from future to past, as illustrated in Fig. 4.2[35]. This is referred to as a bi-directional RNN (Bi-RNN).

In theory, RNNs enable capturing temporal dependencies in the input data. However, there might be numerical issues in the calculation of the gradients in practical applications, such as derivatives tending to 0 or to $+ \infty$, which might make the convergence of the network training challenging. Due to the connection along the time direction, the updating of the model parameters not only depends on the gradient of the loss function at the current time step, but also the gradients of states at all the previous time steps. This algorithm is referred to as backpropagation through time (BPTT). We use the chain rule to derive the gradient of the loss function L_t (e.g., the mean square error between the true output and the prediction from neural networks) with respect to network parameters at time step t. With no loss of generality, we will illustrate the computation with the kernel weights W_{ss}:

$$\frac{\partial L_t}{\partial W_{ss}} = \frac{\partial L_t}{\partial \gamma_t} \frac{\partial \gamma_t}{\partial s_t} \frac{\partial s_t}{\partial W_{ss}}, \qquad (4.4)$$

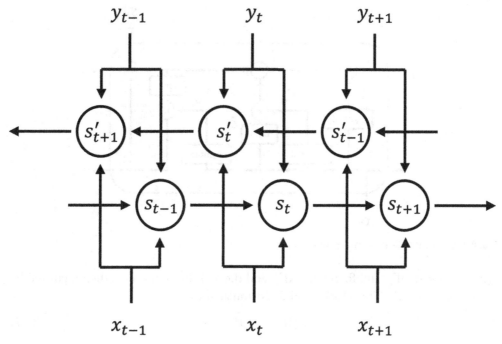

Fig. 4.2 Architecture of Bi-RNN.

and since s_t is a function of both W_{ss} and s_{t-1} (itself being dependent on W_{ss}), the chain rule again gives

$$\frac{\partial s_t}{\partial W_{ss}} = \frac{\partial g}{\partial W_{ss}}\left(W_{ss}s_{t-1} + W_{xs}x_t + b_s\right) + \frac{\partial g}{\partial s_{t-1}}\left(W_{ss}s_{t-1} + W_{xs}x_t + b_s\right)\frac{\partial s_{t-1}}{\partial W_{ss}}, \qquad (4.5)$$

and it can be shown recursively

$$\frac{\partial s_t}{\partial W_{ss}} = \frac{\partial g}{\partial W_{ss}}\left(W_{ss}s_{t-1} + W_{xs}x_t + b_s\right) + \sum_{i=1}^{t-1}\left(\prod_{j=i+1}^{t}\frac{\partial g}{\partial s_{j-1}}\left(W_{ss}s_{j-1} + W_{xs}x_t + b_s\right)\right)\frac{\partial s_i}{\partial W_{ss}}.$$
$$(4.6)$$

In (Eq. 4.6), the recursive product can easily tend to 0 or to $+\infty$ with long time steps (respectively vanishing or exploding gradients), limiting the network capacity to learn long dependency in the former case and making the process instable in the latter one.

To overcome this limitation, variants of RNNs, such as long short-term memory (LSTM)[36] and gated recurrent units (GRU)[37], have been developed by introducing new components (e.g., gate units) in the network architecture. As illustrated in Fig. 4.3, LSTM introduces a mechanism known as cell states with the update gate, forget gate and output gate. The value of the cell state at each time step (c_t) is determined by both the candidate value at current time step $(\tilde{c}_t)$ and the value at the previous time step (c_{t-1})

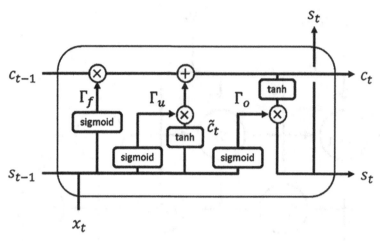

Fig. 4.3 Long short-term memory unit.

with the update (Γ_u) and forget gate (Γ_f), and the activation output is then regulated by the output gate (Γ_o). This process can be formulated as

$$\tilde{c}_t = g\left(W_{cs}s_{t-1} + W_{cx}x_t + b_c\right), \tag{4.7}$$

$$\Gamma_u = \sigma\left(W_{us}s_{t-1} + W_{ux}x_t + b_u\right), \tag{4.8}$$

$$\Gamma_f = \sigma\left(W_{fs}s_{t-1} + W_{fx}x_t + b_f\right), \tag{4.9}$$

$$c_t = \Gamma_u\tilde{c}_t + \Gamma_f c_{t-1}, \tag{4.10}$$

$$\Gamma_o = \sigma\left(W_{os}s_{t-1} + W_{ox}x_t + b_o\right), \tag{4.11}$$

$$s_t = \Gamma_o c_t, \tag{4.12}$$

Where, W and b are trainable weight matrices and bias vectors; and σ is the sigmoid function with output between 0 and 1.

Other variants, such as GRU, are also widely used and provide advantages in some specific tasks in some specific applications such as speech recognition and natural language processing, but their performance is generally similar to the LSTM approach. For this reason, in this study of seismic reservoir characterization, we adopt LSTM. A simplified physical model of wave propagation is the one-dimensional convolution model, where the predicted property not only relates to information above the target (with smaller t) but also deeper, within the wavelet resolution. Therefore, it is justified to adopt a bidirectional LSTM (Bi-LSTM). The proposed model architecture is illustrated in Fig. 4.4, consisting of two Bi-LSTM layers and a fully connected layer with the time-distributed wrapper. The activation function on the final layer can be a softmax when the problem is to predict facies classes, or a linear activation to predict reservoir property such a density or porosity.

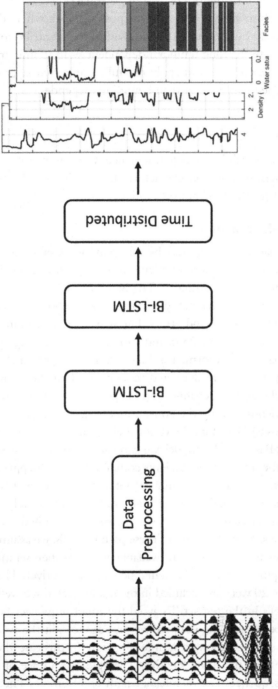

Fig. 4.4 Architecture of neural network used in this study, consisting of two Bi-LSTM layers and a fully connected layer with the time-distributed wrapper.

The softmax function turns logits into probabilities that sum to 1 and outputs a vector that represents the probability distributions of a list of potential outcomes. Note that the framework is flexible enough to allow for a multi-purpose network where different properties are learnt simultaneously by summing their individual loss contributions. The Bi-LSTM layers are used to capture the long temporal dependencies in seismic traces, and by using the time distributed wrapper, we can apply the fully connected layer to every temporal slice of an input independently rather than being interdependent.

4.3 Applications

To demonstrate the applicability and performance, we present two case studies of the proposed RNNs in seismic reservoir characterization, including (1) the synthetic model of Stanford VI-E and (2) the Marmousi2 model.

4.3.1 Stanford VI-E dataset

The Stanford VI-E dataset was created by the Stanford Center for Earth Resources Forecasting Group with the purpose of testing algorithms for reservoir characterization, monitoring and production forecasting[38,39]. The Stanford VI-E reservoir model consists of a three-layer prograding fluvial channel system with $150 \times 200 \times 200$ cells and the thickness of each layer is 80 m, 40 m and 80 m (Fig. 4.5). The dimensions of the cell in the horizontal and vertical direction are 25 m and 1 m, respectively. Stratigraphically, the deltaic deposits (bottom layer) were first formed and then meandering channels (middle layer) and sinuous channels (top layer) were deposited sequentially. Four facies are included in the model: boundary (shale deposits), channel (sand deposits, some of which are saturated with oil), point bar (sand deposits along the convex inner edges of meandering channels) and floodplain (shale deposits). We adopt the original elastic model in terms of P- and S-wave velocities and density. Based on this model, we generate the synthetic pre-stack seismic volumes by convolving the reflection coefficients computed using Zoeppritz equations with a 25 Hz Ricker wavelet. The obtained seismic dataset includes three partial angle-stacks: near, middle, and far stacks corresponding to 12°, 24°, and 36°, respectively (Fig. 4.6). We randomly select 25 traces, referred to as pseudo-wells, from the synthetic geological model to mimic 25 well locations. At the locations of these pseudo-wells, we assume that the original petro-elastic properties and rock facies are available, as well as their seismic response, which are the output and input variables of the neural network, respectively. The pseudo-wells are split into three subsets: 20 wells are included in the training set, three wells in the validation set and two blind wells in the test set. To avoid the input sequences being too long and to increase the number of training samples, we propose to split the well logs and seismic traces into small subsequences along the time direction. In this case, the size of sampling window is 80 samples and the stride length is 1 sample. For a trace of length of 200, we can generate 120 subsequences, and each has a length of 80 samples. Therefore, the number

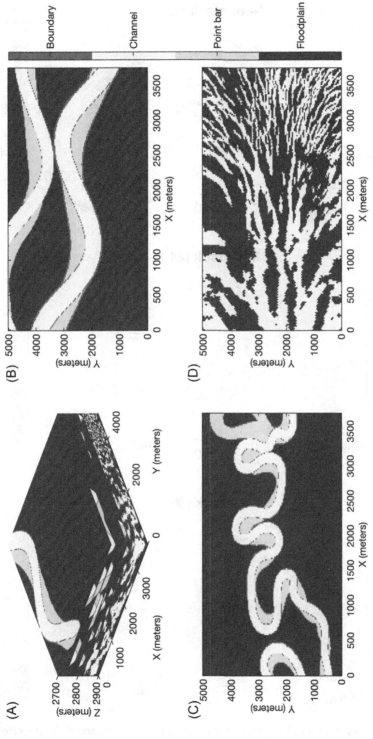

Fig. 4.5 Synthetic reservoir model: (A) 3D facies model of Stanford VI-E, (B) top layer, (C) middle layer, and (D) bottom layer.

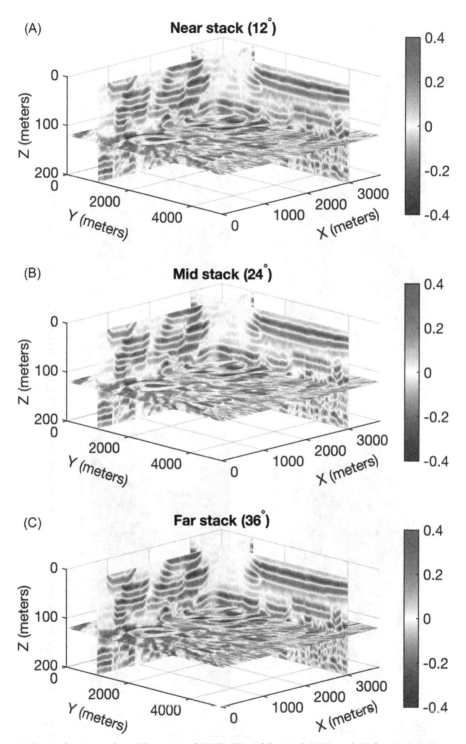

Fig. 4.6 Prestack seismic data: (A) near stack (12°), (B) middle stack (24°), and (C) far stack (36°).

of samples in the training and validation set are $120 \times 20 = 2400$ and $120 \times 3 = 360$, respectively. In practical applications, these numbers depend on the sampling rate of seismic data and the number of wells. Strategies including pseudo–well simulations may be used in case not enough wells are available to train the model[40].

In this case study, we apply the proposed Bi-LSTM model to two common tasks in seismic reservoir characterization: the joint estimation of elastic and petrophysical properties and the prediction of the facies classification. The target output of the first task includes continuous variables. It is a regression problem with the near, middle and far seismic traces as input and the P- and S-wave velocities, density, porosity and water saturation as output. In this case, we use the root mean square error (RMSE) as loss function. The target output of the second task includes discrete variables. The facies prediction is a multiclass classification problem, in which the facies labels are encoded into indicator variables, or according to programming language into one-hot vectors. Considering that the training samples of each facies type are highly unbalanced due to the actual facies proportions, we adopt a weighted categorical cross-entropy as the loss function to avoid facies with fewer training samples being underestimated in the prediction:

$$J = -\frac{1}{N} \sum_{i=1}^{N} \sum_{k=1}^{K} w_k I\left(k, y_i\right) \log P\left(y_i = k \mid x_i\right), \qquad (4.13)$$

where N is the number of training samples; K is the number of classes; x_i and y_i represent the input sequences and output facies type of training sample i, respectively; $I(k, y_i)$ is the indicator function that is equal to 1 if and only if sample i belongs to class k; $P(y_i = k \mid x_i)$ is the output probability of neural network that sample i belongs to class k; w_k is the weight of class k which is proportional to the reciprocal of its frequency in the training samples with normalization. Fig. 4.7 shows the training and validation losses over epochs of the two inversion scenarios using the RMSProp optimizer with a batch size of 128 and a learning rate of 0.0001. The RMSProp optimizer updates network parameters θ as follows:

$$g = \nabla_\theta J\left(\theta\right), \qquad (4.14)$$

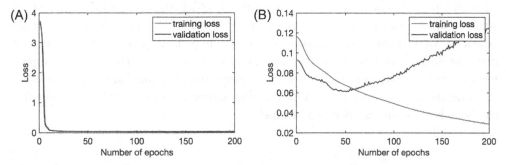

Fig. 4.7 Training history: (A) inversion for petro-elastic properties and (B) inversion for facies types.

$$E\left[g^2\right]_t = \rho E\left[g^2\right]_{t-1} + \left(1-\rho\right)g_t^2, \tag{4.15}$$

$$\theta_{t-1} = \theta_t - \frac{\eta}{\sqrt{E\left[g^2\right]_t + \varepsilon}}, \tag{4.16}$$

Where, the subscript t represents the training epoch step; $E[g^2]$ is the second moment of the gradient g of the loss function with respect to network parameters; ρ is the decaying parameters; η is the initial learning rate; $\in$ is a small constant to make the denominator in (Eq. 4.16) not equal to zero.

Based on the training curves in Fig. 4.7, we choose to stop training around 50 to minimize the possibility of overfitting in both scenarios.

After training, we evaluate the trained models at the two blind well locations. To generate the whole logs, we vertically concatenate the subsequences predicted by the trained network by averaging the overlapped samples. Fig. 4.8 shows the prediction of petrophysical and elastic properties and Fig. 4.9 shows the predicted facies. The predicted reservoir properties show a good agreement with the true models, although the high-frequency details are lost due to the band-limited nature of seismic data. At some locations characterized by high contrasts, the predicted elastic properties deviate from the true values, possibly due to the absence of a low-frequency model. There are some facies misclassifications in particular at the facies boundaries, due to the abrupt transitions that are filtered by the seismic data. The predicted elastic properties and facies models for the entire seismic volume are shown in Figs. 4.10 and 4.11, respectively. Most of the channel reservoirs are correctly predicted.

To quantitatively assess the prediction results, we compute the correlation coefficients between the true and predicted reservoir properties (Table 4.1) and the confusion matrix of facies classification (Fig. 4.12) of the validation dataset. For a fair comparison, the continuous reservoir properties from well logs are filtered to the frequency range of seismic data. Overall, the predicted continuous reservoir properties and discrete facies show a high correlation and accuracy with the true models, despite the difference in resolution. It is interesting to note however that the correlation coefficients S–wave velocity in this case is a bit smaller to other parameters. This is not an intuitive result, especially in comparison to the high correlation coefficients found on density, and we attribute this to the particular rock physics under this model, where density and water saturation were more closely related to acoustic impedance than to S-velocity, and therefore easier to pick up for the network. Fig. 4.13 shows the PDF of elastic properties of each facies. We can see that the elastic properties of channel and point bar facies are close, and therefore they would have similar seismic response. That is why the point bar facies is often misclassified as a channel.

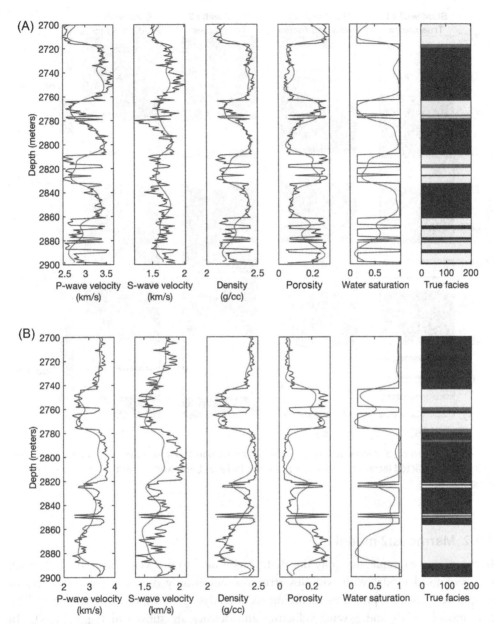

Fig. 4.8 True (*blue lines*) and predicted (*red lines*) petroelastic properties at the blind well locations: (A) blind well #1 and (B) blind well #2.

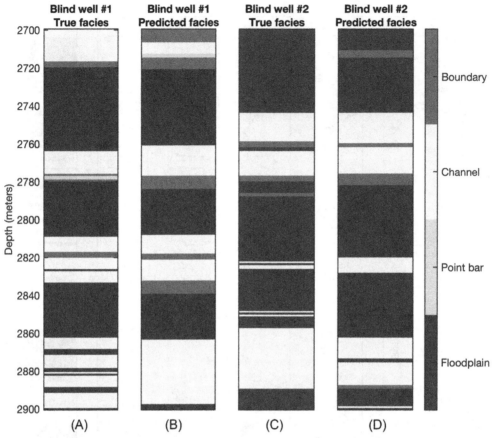

Fig. 4.9 True and predicted petroelastic properties at the blind well locations: (A) true facies at blind well #1, (B) predicted facies at blind well #1, (C) true facies at blind well #2, and (D) predicted facies at blind well #2.

4.3.2 Marmousi2 model

In the second example, we apply the Bi-LSTM model to the Marmousi2 model, which is a geological model with complex structure commonly used for a wide variety of geophysical research problems including AVO analysis and impedance inversion[27]. The true models of P- and S-wave velocities and density are shown in Fig. 4.14A–C. In this example, we compare the performance of the Bi-LSTM model with the Bayesian linearized AVO inversion (BLI) method[5]. We select the central part of the Marmousi2 model as the area of interest consisting of 8000 × 440 grid cells. The pre-stack seismic data are generated by convolving a 25 Hz Ricker wavelet with the reflection coefficients that are computed from the elastic properties (i.e., P- and S-wave velocities and density) using the Zoeppritz equations. The seismic data include near, mid and far stacks with

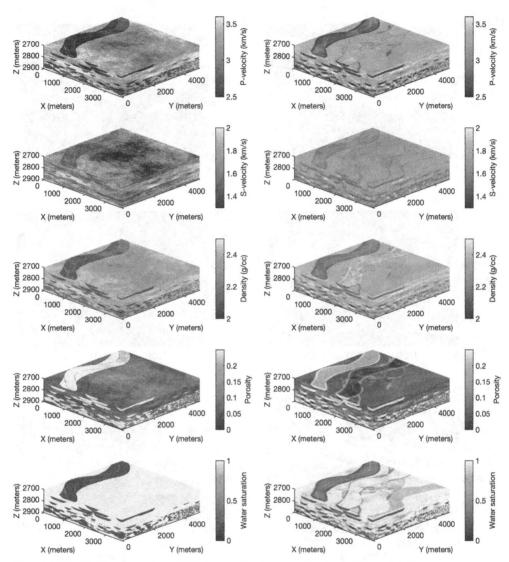

Fig. 4.10 True (left) and predicted (right) petroelastic properties in the 3D model, from top to bottom: P-wave velocity (km/s), S-wave velocity (km/s), density (g/cc), porosity and water saturation.

corresponding incident angles of 15°, 30°, and 45°, respectively (Fig. 4.14D–F). The sampling rate in time is 4 ms and the signal-to-noise ratio is 15 dB. In this example, we randomly extracted 15 traces as the pseudo wells. The well locations are shown as the dash line in Fig. 4.14D. The training data is built according to the same process used in the previous example, but the sequence length is set to 40.

To evaluate the performance of the Bi-LSTM method for elastic properties estimation, we compare its prediction with the results from the method of BLI. Fig. 4.15

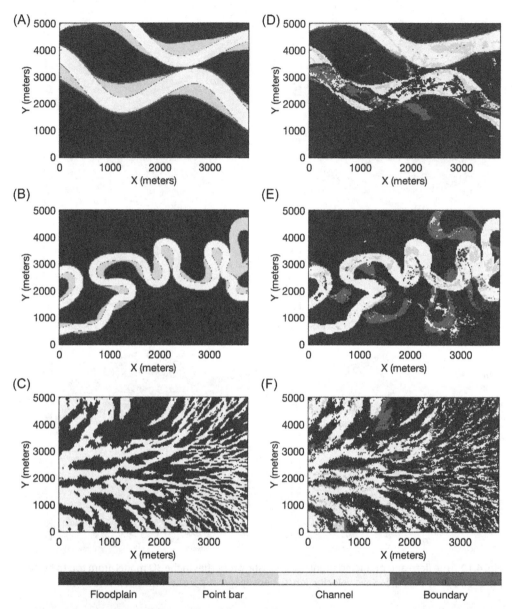

Fig. 4.11 True and predicted facies: (A) true facies of the top layer, (B) true facies of the middle layer, (C) true facies of the bottom layer, (D) predicted facies of the top layer, (E) predicted facies of the middle layer, and (E) predicted facies of the bottom layer.

Table 4.1 Correlation coefficients between the true and predicted reservoir properties for training, validation, and test datasets.

	P-velocity	S-velocity	Density	Porosity	Water saturation
Training	0.882	0.651	0.890	0.887	0.887
Validation	0.841	0.597	0.856	0.834	0.867
Test	0.922	0.695	0.930	0.907	0.913

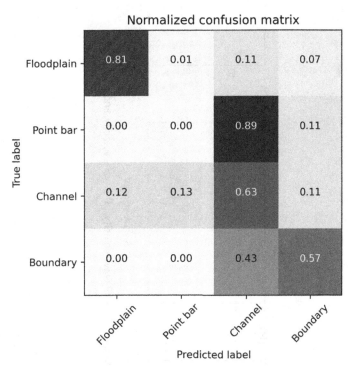

Fig. 4.12 Confusion matrix of facies classification on the validation dataset.

shows the predicted models obtained by the Bi-LSTM and BLI methods. Compared to the true models shown in Fig. 4.14, we can see that the results of the Bi-LSTM show a better agreement with the true models and recover more details than the results of the BLI. The correlation coefficients between the true and predicted reservoir properties of the whole Marmousi2 model by the two methods are summarized

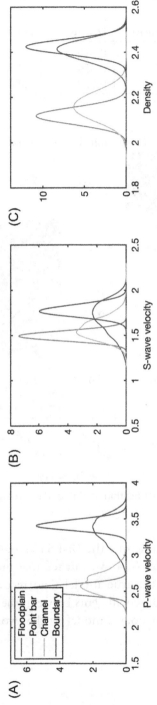

Fig. 4.13 Facies-dependent PDF of elastic properties: (A) P-wave velocity, (B) S-wave velocity, and (C) density.

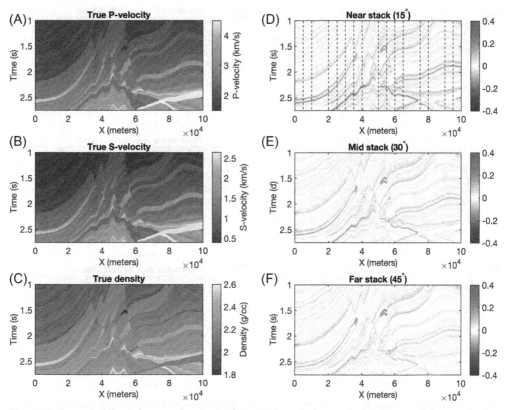

Fig. 4.14 True models and prestack seismic data: (A) true P-wave velocity, (B) true S-wave velocity, (C) true density, (D) near stack (15°), (E) middle stack (30°), and (F) far stack (45°).

in Table 4.2, which quantitatively show the predicted models of the Bi-LSTM has higher accuracy than that of the BLI. In addition, the Bi-LSTM is computationally efficient with GPU. The training of the Bi-LSTM model can converge in 5 minutes and it takes around 10 minutes to perform the prediction on the entire seismic data. However, the main disadvantage of the data–driven Bi-LSTM method is that it usually requires a large training dataset that is often not available in the exploration phase with few drilled wells.

Table 4.2 Correlation coefficients between the true and predicted reservoir properties by the BLI and Bi-LSTM methods.

	P-velocity	S-velocity	Density
BLI	0.921	0.910	0.882
Bi-LSTM	0.958	0.957	0.938

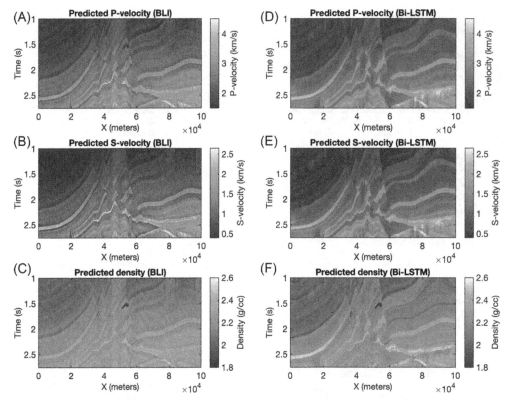

Fig. 4.15 Predicted elastic properties obtained by the BLI and Bi-LSTM methods (A) predicted P-wave velocity by the BLI, (B) predicted S-wave velocity by the BLI, (C) predicted density by the BLI; (D) predicted P-wave velocity by the Bi-LSTM; (F) predicted S-wave velocity by the Bi-LSTM; and (F) predicted density by the Bi-LSTM.

4.4 Conclusion

We proposed a data-driven method for seismic reservoir characterization with RNNs and successfully applied it to two synthetic case studies to investigate its feasibility and performance. The results indicate promising potential for the application of deep neural networks to the inversion of seismic data. The proposed approach is a valid alternative to the conventional model-driven methods, which avoids several pre-processing steps, such as model calibration, and provides a robust and efficient way to efficiently automate the workflow of seismic reservoir characterization. There are still challenges to derive the entire end-to-end workflow, where the neural network can intermediately transform the input seismic traces to reservoir properties without any additional modeling step. In this study, we assumed that the pre-conditioning of seismic data and seismic–well-tie had been previously completed to have consistent input seismic and well log data in the time domain. Future research will investigate the feasibility of deep neural networks to address these limitations and implement the completely data-driven workflow for seismic reservoir characterization based on based deep neural networks. Finally, even if we have tested the algorithms on synthetic reservoirs with a realistic level of heterogeneity, our conclusions were still limited to the relatively simple process that was used to generate data. In real cases, seismic inversion would have to address the problem of properly handling the signal imperfections including multiples, unfiltered converted wave, etc., which in itself is another avenue for future research.

References

1. Doyen P, 2007. Seismic Reservoir Characterization: An Earth Modelling Perspective (Vol. 2, p. 255). Houten: EAGE publications.
2. Grana D, Mukerji T, Doyen P. *Seismic Reservoir Modeling.* New Jersey, USA: Wiley; 2021.
3. Tarantola A, Mukerji T. *Inverse Problem Theory and Methods for Model Parameter Estimation.* Philadelphia, USA: Society for Industrial and Applied Mathematics; 2005.
4. Aster RC, Borchers B, Thurber CH. *Parameter Estimation and Inverse Problems.* Amsterdam, Netherlands: Elsevier; 2018.
5. Buland A, Omre H. Bayesian linearized AVO inversion. *Geophysics* 2003;**68**(1):185–98.
6. Nawaz MA, Curtis A, Shahraeeni MS, Gerea C. Variational Bayesian inversion of seismic attributes jointly for geological facies and petrophysical rock properties. *Geophysics* 2020;**85**(4): MR213–33.
7. de Figueiredo LP, Grana D, Roisenberg M, Rodrigues BB. Gaussian mixture Markov chain Monte Carlo method for linear seismic inversion. *Geophysics* 2019;**84**(3):R463–76.
8. de Figueiredo LP, Grana D, Roisenberg M, Rodrigues BB. Multimodal Markov chain Monte Carlo method for nonlinear petrophysical seismic inversion. *Geophysics* 2019;**84**(5):M1–13.
9. Liu M, Grana D. Stochastic nonlinear inversion of seismic data for the estimation of petroelastic properties using the ensemble smoother and data reparameterization. *Geophysics* 2018;**83**(3): M25–39.
10. Liu M, Grana D. Time-lapse seismic history matching with an iterative ensemble smoother and deep convolutional autoencoder. *Geophysics* 2020;**85**(1):M15–31.
11. Qian F, Yin M, Liu XY, Wang YJ, Lu C, Hu GM. Unsupervised seismic facies analysis via deep convolutional autoencoders. *Geophysics* 2018;**83**(3):A39–43.
12. Saad OM, Chen Y. Deep denoising autoencoder for seismic random noise attenuation. *Geophysics* 2020;**85**(4):V367–76.
13. Zhao T. Seismic facies classification using different deep convolutional neural networks. SEG Technical Program Expanded Abstracts. California, USA: Society of Exploration Geophysicists; 2018. p.2046–50.
14. Liu M, Jervis M, Li W, Nivlet P. Seismic facies classification using supervised convolutional neural networks and semisupervised generative adversarial networks. *Geophysics* 2020;**85**(4):O47–58.
15. Di H, Wang Z, AlRegib G. Seismic fault detection from post-stack amplitude by convolutional neural networks. 80th EAGE Conference and Exhibition 2018. European Association of Geoscientists & Engineers 2018a;**2018**:1–5.
16. Guitton A. 3D convolutional neural networks for fault interpretation. 80th EAGE Conference and Exhibition 2018. European Association of Geoscientists & Engineers 2018;**2018**:1–5.
17. Xiong W, Ji X, Ma Y, Wang Y, AlBinHassan NM, Ali MN et al. Seismic fault detection with convolutional neural network. *Geophysics* 2018;**83**(5):O97–103.
18. Wu X, Liang L, Shi Y, Fomel S. FaultSeg3D: using synthetic data sets to train an end-to-end convolutional neural network for 3D seismic fault segmentation. *Geophysics* 2019a;**84**(3):IM35–45.
19. Wu X, Shi Y, Fomel S, Liang L, Zhang Q, Yusifov AZ. FaultNet3D: predicting fault probabilities, strikes, and dips with a single convolutional neural network. *IEEE Trans Geosci Remote Sens* 2019b;**57**(11):9138–55.

20. Di H, Wang Z, AlRegib G. Deep convolutional neural networks for seismic salt-body delineation. AAPG Annual Convention and Exhibition 2018b.

21. Shi Y, Wu X, Fomel S. SaltSeg: automatic 3D salt segmentation using a deep convolutional neural network. *Interpretation* 2019;**7**(3):SE113–22.

22. Oliveira DA, Ferreira RS, Silva R, Brazil EV. Interpolating seismic data with conditional generative adversarial networks. *IEEE Geosci Remote Sens Lett* 2018;**15**(12):1952–6.

23. Laloy E, Linde N, Ruffino C, Hérault R, Gasso G, Jacques D. Gradient-based deterministic inversion of geophysical data with generative adversarial networks: Is it feasible? *Comput Geosci* 2019;**133**:104333.

24. Liu M, Grana D. Accelerating geostatistical seismic inversion using TensorFlow: a heterogeneous distributed deep learning framework. *Comput Geosci* 2019;**124**:37–45.

25. Pan S, Chen K, Chen J, Qin Z, Cui Q, Li J. A partial convolution-based deep-learning network for seismic data regularization1. *Comput Geosci* 2020;**145**:104609.

26. Alfarraj M, AlRegib G. Semisupervised sequence modeling for elastic impedance inversion. *Interpretation* 2019;**7**(3):SE237–49.

27. Martin GS, Wiley R, Marfurt KJ. Marmousi2: an elastic upgrade for Marmousi. *The Leading Edge* 2006;**25**(2):156–66.

28. Mustafa A, Alfarraj M, AlRegib G. Estimation of acoustic impedance from seismic data using temporal convolutional network. *SEG Technical Program Expanded Abstracts.* Texas, USA: Society of Exploration Geophysicists; 2019. p. 2554–8.

29. Mustafa A, Alfarraj M, AlRegib G. Spatiotemporal modeling of seismic images for acoustic impedance estimation. *SEG Technical Program Expanded Abstracts.* Texas, USA: Society of Exploration Geophysicists; 2020. p. 1735–9.

30. Das V, Pollack A, Wollner U, Mukerji T. Convolutional neural network for seismic impedance inversion. *Geophysics* 2019;**84**(6):R869–80.

31. Di H, Chen X, Maniar H, Abubakar A. Semi-supervised seismic and well log integration for reservoir property estimation. *SEG Technical Program Expanded Abstracts* 2020. p. 2166–70.

32. Sun H, Demanet L. Extrapolated full-waveform inversion with deep learning. *Geophysics* 2020;**85**(3):R275–88.

33. Zhang ZD, Alkhalifah T. High-resolution reservoir characterization using deep learning-aided elastic full-waveform inversion: The North Sea field data example. *Geophysics* 2020;**85**(4):WA137–46.

34. Goodfellow I, Bengio Y, Courville A, Bengio Y. Deep Learning 12016. MIT press Cambridge:2.

35. Schuster M, Paliwal KK. Bidirectional recurrent neural networks. *IEEE Trans Signal Process* 1997;**45**(11):2673–81.

36. Hochreiter S, Schmidhuber J. Long short-term memory. *Neural Comput* 1997;**9**(8):1735–80.

37. Cho K, Van Merriënboer B, Gulcehre C, Bahdanau D, Bougares F, Schwenk H, et al. 2014. Learning phrase representations using RNN encoder-decoder for statistical machine translation. arXiv preprint arXiv:1406.1078.

38. Castro S, Caers J, Mukerji T. The Stanford VI reservoir 18th Annual Report Stanford Center for Reservoir Forecasting. California, USA: Stanford University; 2005.

39. Lee J, Mukerji T. The Stanford VI-E reservoir: A synthetic data set for joint seismic-EM time-lapse monitoring algorithms. 25th Annual Report: Technical Report 2012. Stanford Center for Reservoir Forecasting, Stanford University, Stanford, CA.

40. Joseph C, Fournier F, Vernassa S. Pseudo-well methodology: a guiding tool for lithoseismic interpretation. SEG Technical Program Expanded Abstracts. Texas, USA: Society of Exploration Geophysicists; 1999. p. 938–41.

CHAPTER 5

Convolutional neural networks: core interpretation with instance segmentation models

Rafael Pires de Lima[a], Fnu Suriamin[b]
[a]Geological Survey of Brazil, São Paulo, Brazil
[b]Oklahoma Geological Survey, Norman, OK, United States

Abstract

Core acquisition is an important step in many geoscientific research projects, ranging from oil and gas exploration to the study of oceanic crust. Cores are ideally stored in climate-controlled core warehouse as they are expensive to obtain and contain important information about the subsurface. Thanks to storing and preservation procedures, many miles of cores from different regions are available for analysis in different parts of the world. However, this vast collection of cores is also a source of disperse information that is sometimes overlooked. Recently, geoscientists are exploring machine learning techniques to aid in core interpretation. Instance segmentation, a computer vision task that aims to individually identify instances on an image, shows promising results for the interpretation of cores. Instance segmentation models use standard core photographs as input and generate an output that directly matches the photograph. We show that instance segmentation models built with convolutional neural networks have the potential to greatly accelerate core interpretation.

Keywords

Convolutional neural networks; Core intepretation; Machine learning

5.1 Introduction

Among a wide range of tools and data used for the analysis of subsurface resources, drill-hole cores and cuttings are some of the few data that can be considered as the ground truth of the subsurface. Moreover, cores are crucial to verify the hypotheses developed from the geophysical and well logging studies, helping researchers understand the advantages and drawbacks of different propositions. Core descriptions can be useful to identify key lithofacies and facies associations, evaluate facies stacking and interpret depositional environments as well as their history and evolution[1], evaluate the relationships among porosity, permeability, and lithofacies,[2] and help operators to identify optimal zones for designing completions. Although the standards used in geological description, such as grain size and grain roundness are based on quantitative characteristics, traditional core interpretation includes a strong qualitative and subjective component. This subjective component occurs because of the interaction of the geologist with the rock

Advances in Subsurface Data Analytics
DOI: https://doi.org/10.1016/B978-0-12-822295-9.00004-2
117

samples; most of the interpretation is based on visual observations of color, structure, texture, and biota in core as well as ready-to-analyze physical and chemical properties. Thus, different geologists describing the same data might reach different interpretations. The variations in interpretation tend to be minor, but they can accumulate and lead workers to different lithofacies interpretations and consequently different conclusions. Moreover, geological concepts can change over time and the same core sample can possibly have its interpretation changed as well.

Aware of the importance of cores, many companies and research institutions maintain core libraries. For example, the Oklahoma Petroleum Information Center (OPIC) holds more than 9,200 records, summing over 161 km (100 miles) of cores drilled around Oklahoma[3]. The core libraries of the Geological Survey of Brazil estimated in 2015 to have more than 350 km (271 miles) of core stored in different parts of the country, with an estimated cost for the development of such core libraries of over $1 billion dollars[4]. Australian AuScope stores imagery of more than one million meters of core, a project with an estimated implementation cost of over $1.7 billion Australian dollars[5]. Despite the effort in storing cores, the access to such cores and their interpretation is not always easy. Quick access to core interpretations, perhaps interpretations conducted with different objectives or by different experts, could leverage the geological knowledge of a region, facilitating research and reducing risks for operators. Aiming to investigate the options to standardize and to accelerate core interpretation for lithofacies identification, we evaluate applications machine learning as an aid for geologists on their visual-recognition task.

Machine learning has been widely used in different geoscience fields and applications aiming to classify image data into rock types are increasing. For example, Valentín et al.[6] used convolutional neural networks (CNN) to classify ultrasonic and microresistivity borehole image logs into four different lithofacies, Ran et al.[7] used CNN to classify outcrop photographs into six different rock types, Pires de Lima et al.[8] used CNN to classify images of core into 11 lithofacies, Baraboshkin et al.[9] used CNN to classify approximately 2,000 m of cores into six different lithotypes, Pires de Lima et al.[10] used CNN to classify petrographic thin sections, hand samples, images of cores, and microfossil images, Liu et al.[11] used CNN to classify hand samples into nine different rock types. As these examples show, CNN are very useful tools for geoscientific image analysis.

CNN are designed to process data that is input in the form of multiple arrays. In fact, many geoscience data are in the form of multiple arrays: 1D for a well log or a single seismic trace; 2D for images such as photographs; and 3D for seismic volumes. The defining building block of CNN is the convolutional layer, each convolutional layer containing one or more filters. The convolutional filters act much like any other signal or image convolutional filter on which the output is computed based on a locally weighted sum of the input data. LeCun et al.[12] observed that CNN can be traced back to Fukushima's[13] neocognitron, an architecture that was much similar to CNN, but lacked a supervised learning algorithm to update the convolutional filter weights. The lack of a supervised learning was addressed by LeCun et al.[14] with the use of backpropagation for the task of classifying handwritten digits. With back propagation, the weights of the convolutional filters can be initialized randomly and automatically updated during the training process. The updates are performed in such a way that the difference, or loss, between the models' prediction and the desired output is reduced. The loss is an objective function that measures the error (or distance) between the output of a model and the desired output, or target.

Despite the early applications, CNN's major increase in popularity was due to the ImageNet competition in 2012. Krizhevsky et al.'s[15] AlexNet CNN model was applied to a data of about a million images that contained 1,000 different classes[16] with the objective to perform image classification and achieved results far superior than the best competing approaches in 2012. Such performance led CNN to be the current dominant approach for almost all recognition and detection tasks[12]. After AlexNet's performance in the image classification task, a task on which the objective is to categorize an image according to its content, a vast collection of CNN models appeared (e.g.,[17–21]). Image classification is perhaps the most common task previous studies investigated using geoscientific images[6–8,10]. However, research in computer vision field advanced to address different tasks, such as object localization, semantic segmentation, and instance segmentation. Here we use the same terminology as He et al.[22]. In object detection, the goal is to classify individual objects and localize each object using a bounding box. Liu et al.[11] study is one example of the use of object detection using CNN with geoscience images. Semantic segmentation's goal is to classify each pixel into a fixed set of categories, regardless of object instances. Instance segmentation is a combination of object detection and semantic segmentation, where the goal is to detect all objects in an image and perform the segmentation of each instance. The current instance segmentation task was introduced by Hariharan et al.[23] and became popular due to COCO (Common Objects in Context)[24] dataset publication. In this chapter, we show how instance segmentation can be used to accelerate core-based lithofacies interpretation, using standard photographs of core.

With few exceptions, for example deep sea drilling[25], vast collections of core images are not readily available to most geoscientists even today. Further digitization core samples in the form of images will not only facilitate access to data for traditional analysis, but will also provide key elements for the implementation of innovative machine learning algorithms for geoscience. The methodology presented here has the potential to organize many miles or kilometers of slabbed cores into a reliable and coherent system easily accessible to a variety of users. In this chapter, we show how instance segmentation algorithms can be used for the automatization of core lithofacies classification. We begin with an overview of the methodology, which includes data description and some geological background. Then, we apply the instance segmentation algorithms to our core data set, providing metrics for the analysis of the performance of the results, analyzing the method's advantages and limitations. We conclude with a summary of our findings and suggestions.

5.2 Methods

Machine learning techniques are currently well-disseminated in many fields of science and deep learning methods are gaining popularity. LeCun et al.[12] presented a review on deep learning, explaining the key components of the technique as well as how it was used for speech recognition[26], natural language processing[27], and computer vision[15,28], among other applications. The advances in the computer vision field are particularly interesting for the application of deep learning techniques in geoscience. We use standard cores photographs to define the lithofacies based on differences in lithology, texture, primary sedimentary structure, composition, and bioturbation. We then use the input core photographs and the output target lithofacies interpretation, as data for a supervised instance segmentation model. In other words, the model uses standard core photographs as input with no preprocessing other than rescaling and outputs lithofacies defined by polygons. The next subsection provides more details about the data and the geological settings. The following subsection explains the instance segmentation model we use, as well as how performance is evaluated.

5.2.1 Geological setting and data

The Meramecian-aged rocks that form unconventional reservoirs in the STACK play (Sooner Trend Anadarko [Basin] Canadian Kingfisher) area of the Anadarko Basin consist of regionally extensive low permeability rocks. Based on detailed description of the cores, Suriamin[29] concluded the Meramecian-aged rocks can be divided into eight lithofacies, including skeletal wackestone-packstone, chert – cherty breccia, structureless siltstone, crosslaminated siltstone, laminated siltstone, bioturbated siltstone, glauconitic siltstone-sandstone, and structureless sandstone (Table 5.1). These lithofacies were deposited on a shallow marine setting with dominant wave-influence. We conducted detailed core description and lithofacies definition using 260 meters (~850 ft) of core from
five wells located in Kingfisher, Blaine, and Canadian Counties, Oklahoma, United States. For this study, the structureless siltstone and structureless sandstone were combined into one lithofacies as they are challenging to define using only core photography without thin section analysis. Likewise, crosslaminated siltstone and laminated siltstone are combined into laminated siltstone as the differences in photographs are minor. Additionally, the presence of glauconitic siltstone-sandstone is very limited in the cores (~3 ft or 0.9 m). Therefore, glauconitic siltstone-sandstone lithofacies was excluded from the dataset. Similarly, defective cores' interval such as rubble section (due to extensive fractures or core plug holes) were also not included to be targets to the model. Fig. 5.1 shows one example of the interpretation of two core photographs. We interpreted a total of 71 photographs with dimensions ranging from 1,307 by 4,268 to 2,818 by 4,033 pixels.

Table 5.1 Lithofacies Characteristics

No.	Lithofacies	Color	Thickness	Texture, sorting & grain size	Thin section petrography	Sedimentary structures	Bioturbation
1	Skeletal wackestone-packstone	White to very light gray	Bed thickness varies from ~2.0 – 238 cm (~0.8 – 94 in.).	Very coarse sand, angular to sub-angular, moderately to poorly sorted.	Peloids, quartz, crinoids, brachiopods, bryozoan, and monoaxon sponge spicule. Well-cemented grains. Blocky and poikilotopic calcite cements. Dolomite, fluid inclusion, and quartz overgrowths.	Wispy laminations	Not observed
2	Chert – cherty breccia	Grayish black to dark gray	Bed thickness varies from ~0.5 – 13 cm (0.2 – 5 in.).	Very fine sand to pebble, angular to sub-angular, moderately sorted	Microcrystalline quartz. Calcite cement, sponge spicules, organic materials, and possible spores and pollen. Rhombic dolomite crystals, baroque dolomite, ferroan dolomites, detrital quartz, and opaque minerals are rare. Ferroan dolomite as overgrowth cement.	Wispy laminations	Unidentifiable bioturbations
3	Structureless siltstone	Light to medium gray	Bed thickness varies from ~0.02 – 2.7 m (~0.07 – 9 ft).	Medium to coarse silt, angular to sub-angular, well sorted.	Detrital monocrystalline quartz and calcite cement. Peloids, crinoids and brachiopods. Pyrite, overgrowth quartz cements, calcite cement, or grain replacement. Anhedral and euhedral dolomite, ferroan dolomite, twin potassium feldspar, greenish clay clasts, organic-rich clasts, and Fe-dolomites	Faint laminations	Unidentifiable bioturbations

(Continued)

Table 5.1 Cont'd

No.	Lithofacies	Color	Thickness	Texture, sorting & grain size	Thin section petrography	Sedimentary structures	Bioturbation
4	Cross-laminated siltstone	Light to medium gray	Bed thickness varies from ~0.02 – 0.6 m (0.07 – 2 ft).	Medium to coarse silt, angular to sub-angular, moderately sorted	Monocrystalline quartz grains and microcrystalline texture with vacuoles inclusion. Authigenic quartz overgrowth, a diffuse boundary and a thinly coated clay mineral. Calcite cement, peloids, and clay minerals. Fe-dolomite, rhombic dolomite crystals, crinoids, twin feldspar, opaque minerals, greenish clay clasts, and muscovite.	Mm-to-cm-scale low-angle planar laminations, ripple laminations, and hummocky cross-stratification.	*Thalassinoides* and horizontal spreiten.
5	Laminated siltstone	Light to medium gray	Bed thickness varies from ~0.3 - 750 cm (~0.01 – 24 ft).	Medium to coarse silt, angular to sub-angular, moderately sorted	Coarse calcite cement, carbonate grains, peloids, silt-sized detrital quartz grains, greenish clay clasts, brachiopods, and sponge spicules. Occasionally silt-sized detrital quartz grains with clay matrix, quartz overgrowth, fluid inclusions. Oncoid and grapestone, Fe-dolomite, and pyrite.	Parallel, occasionally discontinuous, wavy, or wispy laminations	*Phycosiphon*

No.	Lithofacies	Color	Thickness	Texture, sorting & grain size	Thin section petrography	Sedimentary structures	Bioturbation
6	Bioturbated siltstone	Dark gray to black	Bed thickness varies from ~0.003 – 11 m (~0.01 – 36 ft).	Very fine to coarse silt, angular to sub-angular, moderately sorted	Clay-rich, silt-size detrital monocrystalline quartz grains with microinclusion minerals or fluid inclusions. Occasionally calcite-rich and detrital quartz-rich. Calcite cement, micrite, organic materials, pyrite, titanium oxide minerals, dolomite, Fe-dolomite, greenish clay clasts, rock fragments, calcified spicules, and monoaxon sponge spicules.	Wavy and wispy lamination	Phycosiphon, Chondrites, Skolithos, Thalassinoides, Teichicnus, Planolites, vertical and horizontal spreiten, and Bergaueria
7	Glauconitic siltstone and sandstone.	Greenish gray	Bed thickness varies from ~0.1 – 1.2 m (~0.3 – 4 ft).	Coarse silt – very fine sand, angular-to-sub-angular, moderately sorted	Glauconite. Rhombic dolomite crystals, detrital angular to sub-angular quartz, organic materials, opaque minerals, and clay. Well-compacted grains. Plastically deformed glauconite. Sutured grain contact. Rhombic dolomite crystals, baroque dolomite, and possible phospathic debris.	Structureless	Horizontal spreiten
8	Structureless sandstone	Light to medium gray	Bed thickness varies from ~0.6 – 21 m (~2 – 69 ft).	Very fine sand, angular-to-sub-angular, moderately to well-sorted.	Well-cemented quartz-rich sandstone. Quartz grains with diffuse boundary and fluid inclusions along authigenic quartz overgrowths. Calcite grains, calcite cement, Fe-dolomite, twinned plagioclase feldspar, and clay minerals.	Structureless	Skolithos and Planolites

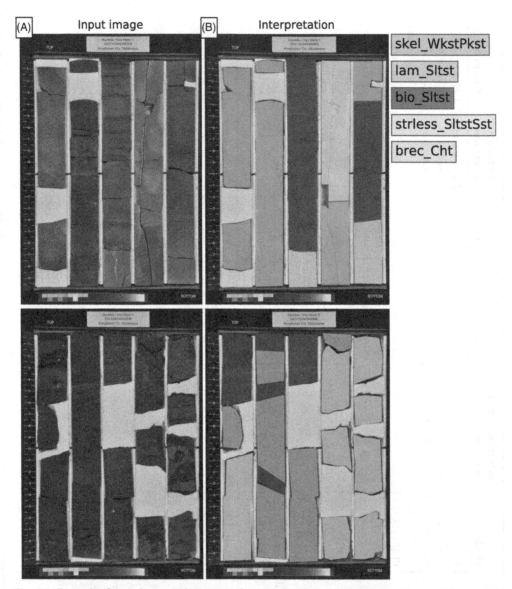

Fig. 5.1 *Example of core photograph (input to the model) and the interpretation (target to the model).* Each polygon is an instance and the model's objective is to segment each one of the instances. The abbreviations used for the lithofacies are used for the other figures as well. skel_ WkstPkst is used as short for skeletal wackestone-packstone, lam_Sltst as short for laminated siltstone, bio_Sltst as short for bioturbated siltstone, strless_SltstSst as short for structureless siltstone/sandstone, brec_Cht as short for chert – cherty breccia.

5.2.2 Instance segmentation

He et al.[22] observed that results for object detection and semantic segmentation were improved over a short period of time. The objective of object detection systems is to fit a bounding box, that is, a rectangle, as close as possible to the provided bounding box target. Huang et al.[30] performed a study on the accuracy and speed of object detection methods. In contrast to object detection, the objective of semantic segmentation is to classify every unit (a pixel in the case of a photograph) into one of the target categories. Improvements in object detection and semantic segmentation are credited to the development and popularization of powerful baseline region-based convolutional neural networks (R-CNN) systems, such as the fast R-CNN[31] and faster R-CNN[32] for object detection, and fully convolutional network (FCN)[33] and U-net[34] frameworks for semantic segmentation. He et al.'s[22] Mask R-CNN addresses the instance segmentation task, where the goal is to detect all objects in an image while also precisely segmenting each instance.

Mask R-CNN was inspired in faster R-CNN that by itself is an improvement on fast R-CNN. Thus, we briefly review faster R-CNN as a means to describe Mask R-CNN. Faster R-CNN uses a two-stage process for object localization, consisting of a class-agnostic proposal and a class-specific detection stage. In the first stage, the input is processed by a CNN, often called "backbone," that outputs a feature map used by the region proposal network (RPN), a small fully convolutional network. The backbone CNN model can have different architectures, but the objective is to process the input and to extract feature maps that contain the most important information of the data. We provide details of the backbone used in this study later in this section. In the second stage of Faster R-CNN, the region proposals are used to crop features from the same intermediate feature map. The cropped features are subsequently fed to the remainder of the feature extractor in order to predict a class and class-specific bounding box refinement for each proposal. The remainder of the feature extractor, that is, the classification branch and the bounding-box regression, is called "head." We refer readers to Huang et al.[30] for a comprehensive comparison between Faster R-CNN and other object detection frameworks. Mask R-CNN adds a branch to the head for predicting segmentation masks on each region of interest (RoI), which runs in parallel with the existing branches for classification and bounding box regression (Fig. 5.2). The mask branch is a small FCN applied to each RoI and predicts a pixel-by-pixel segmentation mask. Mask R-CNN also proposed a layer called RoIAlign to improve the mapping of RoIs to the regions of the input image, compensating location misalignment between the input image and the lower resolution feature maps. Chen et al.[35] observed that the majority of current instance segmentation frameworks are methods that first detect object bounding boxes, and then crop and segment these regions, as popularized by Mask R-CNN.

The metrics for evaluation of instance segmentation need to consider both the class of the instance as well as its dimensions. Mean average precision (mAP) is a common metric used to evaluate the performance of object detection and instance segmentation

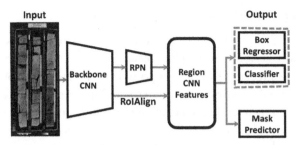

Input

Output

Backbone CNN

RPN

RoIAlign

Region CNN Features

Box Regressor

Classifier

Mask Predictor

Fig. 5.2 *Mask R-CNN representation.* A photograph of a core is used as input to the network. The network processes the image with the CNN backbone to extract region proposals that are then used by the head on the final section of the model to classify the data, as well as to predict bounding boxes and masks.

tools. To better define mAP, we briefly review precision, recall, and intersection over union (Iou). IoU is a ratio of the area of overlap to the area of union (Fig. 5.3). Precision and recall are defined as:

$$Precision = \frac{TP}{TP + FP}$$

(5.1.)

$$Recall = \frac{TP}{TP + FN}$$

(5.2)

Where, TP are the true positives, FP are the false positives, and FN are the false negatives. The equations show that recall measures the ability of the model to find all relevant instances in the dataset, while precision measures the proportion of those instances that are in fact correctly selected. The equations also show that there is a trade-off between precision and recall due to the FP and FN. For most classification-related tasks, results have the form of a confidence level, where larger values indicate greater confidence

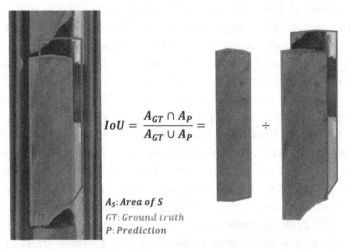

$$IoU = \frac{A_{GT} \cap A_P}{A_{GT} \cup A_P} =$$

$\div$

A_S: *Area of S*
GT: Ground truth
P: Prediction

Fig. 5.3 *Intersection over union (IoU).* IoU is the ratio of the area of overlap to the area of union. It ranges from 0, when there is no alignment between the analyzed shapes, to 1.0, when they perfectly overlap.

that the sample should be classified as a specific class. Thus, a precision–recall curve is computed from a method's ranked output, where the rank depends on the confidence level. Finally, the average precision (AP) is calculated as the area under the curve of the precision-recall curve. For multiclass problems, we compute AP separately for each class, then average over classes to compute the mean average precision (mAP). A detection is a true positive if it has IoU greater than a defined threshold value. For example, mAP@0.5 is the mAP when the IoU threshold is equal to 0.5. Here we use COCO's definition for mAP and AP. The COCO mAP averages the AP for each class at IoUs starting from 0.50 to 0.95 with increments of 0.05. Everingham et al.[36] provided more information about metrics for object detection and instance segmentation tasks.

In this study, we use the Mask R-CNN with a ResNet50 backbone to perform instance segmentation using photographs of core – an analogy to core interpretation. ResNets are an architecture introduced by He et al.[17] and widely used in computer vision tasks. ResNets use shortcut identity connection in which the outputs of a previous layer are added to the outputs of upcoming layers. ResNet50 is an indication of a model composed of 50 layers. We make experiments using two ResNet architectures, the first one extracts features from the last convolutional layer of the 4th stage of the network, while the second extracts feature maps at different scales within the network. Such multi-scale network architecture is commonly denominated Feature Pyramid Network (FPN)[37]. Thus, the results from the models are identified as ResNet50-C4 and ResNet50-FPN. We use Detectron2[38] Mask R-CNN implementation and framework, and we execute the experiment using Colaboratory (or Colab, https://research. google.com/colaboratory/faq.html, 2020)[43]. Detectron2 provides a collection of baselines trained models and we fine-tune such models for core instance segmentation. Fine-tuning is a transfer learning technique, repurposing a model trained to perform a primary task "A" to perform a secondary task "B," widely used in many fields of science and frequently outperforms models trained from scratch[39–42]. During fine-tuning, the first two layers of the models remain frozen, meaning the backpropagation algorithm does not modify the weights of such layers. Samples are augmented on-the-fly during training using simple resizing (such that the short side is never larger than 1,000 pixels) and horizontal flipping techniques. We train the models with stochastic gradient descent (SGD) method, with a learning rate of 10^{-3} and 0.9 momentum, and a batch size of eight images for 2,000 iterations. SGD is a stochastic approximation of gradient descent, a simple, yet efficient and widely used optimization algorithm that minimizes the loss in an iterative fashion. Detectron2's Mask R-CNN implementation uses a multitask loss that consider the classification, bounding box, and mask losses as described by He et al.[22]. The classification loss measures how close the model is to predicting the correct class, the bounding box loss measures the distance between the labeled and the predicted bounding box, and the mask loss penalizes wrong per-pixel classifications.

Ideally, data used in machine learning applications is divided into training, validation, and test sets, all sets with samples assumed to be extracted from the same distribution.

Table 5.2 Data split and number of instances (total and by lithofacies.

Fold	Set	# Photographs	# Instances (total)	# skel_ WkstPkst	# lam_ Sltst	# bio_ Sltst	# strless_ SltstSst	# brec_ Cht
1	Training	56	543	26	111	253	132	21
	Validation	15	113	0	16	56	41	0
2	Training	57	493	18	98	234	139	4
	Validation	14	163	8	29	75	34	17
3	Training	57	551	24	103	261	142	21
	Validation	14	105	2	24	48	31	0
4	Training	57	543	20	102	243	157	21
	Validation	14	113	6	25	66	16	0
5	Training	57	494	16	94	245	122	17
	Validation	14	162	10	33	64	51	4

Training data is the data effectively used by the model during the optimization stage, while the validation data is used to tune hyperparameters, and the test set is reserved to be used only on the final stages to evaluate model's performance. As we have a relatively low number of samples (compare, for example, our 71 samples with COCO's more than 2,00,000 samples), the experiments are conducted using five-fold cross validation. In k-fold cross-validation, the models are trained k times with different subsets of the data, providing a more realistic information of model's performance in cases which it is difficult to ensure the split of data into three different sets (training, validation, and test) with same distribution. In each one of the folds, the data is divided into training and validation sets. We randomly select 20% of the photographs to be part of the validation, without making any considerations or assumptions about the lithofacies (or instances) interpreted for the photographs. Table 5.2 shows the number of photographs and instances for each one of the folds, for both training and validation sets, as well as the number of instances of each one of the lithofacies.

5.3 Results

Colab resources vary depending on usage by the users, thus the graphic processing units (GPUs) available for the experiments vary over time. Table 5.3 shows the necessary time to train and evaluate the models in one-fold for different graphical processing units (GPUs). The results are rounded to the nearest half-hour (e.g., an execution of three hours and 40 minutes is rounded to 3.5 h). Results in Table 5.3 show that the time the models take to execute 2,000 training steps in Colab resources is roughly three hours in average.

Table 5.3 Training and evaluation execution time by fold rounded to the half-hour mark.

GPU	Model	Time (h)
Tesla T4	ResNet50–FPN	2.5
Tesla K80	ResNet50–FPN	3.5
Tesla P100	ResNet50	2.5
Tesla K80	ResNet	3.5

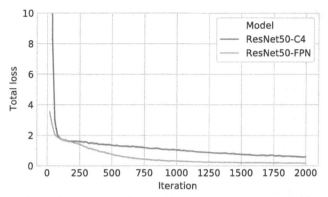

Fig. 5.4 *Training loss for the experiments.* The loss is stored every 20th step. The figure shows the central tendency and the standard deviation confidence interval of the training loss for each model, aggregated by fold. The results are for each fold are very similar, thus the confidence interval might not be visible at this scale.

Fig. 5.4 shows that both models' training loss reduction is very similar for each one of the folds. Both ResNet50-C4 and ResNet50-FPN greatly reduce the training loss in the first 200 training steps. Then the loss of ResNet50-FPN approaches zero faster than ResNet50-C4 for the rest of the training and have only marginal improvements after training step 1,000. After a fast improvement in the first 200 training steps, ResNet50-C4 lags behind results of ResNet50-FPN, although ResNet50's loss seems to constantly get closer to zero and loss plateauing is not as clear as in ResNet50-FPN. The training loss helps understand if the backpropagation algorithm is working as expected (as the objective is to reduce the loss) and if the model's architecture affects the results, however it cannot inform on overfitting. Thus, many machine learning practitioners observe the validation loss during training to evaluate overfitting. We chose to present the training loss by itself and to evaluate training and validation mAPs. Although the loss is an important measure, the mAPs are more interpretable and are the metrics more commonly used in computer vision final evaluation[24]. Fig. 5.5 shows the mAP and mAP@50 for training and validation sets computed at every 200 steps during training, the larger step interval is chosen due to computational cost. Similar to the loss in Fig. 5.4, the metrics in Fig. 5.5 show that the ResNet50-FPN quickly improves on the results computed on the training set, while ResNet50-C4 underperforms. The mAP for both models (Fig. 5.5A) slowly increases throughout the training, with ResNet50-FPN having the better performance. On the mAP@50 (Fig. 5.5B), a more accommodating metric, the ResNet50-FPN seems to achieve perfect scores on the training set at step 1,000, while ResNet50-c4 continues improving and approaches ResNet50-FPN results at the final training steps. However, perhaps the most important information in Fig. 5.5 are the results of the validation set. There is a large difference for both mAP and mAP@50 between training and validation results. The training metrics are constantly

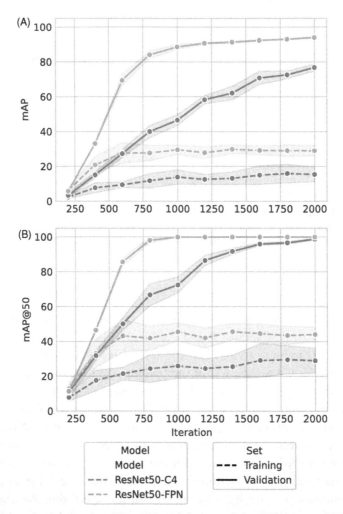

Fig. 5.5 *Training and validation sets APs computed during training.* (A) shows mAP for training and validation sets for ResNet50-C4 and ResNet50-FPN. (B) shows mAP@50 for training and validation sets for ResNet50-C4 and ResNet50-FPN. The metrics are computed at each 200th step and marked as circles in the figures. The lines show the central tendency shaded by the standard deviation confidence interval loss for each model, aggregated by fold.

better than the validation metrics, which indicate overfitting of the models. After step 250 training results tend to improve faster than validation for both models and both mAP and mAP@50 metrics. Fig 5.5A shows that ResNet50-FPN validation mAP seems to stay somewhat stable after step 1,000, with a slightly more concentrated standard deviation of the results, while ResNet50-C4 validation seems to have a higher oscillation on a lower performance level. Results are similar in Fig 5.5B, however mAP@50 results seem to have a larger variation (wider standard deviation shade) than mAP.

Fig. 5.6 shows a summary of the trained models. Although both models overfit the training data, results of ResNet50-FPN tend to be better on the validation set for almost all folds, across all classes. The performance of instance segmentation of bio-turbated siltstone in the validation set seems to be comparable for both models with

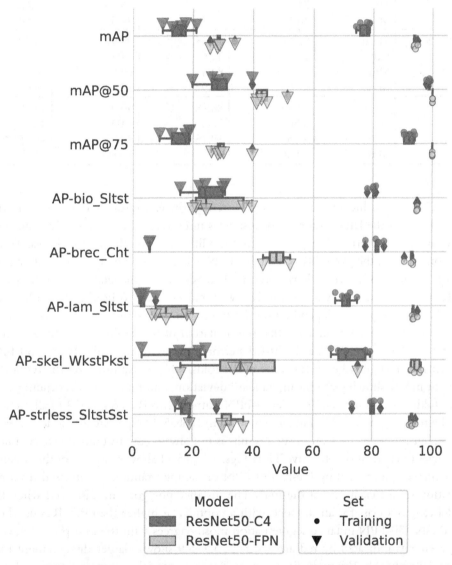

Fig. 5.6 *Training and validation sets APs evaluation.* Markers for each one of the folds, models, and sets are overlain on top of their boxplots. Colors indicate model, while marker style indicates training or validations set. Multi-class averaged results mAP, mAP@75, and mAP@50 are displayed on the first rows of the figure, while class-by-class results are displayed on the bottom rows. Training variance is small, thus boxplots are reduced to lines in several cases. Gray diamond-shaped markers indicate boxplot outliers.

Table 5.4 Summarized metrics for ResNet50-C4 and ResNet50-FPN.

		ResNet50-C4		ResNet50-FPN	
		Training	*Validation*	*Training*	*Validation*
mAP	Mean	76.84	15.45	94.01	29.05
	Std	2.02	4.26	0.52	2.99
	Min	74.05	9.82	93.56	25.68
	Max	78.64	21.01	94.87	33.88
mAP@75	Mean	92.11	15.35	99.97	30.72
	Std	1.72	4.36	0.03	5.20
	Min	90.33	8.84	99.92	26.20
	Max	94.00	18.92	100.00	39.64
mAP@50	Mean	98.72	28.98	99.97	44.08
	Std	0.56	7.26	0.03	4.26
	Min	97.81	19.75	99.92	40.95
	Max	99.26	39.58	100.00	51.27

ResNet50-FPN achieving slightly better performance. Bioturbated siltstone is the lithofacies with the largest number of instances interpreted. In contrast, there are only a few samples of chert – cherty breccia and the lithofacies is selected for the validation set in only two of the five folds. ResNet50-FPN is significantly better in identifying the low occurrences of chert – cherty breccia than ResNet50. Skeletal wackestone-packstone, another lithofacies with relatively few instances, seem to be the class/lithofacies with the largest variance depending on the fold, with one of the folds achieving an AP of more than 70. Results indicate that segmentation of the laminated siltstone is challenging for both models. ResNet50-C4 performance is weak, while ResNet50-FPN performance is slightly better for some of the folds. Table 5.4 summarizes the results for different mAPs, showing the mean, standard deviation, minimum, and maximum values across folds. Results indicate ResNet50-FPN outperforms ResNet50-C4 in all metrics.

The metrics presented in Figs. 5.3–5.6 and in Table 5.4 help evaluate models' performance, but the interpretation of such results is not always easy to translate to the initial lithofacies interpretation objective. Thus, Figs. 5.7 to 5.11 show examples of the instance segmentation performed by ResNet50-FPN on samples randomly extracted from the validation set for each one of the folds. The instance polygons are displayed when the model assigns a class for an instance with a confidence higher than 0.7. Results show that ResNet50-FPN instance segmentation mostly agrees with the interpretation, with disagreements indicated by red arrows. Figs. 5.7–5.9 show a larger disagreement than Figs. 5.10 and 5.11. Although disagreement between model output and existing interpretations shown in Figs. 5.10 and 5.11 is minor, with the main differences in regions not segmented by the model, the core photographs sampled only structureless siltstone and structureless sandstone that were combined into a single lithofacies (structureless

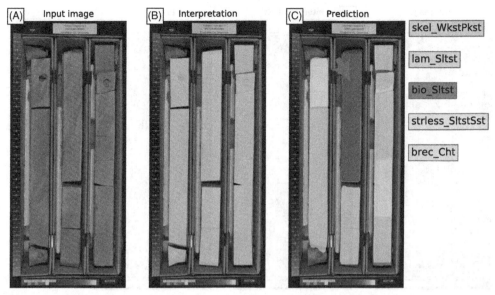

Fig. 5.7 *Example of instance segmentation on a sample from the validation set from the first fold.* (A) shows the input image, a photograph not used in training. (B) shows the lithofacies interpretation. (C) shows the instance segmentation results from trained ResNet50-FPN. The red arrow points to a location of disagreement between the interpretation and the model output. The photograph was randomly selected from the validation set. Note the evident saw marks on (A).

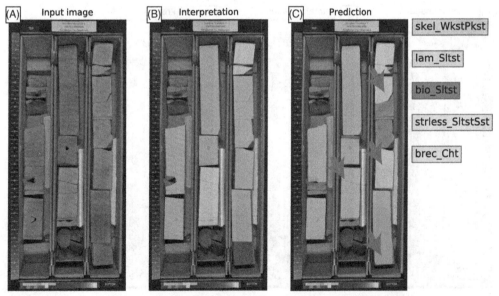

Fig. 5.8 *Example of instance segmentation on a sample from the validation set from the second fold.* (A) shows the input image, a photograph not used in training. (B) shows the lithofacies interpretation. (C) shows the instance segmentation results from trained ResNet50-FPN. The red arrow points to a location of disagreement between the interpretation and the model output. The photograph was randomly selected from the validation set. Note the rubble section is not interpreted in the top part of the box on the left.

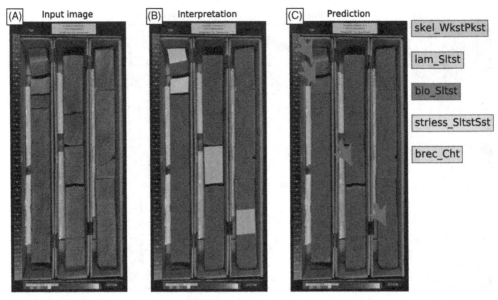

Fig. 5.9 *Example of instance segmentation on a sample from the validation set from the third fold.* (A) shows the input image, a photograph not used in training. (B) shows the lithofacies interpretation. (C) shows the instance segmentation results from trained ResNet50-FPN. The red arrow points to a location of disagreement between the interpretation and the model output. The photograph was randomly selected from the validation set.

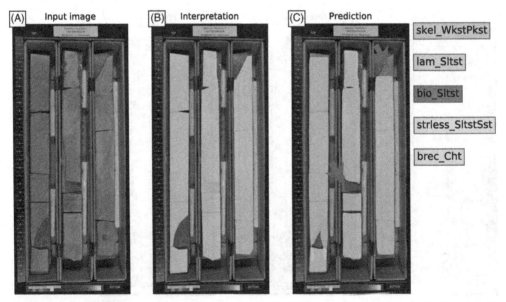

Fig. 5.10 *Example of instance segmentation on a sample from the validation set from the fourth fold.* (A) shows the input image, a photograph not used in training. (B) shows the lithofacies interpretation. (C) shows the instance segmentation results from trained ResNet50-FPN. The red arrow points to a location of disagreement between the interpretation and the model output. The photograph was randomly selected from the validation set.

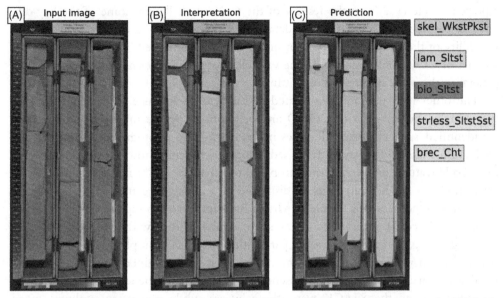

Fig. 5.11 *Example of instance segmentation on a sample from the validation set from the fifth fold.* (A) shows the input image, a photograph not used in training. (B) shows the lithofacies interpretation. (C) shows the instance segmentation results from trained ResNet50-FPN. The red arrow points to a location of disagreement between the interpretation and the model output. The photograph was randomly selected from the validation set.

siltstone/sandstone). The photograph used as input for the results presented in Fig. 5.7 show only the structureless siltstone/sandstone; yet the model assigned bioturbated siltstone to a small portion of the core at the top part of the center box. Curiously, that piece of core shows a vertical stripe, a man–made stain. Fig. 5.8 shows the model confusing laminated siltstone with structureless siltstone/sandstone. Fig. 5.9 shows the model confusing the interpretation of structureless siltstone/sandstone and laminated siltstone as bioturbated siltstone. The difference in color in the pieces of core in Fig. 5.9 is small, which might indicate the model is not too sensitive to laminations.

5.4 Discussion

The results in the previous section show that the computer vision instance segmentation task can be adapted to facilitate core lithofacies interpretation. All examples extracted from the validation set show appropriate interpretation of the core photograph, with few disagreements between the geologists' interpretation and model output, especially for ResNet50–FPN instance segmentation. The multi-scale ResNet50-FPN is more effective to segment the photograph across all lithofacies in comparison to the single-scale ResNet50-C4 approach. Although the performance of the models in the validation set seems below par, they are comparable to results obtained by Mask R-CNN reported by He et al.[22] on the test set of the COCO challenge. Some of the key differences between ours and COCO datasets are the number of samples available, the variability of the data, and the type of instance to be segmented. The sheer number of samples of COCO combined with the variability of the samples helps models to better generalize the data and prevent overfitting.

The other main difference is related to the type of instance to be segmented by the models. The instances in COCO are well-defined by the images on which they are identified, such characteristics does not occur in core photographs. Geologists can distinguish man-made features from sedimentary structures; however, sedimentary or man-made features on photograph can blend together for CNN models. The curvilinear features on Fig. 5.7 can easily be dismissed by a geologist as man-made saw marks, however a model sees that feature as part of the data it tries to segment. That does not mean a CNN could not learn to ignore such features, note how the model is very precise in ignoring everything that is not part of the core in all examples and even ignoring rubble section in Fig. 5.8, but that the model needs to be trained on enough data to be able to learn which features are useful and which should be suppressed. Moreover, the differences in the STACK play lithofacies are subtle, and often geologists need additional data, such as thin sections, to augment with cores to better identify different sedimentation patterns. The other observable drawback of the methodology is overfitting, with the performance on the training sets being superior than the performance on the validation sets. An increased number of samples could likely help prevent overfitting. Another possibility would be to use backbones with fewer convolutional layers. The combination of a large number of model's parameters with a limited number of training samples facilitates overfitting as the model has might have enough capacity to memorize the data.

This work uses instance segmentation models, in contrast to the classification of cropped images of cores (e.g.,[8,9]) or object detection of rock samples[11]. Photographs used as input for object detection, segmentation, and instance segmentation models not necessarily need to be pre-processed, contrary to the classification of cropped images of core photographs. Directly using standard photographs of core as input to a model with barely any preprocessing has as advantage a direct relationship with data commonly used by geologists, which makes it easier for workers to quickly go back-and-forth between model's results and the raw data for quality control and analysis. Although pre-processing of photographs might prove to be somewhat cumbersome in some applications, such pre-processing might help direct the model to focus strictly on the main part of the image to be analyzed (i.e., only the core, not the box or other objects) and might be useful depending on the project. Moreover, the addition of more training data should be helpful to train more robust models – note our entire dataset is comprised of 71 photographs, whereas usually CNN applications rely on hundreds or thousands of samples. Another promising path is the addition of supplementary information correctly localized on the photographs, such as X-ray fluorescence or gamma-ray measurements. Such measurements can help the models differentiate between lithofacies that are distinct, yet visually similar.

5.5 Conclusion

This work shows that the geological task of core lithofacies interpretation can be aided by the use of instance segmentation CNN models. Particularly, we make use of Mask R-CNN with different backbones and observe that instance segmentation performed by such models are not far from a geologist interpretation, yet the models lack intuition that can lead to interpretation mistakes. Nonetheless, results are robust enough and indicate that they could potentially facilitate interpretation or re-interpretation of large core collections, accelerating an important task for many geoscience projects. We foresee that an increase in the digitalization of core collections and their interpretation could potentially lead to more diverse datasets, increasing the performance of models trained on such datasets.

5.6 Acknowledgments

Pires de Lima thanks the Geological Survey of Brazil for permission to develop and publish this study. We thank David Duarte for his help reviewing the manuscript as well as his suggestions for improvements. Most of the code we use for the experiments and analysis can be accessed at https://github.com/raplima/2020_cores_auto.

References

1. Grauch VJS, Skipp GL, Thomas JV, Davis JK, Benson ME, 2015. Sample Descriptions and Geophysical Logs for Cored Well BP-3-USGS, Great Sand Dunes National Park and Preserve, Alamosa County, Colorado, Data Series. Reston, VA. doi: 10.3133/ds918
2. Suriamin F, Pranter MJ. Stratigraphic and lithofacies control on pore characteristics of Mississippian limestone and chert reservoirs of north-central Oklahoma. *Interpretation* 2018:1–66. doi:10.1190/int-2017-0204.1.
3. Oklahoma Geological Survey, 2020. Oil and gas databases [WWW Document]. http://www.ou.edu/ogs/data/oil-gas. [Accessed 6 January 2020].
4. Geological Survey of Brazil, 2020. Rede de Litotecas [WWW Document]. http://www.cprm.gov.br/publique/Redes-Institucionais/Rede-de-Litotecas/Sobre-5637.html. [Accessed 6 January 2020].
5. AuScope, 2019. National Virtual Core Library — AuScope [WWW Document]. Available from: https://www.auscope.org.au/nvcl. [Accessed 6 January 2020].
6. Valentín MB, Bom CR, Coelho JM, Correia MD, de Albuquerque MP, de Albuquerque MP et al. A deep residual convolutional neural network for automatic lithological facies identification of Brazilian pre-salt oilfield wellbore image logs. *J Pet Sci Eng* 2019. doi:10.1016/J.PETROL.2019.04.030.
7. Ran X, Xue L, Zhang Y, Liu Z, Sang X, He J. Rock classification from field image patches analyzed using a deep convolutional neural network. *Mathematics* 2019;**7**:755. doi:10.3390/math7080755.
8. Pires de Lima R, Suriamin F, Marfurt KJ, Pranter MJ, 2019b. Convolutional neural networks as aid in core lithofacies classification. Interpretation 7, SF27–SF40. doi:10.1190/INT-2018-0245.1.
9. Baraboshkin EE, Ismailova LS, Orlov DM, Zhukovskaya EA, Kalmykov GA, Khotylev OV et al. Deep convolutions for in-depth automated rock typing. *Comput Geosci* 2020;**135**:104330. doi:10.1016/j.cageo.2019.104330.

10. Pires de Lima R, Bonar A, Coronado DD, Marfurt K, Nicholson C. Deep convolutional neural networks as a geological image classification tool. *Sediment. Rec.* 2019a;**17**:4–9. doi:10.210/sedred.2019.2.
11. Liu X, Wang H, Jing H, Shao A, Wang L. Research on intelligent identification of rock types based on faster R-CNN method. *IEEE Access* 2020;**8**:21804–12. doi:10.1109/ACCESS.2020.2968515.
12. LeCun Y, Bengio Y, Hinton G. Deep learning. *Nature* 2015;**521**:436–44. doi:10.1038/nature14539.
13. Fukushima K. Neocognitron: a self-organizing neural network model for a mechanism of pattern recognition unaffected by shift in position. *Biol Cybern* 1980;**36**:193–202. doi:10.1007/BF00344251.
14. LeCun Y, Boser BE, Denker JS, Henderson D, Howard RE, Hubbard WE et al. Handwritten digit recognition with a back-propagation network. In:D.S. Touretzky (Ed.), Advances in Neural Information Processing Systems 2. Morgan-Kaufmann:1990:396–404.
15. Krizhevsky A, Sutskever I, Hinton GE. ImageNet classification with deep convolutional neural networks, in: Proceedings of the 25th International Conference on Neural Information Processing Systems - Volume 1, NIPS'12. Curran Associates Inc.,USA. 2012:1097–105.
16. Russakovsky O, Deng J, Su H, Krause J, Satheesh S, Ma S et al. ImageNet large scale visual recognition challenge. *Int. J. Comput. Vis.* 2015;**115**:211–52. doi:10.1007/s11263-015-0816-y.
17. He K, Zhang X, Ren S, Sun J. Identity mappings in deep residual networks. in: B. Leibe, J. Matas, N. Sebe, M. Welling. (Eds.), Computer Vision – ECCV 2016. Amsterdam, The Netherlands: Springer International Publishing; 2016. p. 630–645.
18. Howard AG, Zhu M, Chen B, Kalenichenko D, Wang W, Weyand T, et al., 2017. Mobilenets: efficient convolutional neural networks for mobile vision applications.
19. Huang G, Liu Z, Maaten LVD, Weinberger KQ. Densely Connected Convolutional Networks, in: Computer Vision – ECCV 2016. Honolulu, HI, USA; 2016. p. 2261–2269. https://doi.org/10.1109/CVPR.2017.243.
20. Simonyan K, Zisserman A. Very deep convolutional networks for large-scale image recognition, in: Y. Bengio, Y. LeCun, (Eds.), 3rd International Conference on Learning Representations (ICLR). San Diego, CA, USA 2015.
21. Szegedy C, Vanhoucke V, Ioffe S, Shlens J, Wojna Z, 2015. Rethinking the inception architecture for computer vision.
22. He K, Gkioxari G, Dollár P, Girshick R. Mask R-CNN 2017 IEEE International Conference on Computer Vision (ICCV) Venice, Italy 2017:2980–8. doi:10.1109/ICCV.2017.322.
23. Hariharan B, Arbeláez P, Girshick R, Malik J, 2014. Simultaneous Detection and Segmentation BT - Computer Vision – ECCV 2014, In: Fleet D, Pajdla T, Schiele B, Tuytelaars T. (Eds.), Springer International Publishing, Cham, pp. 297–312.
24. Lin T-Y, Maire M, Belongie S, Hays J, Perona P, Ramanan D, et al., 2014. Microsoft COCO: Common Objects in Context BT - Computer Vision – ECCV 2014, In: Fleet D, Pajdla T, Schiele B, Tuytelaars T (Eds.), Springer International Publishing, Cham, pp. 740–55.
25. Project DSD. Archive of core and site/hole data and photographs from the deep sea drilling project (DSDP) [WWW Document]. *NOAA Natl. Centers Environ. Inf.* 1989.
26. Hinton G, Deng L, Yu D, Dahl GE, Mohamed A, Jaitly N et al. Deep neural networks for acoustic modeling in speech recognition: the shared views of four research groups. *IEEE Signal Process Mag* 2012;**29**:82–97. doi:10.1109/MSP.2012.2205597.
27. Collobert R, Weston J, Bottou L, Karlen M, Kavukcuoglu K, Kuksa P. Natural language processing (almost) from scratch. *J. Mach. Learn. Res.* 2011;**12**:2493–537.
28. Szegedy C, Liu W, Jia Y, Sermanet P, Reed SE, Anguelov D, et al., 2014. Going deeper with convolutions.
29. Suriamin F, 2020. Integrated Reservoir Characterization of a Mixed Siliciclastic-Carbonate Reservoirs, Mississippian Strata of Northern and Central Oklahoma. University of Oklahoma. Available from: https://shareok.org/handle/11244/324356.
30. Huang J, Rathod V, Sun C, Zhu M, Korattikara A, Fathi A et al. Speed/accuracy trade-offs for modern convolutional object detectors 2017 IEEE Conference on Computer Vision and Pattern Recognition (CVPR)Honolulu, HI, USA 2017:3296–7. doi:10.1109/CVPR.2017.351.
31. Girshick R. Fast R-CNN 2015 IEEE International Conference on Computer Vision (ICCV) Santiago, Chile 2015:1440–8. doi:10.1109/ICCV.2015.169.

32. Ren S, He K, Girshick R, Sun J. Faster R-CNN: towards real-time object detection with region proposal networks. Cortes C, Lawrence ND, Lee DD, Sugiyama M, Garnett R. Advances in Neural Information Processing Systems 28, Curran Associates, Inc. Montreal, Quebec, Canada 2015:91–9.

33. Long J, Shelhamer E, Darrell T. Fully convolutional networks for semantic segmentation 2015 IEEE Conference on Computer Vision and Pattern Recognition (CVPR) Boston, MA, USA 2015:3431–40. https://doi.org/10.1109/CVPR.2015.7298965.

34. Ronneberger O, Fischer P, Brox T. U-Net: convolutional networks for biomedical image segmentation. Navab N, Hornegger J, Wells WM, Frangi AF, Medical Image Computing and Computer-Assisted Intervention – MICCAI 2015. Springer International Publishing Cham 2015:234–41.

35. Chen X, Girshick R, He K, Dollar P. TensorMask: a foundation for dense object segmentation The IEEE International Conference on Computer Vision (ICCV) Seoul, Korea 2019.

36. Everingham M, Van Gool L, Williams CKI, Winn J, Zisserman A. The Pascal visual object classes (VOC) challenge. *Int. J. Comput. Vis.* 2010;**88**:303–38. doi:10.1007/s11263-009-0275-4.

37. Lin T, Dollár P, Girshick R, He K, Hariharan B, Belongie S. Feature pyramid networks for object detection 2017 IEEE Conference on Computer Vision and Pattern Recognition (CVPR). Honolulu, HI, USA; 2017. p. 936–44. doi:10.1109/CVPR.2017.106.

38. Wu Y, Kirillov A, Massa F, Lo W-Y, Girshick R, 2019. Detectron2 [WWW Document]. https://github.com/facebookresearch/detectron2. [Accessed 6 April 2020].

39. Huot F, Biondi B, Beroza G. Jump-starting neural network training for seismic problems, in: SEG Technical Program Expanded Abstracts 2018. San Antonio, Texas, USA: Society of Exploration Geophysicists; 2018:2191–5. doi:10.1190/segam2018-2998567.1.

40. Pan SJ, Yang Q. A survey on transfer learning. *IEEE Trans Knowl Data Eng* 2010;**22**:1345–59. doi:10.1109/TKDE.2009.191.

41. Razavian AS, Azizpour H, Sullivan J, Carlsson S. CNN features off-the-shelf: an astounding baseline for recognition 2014 IEEE Conference on Computer Vision and Pattern Recognition Workshops 2014:512–9. https://doi.org/10.1109/CVPRW.2014.131.

42. Yosinski J, Clune J, Bengio Y, Lipson H. How transferable are features in deep neural networks? *Adv. Neural Inf. Process. Syst.* 2014;**27**:3320–8.

43. Colaboratory – Google [WWW Document], 2020. https://research.google.com/colaboratory/faq.html. [Accessed 6 April 2020].

CHAPTER 6

Convolutional neural networks for fault interpretation – case study examples around the world

Hugo Garcia[a], Peter Szafian[b], Chris Han[b], Ryan Williams[b], James Lowell[b]
[a]Geoteric, United States
[b]Geoteric, United Kingdom

Abstract

Machine learning is now commonly used across many industry sectors, for example, in the medical sector machine learning has been used to learn how to segment blood vessels from medical images by observing clinicians' manual blood vessel depiction (one deep learning network for each clinician). The oil and gas industry has finally caught up with recent advances in machine learning technology. Operators are turning towards machine learning to efficiently interpret geological systems, improve safety, and maximize investment and return of investment in oil and gas exploration. But what about the tasks that require not just knowledge but understanding of the geologic setting, structural history of a basin, and stress regime, etc.? Would a neural network be able to not just recognize a seismic character to identify a fault but also understand the local and regional context that is required to generate a fault framework? This study deals with automated fault interpretation in seismic data using machine learning. In this chapter, we will show some of the machine learning-guided fault interpretation results from 3D seismic data sets around the world. In terms of results, the interpretation delivered by the networks is orders of magnitude better than the conventional seismic attribute-based workflows. This is especially true for the 3D networks where the true 3D analysis of the data considerably improves not only fault identification but lateral and vertical continuity. Some of the convolutional neural networks' (CNN) limitations (quality of initial training, representativity of particular structural styles in the CNN's knowledge, etc.) can be mitigated with training, by teaching the network to better recognize structural information, ignore stratigraphic features, artifacts in the data and/or noise. The results can be achieved at a fraction of the time (compared with manual interpretation) with the interpreter always at the center of the process.

Keywords

Artificial intelligence; Convolution neural networks; 2D networks; 3D networks; Fault interpretation; Machine learning

6.1 Introduction

Despite the advancements in seismic acquisition and processing, and the computational resources available for the geoscientists, seismic interpretation is still, in its essence, a 2D exercise. 3D seismic data sets are now widely used in the oil and gas industry, and these

Advances in Subsurface Data Analytics
DOI: https://doi.org/10.1016/B978-0-12-822295-9.00005-4

141

are the interpreter's main resource for interpreting the subsurface but the actual work is still mostly done using 2D sections (inline and crosslines). The Society of Exploration Geophysicists, in its website, says this about structural interpretation "A 3-D structural interpretation session may begin with viewing selected inline and crossline sections to acquire a regional understanding of the subsurface geology. Other orientations, such as vertical sections along a dominant dip direction, may also be needed to determine the structural pattern. Time slices then are studied to check the structural pattern."[23] The main product of the structural interpretation is structural maps. In the American Association of Petroleum Geologists' website, it is stated that "Structural seismic interpretation is directed toward the creation of structural maps of the subsurface (…)."[22] Structural maps are, in their essence a 2.5D section (the "half" dimension, time or depth, being displayed with color)." Within an exploration workflow we only return to a true 3-dimensional analysis when a geological model is built.

Does this mean that the interpreter is limited to a 2D world? No, this reflects how, historically, the interpretation workflow was built. With the recent developments in the interpretation software and workflows, the interpreter has the necessary tools to deliver a true 3D interpretation early in a project life cycle. Having this information earlier allows the interpreter to start thinking spatially when looking at the faults, how they relate to each other, and implications on reservoir geometry, compartmentalization, lateral and vertical connectivity: all of this instead of a systematic, line-by-line fault picking. In the last few decades, several automations have been introduced in the interpretation workflow to aid the geoscientist. Auto-trackers analyze the data spatially, following a seismic reflector to generate a surface. With the increase in computational power, this can now be extended to the full volume, with every reflector being tracked. Other tools like geobody interpretation allow for a true 3D data segmentation and extraction of geological features.[1,2] In terms of structural interpretation, an important milestone towards a 3D interpretation was the introduction of seismic-derived edge attributes.[3-7] These attributes analyze the data, highlighting the discontinuities, "filtering" everything else. Although the edge attributes were a big development in structural interpretation and are now an integral part of the workflow, they have several limitations. They are very sensitive to noise, data quality (amplitude and frequency content) has a big impact in the attribute performance and because it is a window-based analysis, the footprint is imbedded in the results (stairstep effect). But the biggest disadvantage is that the edge attributes detect edges in the seismic data as opposed to faults. This is an important distinction, as there can be several stratigraphic edges that will be imaged by the attributes alongside with the faults. There are no mechanisms to either prevent the attributes from identifying the stratigraphic edges or separate those responses from the fault responses in the output volume. These limitations have limited the usefulness of edge attributes as a skeleton for automated fault stick extraction and automated fault surface generation. Despite this, edge attributes are now the pillars of the traditional seismic interpretation.

The next jump in fault interpretation is associated to the development (and industry acceptance) of convolutional neural networks.[8-10] Machine learning, and particularly deep learning, solve most of the edge attribute limitations. The quality of the input seismic data is still an important factor; however, the ability to train the network allows for the minimization of the data quality impact on the results. The training process mimics what a human does when presented with a data set of the same quality. This is a big change in relation to all other tools available to the interpreter and allows for the generation of a true 3D analysis of the data, with an output volume that shows only structural information and that can be used, directly, to produce fault planes. In this chapter we will show several case studies covering the Beagle Sub-basin, Australia (offshore), East Shetland Basin, UK (offshore), Main Pass, Gulf of Mexico, USA (offshore) and Cooper-Eromanga Basin, Australia (onshore), to illustrate the benefits of convolutional neural networks in structural interpretation.

6.2 Machine learning algorithms

Geoscientists' interpretation reasoning is influenced by experiences and lessons learned over time. Initially, their decision-making is based on what they have been taught during their studies, with competence growing through industry experience.

Artificial neural networks were inspired by biological learning processes and structures. Over recent years, advances in artificial intelligence have led to deeper neural networks (known as deep learning) that are cable of solving problems with diverse, unstructured, and inter-connected data. Deep learning adopts the same theory of being taught through example, making it an exciting technology for seismic interpretation. A deep learning network's decisions carry with it the quality and values that were used to train (teach) it and as the network sees more data (examples) its accuracy continues to improve. The accuracy of a deep learning network is a measure of how often the network classifies the answer correctly. More specifically, it is the total number of true positives and true negatives divided by total number of true positives, true negatives, false positives, and false negatives.

Unfortunately, the power of deep learning networks does come at a price, and that is the amount of labeled data required to understand and predict the location of faults within seismic data. Another problem is the quality of labeled data itself. Different geoscientists could well depict the location of faults slightly differently. This inter-observer variation can produce uncertainty within the labeled training data, and in turn reduce the prediction accuracy of a deep learning network. Imprecise or inconsistent labeling, produces adversarial gradients in the training space, forcing the network to try and incorporate the incorrect results into the produced outcomes. Such inaccuracy can result in the deep learning network incorrectly predicting faults when there are none and missing actual fault locations. One method to test the accuracy of the network is compare against synthetically generated seismic datasets such as SEAM.

To reduce the need for large amounts of manually labeled interpreted data, and to reduce the potential errors caused by inter-observer variation, the generation and use of synthetic data has been shown to be an effective method to train 2D and 3D convolutional neural network (CNN) to identify faults with seismic data. For example, Huang et al.[11] proposed to train a 3D CNN on fault attributes generated from synthetic seismic cubes with simple fault configurations. Pochet et al.[12] trained a 2D CNN on patches of a synthetic data set with simple fault geometries and obtained encouraging results when testing on actual sections offshore Netherlands. The results from the case studies shown below have all been initially trained on synthetic seismic cubes with a range of fault configurations.

A CNN is a deep learning neural network, that is designed to recognize visual patterns directly from 2D pixel images or 3D voxel volumes. This is due to the deep learning architecture being capable of capturing semantic contextual information as well recovering spatial information.

CNNs are built from locally connected layers, such as convolution, pooling and up-sampling. There are many variants of the CNN architectures such as LeNet, AlexNet, VGG, Inception ResNet. A popular network architecture for seismic fault prediction is a U–Net as shown below, Fig. 6.1.

Even a CNN that has been trained on hundreds of thousands of synthetically generated fault signature examples, will at some point, be presented with unfamiliar data with characteristics that differ from training. To maintain network accuracy, it is therefore necessary to allow an interpreter to update the network to the nuances of that unseen data. This can be achieved by allowing real-time fault training with the interpreter at the heart of the process.

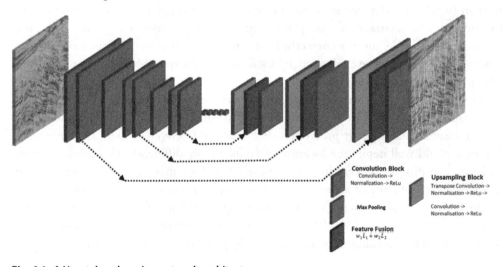

Fig. 6.1 *A U-net deep learning network architecture.*

On the case studies that we are going to show two types of networks were applied: 2D networks and 3D networks. The 2D networks analyze the data on the inline direction and crossline direction separately. This is, to a certain extent, mimicking how humans interpret. One of the biggest limitations of the 2D networks is its difficulty to identify curvilinear fault or fault that change azimuth but keeping the analysis limited to slices reduces the computation resources needed, allowing for the networks to run on a typical workstation. The 3D networks, on the other hand, analyze the data spatially, doing a true 3D, voxel by voxel interpretation. These are much more computationally intensive and so benefits from the hardware scaling capabilities of a cloud environment. Comparatively to the 2D networks, the results are more continuous, are able to capture changes in fault curvature and azimuth. The 3D nature of the network allows it to see through localized noise and data quality issues, often producing a superior interpretation.

6.3 Data and workflow

We are going to demonstrate and discuss the results of the networks using four different data sets. The case studies were chosen to show the performance of the networks in different geographical locations, different data quality, different scale of analysis (basin wide *vs.* reservoir scale) and offshore *vs.* onshore. All case studies used public access data.

The case studies are:

Beagle Sub–Basin, Australia – This is a basin-wide analysis. The data set is from the Beagle Sub-basin, Northwest shelf of Australia. It is an offshore data set, covering 4,466 km^2, with a size of approximately 100 GB. Data courtesy of Geoscience Australia (https://nopims.dmp.wa.gov.au/nopims).

East Shetland Basin, UK – This is a well-scale analysis. The data set is from the Tern-Eider Ridge, East Shetland Basin, Northern North Sea covering 720 km^2, with a size of approximately 11 GB. Data courtesy of the UK National Data Repository (https://ndr.ogauthority.co.uk/)

Main Pass, Gulf of Mexico, USA – Regional-scale analysis. The data set is from the Main Pass area, shallow water, Gulf of Mexico. It is an offshore data set, covering 270 sq. miles (~700 km^2), with a size of, approximately 10 GB. Data courtesy of the Bureau of Ocean Energy Management (https://walrus.wr.usgs.gov/namss/search/).

Cooper-Eromanga Basin, Australia – Regional-scale analysis. The data set is from the Cooper-Eromanga Basin, South Australia. It is an onshore data set, covering 165 km^2, with a size of, approximately 10 GB. Data courtesy of Government of South Australia (https://map.sarig.sa.gov.au/).

The workflow applied in each case study is slightly different as it is dependent on the data quality and objectives, but they have several steps in common.

Below is a generic workflow used in Geoteric^{TM} software; any deviation from this workflow is discussed during the individual case studies:

Sesimic data conditioning – A noise attenuation workflow is sometimes applied to the data before the AI analysis. This is to improve signal-to-noise ratio of the data to improve the AI results. A noise attenuation workflow is designed based on the objectives. All filters available are structurally oriented, they are applied on a plane that follow the local dip and azimuth of the reflectors (witin the window of analysis), as opposed to the X-Y plane. Each filter is designed to attenuate a specific type of noise: coherent, random, or aggressive. The coherent and random noise attenuating filters are relative amplitude and edge preserving. The aggressive noise filter, simple mean within the window of analysis, is not relative amplitude preserving and can smear across fault planes. The aggressiveness of the workflow depends on the number of filters used (in cascade) and the size of the filter sizes. A typical workflow would be a first pass of finite-impulse-response, median hybrid filter[13,14] to target coherent noise. This would be followed by a pass of the edge-adaptive, anisotropic tensor diffusion filter[15] to target random noise.

Attribute-based edge detection - This workflow is applied to generate a base line for fault identification comparison. The workflow consists of the combination (weighted average) of three edge detection seismic attributes: Tensor, SO Semblance and Dip. The Tensor attribute is based on a local structural tensor which is analyzed to find the dominant direction of the reflectors and their structure, it images faults that are expressed by lateral changes of amplitude. The SO Semblance is based on a measure of the Structurally Oriented Semblance. The Dip attribute computes the orientation of local structure within seismic data, the orientation is quantified in terms of Dip angles, and these are calculated at each point within the source data volume.

This is more for the benefit of the interpreter as the AI does not incorporate this result in the analysis.

Convolution neural network design and analysis – A deep learning network (2D and 3D) is trained with 1,000s of structurally accurate fault interpretations that have been mathematically modeled. This synthetically generated (and labeled) data provides our deep learning network with a strong foundation to predict the location of faults in unseen seismic data. The out-the-box foundation network can be run directly on 3D reflection seismic data without the need for additional interpreter-provided training examples. No other attribute or information (tectonic regime, geological setting, depositional environment, fault type, etc.) is necessary. With the foundation network being trained using labeled data (albeit mathematically generated with automatically generated interpretation labels), the network is classed as being supervised learning. The network output is a confidence volume. This volume ranges from 0 to 1 and depicts the probability the network gives of fault presence at each location.

Model training – Training is a process to calibrate or specialize a deep learning network for a data set or type of data set. The result of the training process is a network which should perform better than the input network, on data which is similar to the training set (which is vastly smaller than the original network training data, that is, tens of examples, instead of many thousands).

Training is the process by which the interpreter influences the network. The training is made by showing the network the interpreter's own interpretation via fault sticks. The teaching is made by positive and/or negative reinforcement. The way the network learns is by comparing its own interpretation with the interpretation done by the interpreter. If the same fault is interpreted by the interpreter and the CNN, then the network has confirmation that it did a good interpretation (positive reinforcement). If the CNN predicted a fault, but the interpreter did not interpret a fault stick in the same location, then the network knows it made a mistake (negative reinforcement) if the interpreter puts a fault stick in a location where there is no interpretation from the CNN them that is new knowledge to be learned by the network. With this method we can analzye the CNN model performance during model training.

Quality control – For quality control, we use two approaches; comparing the network results with the input seismic (either slice-by-slice or volumetrically by combining both volumes) and comparing the network results with the frequency decomposition results. The frequency decomposition approach has the advantage of comparing two results derived from two completely different mathematical processes. This comparison is also done volumetrically. Our current study does not quantify model performance, using metrrics, such as accuracy and error.

Re-analysis of the data – The trained network is used to analyze the seismic data again with a new interpretation being generated. If the results are still not satisfactory a new trained network is generated, and the data analyzed once more. This is an iterative process until quality of the results is within the objectives of the interpreter.

Integration of results – The results of the AI final run are integrated with the results from other workflows. The integration can be made with confidence volume (the volume acts like any other edge attribute) or fault sticks/fault surfaces can be automatically extracted using it as input.

6.4 Case studies

6.4.1 Beagle Sub-Basin, Australia

The Canning TQ3D seismic survey was acquired offshore NW Australia, on the Beagle Platform, positioned between the Beagle Sub-Basin and the Exmouth Plateau. The Platform is part of a regional NE-oriented structural high which forms the NW margin of the Beagle Sub-Basin, a Late Triassic-Early Cretaceous failed intracratonic rift that developed into a passive margin. The area consists of a series of north trending horsts and graben structures bifurcated by NE-striking faults found SW of NW-striking transform faults.[19,20] Four main populations of faults have been identified from seismic data.[16]

1. Late Triassic-Early Cretaceous N-S striking normal faults
2. Late Triassic–Early Cretaceous NE–SW-striking normal faults
3. Cretaceous polygonal faults
4. Neogene-present N-S and NE-SW trending en-echelon conjugate fault arrays

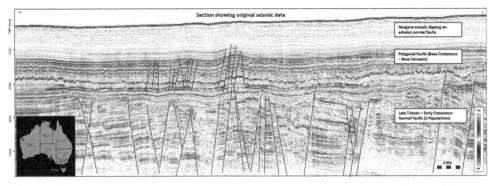

Fig. 6.2 *Vertical section of the Canning TQ3D dataset.* Dotted lines are examples of some of the manually interpreted fault. A thorough manual fault interpretation was not attempted.

In unexplored settings where the petroleum system is not well defined, it is crucial to understand the elements of structural geology to discover successful hydrocarbon plays. The 3D survey was intended for relatively frontier exploration purposes since there are only two wildcat wells drilled within the area. This is a high-quality data set with a very good signal-to-noise ratio. For this reason, no data conditioning workflows were applied prior to running the AI workflows.

As can be seen in Fig. 6.2, the biggest challenge is not identifying the faults, they are clearly visible in the seismic. The interpretation challenges are related to the density and complexity of the faulting, the decrease in the signal-to-noise ratio with depth, and the sheer size of the area to interpret and the total number of faults to map. With the typical oil and gas industry project time constraints, it would be difficult to map all the relevant faults. Data analysis to understand the tectonic evolution, potential play fairways and identify prospects within such a large area can be a very time-consuming process; therefore, the application of AI technology can be of great assistance to the interpreter.

The objectives of this case study were:
- Understand regional structural trends, which provide indication of the timing of tectonic events and formation of suitable structural hydrocarbon traps.
- Identification of specific compartmentalized fault blocks which could constitute potential prospects.
- The detail of fault closures that indicate a possible compartment may be breached.

The AI result for the full dataset was processed using a 3D CNN, Fig. 6.3. The results highlighted unseen structural details at both regional and local scales. Clear faults were detected with high confidence values throughout.

Quality control of the AI results was made by combining the output volume from the network analysis with the frequency color blend volume. Frequency decomposition and RGB color blending is a very well-established technique for stratigraphic imaging. Analysis of the interplay of three frequency restricted volumes allows us to see stratigraphic features that otherwise would be masked in the full spectrum data. The

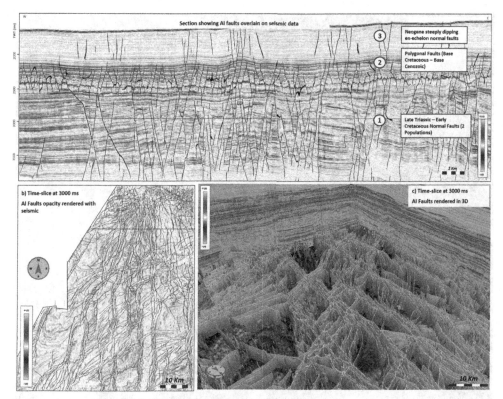

Fig. 6.3 *Results of CNN applied to the Canning TQ3D data set:* (A) section showing the fault interpreted by the network overlaid with the seismic, (B) faults interpreted by the network overlaid with the seismic at time slice 3,000 ms, (C) 3D rendering of the faults interpreted by the network (*yellow and green*) top of the rendered faults at time 3,000 ms. Modified after [21].

frequency volumes that are selected to calculate the blend depend on the objective, bandwidth of the data, what frequencies highlight the target zones, etc. One significant advantage of frequency decomposition is that it provides both stratigraphic and structural information. By comparing the AI results with the results of a well-established workflow such as frequency decomposition color blending (which uses a completely different mathematical process) the confidence on the AI fault interpretation increases. This is also a very fast process, and the QC can be made volumetrically, Fig. 6.4.

Two aspects are taken into consideration when comparing the AI network's interpretation and the frequency decomposition blend. One is a direct comparison of the faults identified by the two methods. In the frequency decomposition blend, discontinuities are imaged as dark lineations (simultaneous low magnitude response for the three frequency volumes, Fig. 6.4 blue arrow) if this aligns with the faults mapped by the network our confidence in the CNN interpretation increases. The second aspect is the sharp lateral changes in the frequency responses displayed as color changes in the RGB

Fig. 6.4 *Results of the CNN analysis is combined with the results from the frequency decomposition.* The combination is a volume that allows for a rapid QC of the faults interpreted by the network. Arrows indicate different types of lineations.

blend. Such sharp lateral transitions are commonly associated with the presence of faults, Fig. 6.4 yellow arrow (note: sharp changes in thickness can also produce sharp lateral changes in frequency). If the frequency response changes are gradual (Fig. 6.4 white arrow), they can be associated with either gradual thickness or stratigraphic transitions or a combination. If we have a dark lineation in the frequency decomposition blend that is not associated with a fault mapped by the network (Fig. 6.4 blue arrow) or a gradual lateral frequency change that is associated with a fault mapped by the network (Fig. 6.4 green arrow), then this are the areas that could be further investigated to understand why the network is not recognizing the feature as a fault.

The results can be further processed to fully characterize the AI-interpreted faults. For example, the fault azimuth was calculated for the faults interpreted by the network providing a rapid understanding of different families of faults. In Fig. 6.5 we compared the AI interpreted faults with the structural map form[16]. Fig. 6.5A) shows the summary fault architecture illustrating characteristic orthorhombic fault pattern, this is the structural interpretation made by[16], in Fig. 6.5B) we have the same section, with the faults interpretated by the AI, rendered in 3D. The tectonic fabric interpreted is almost identical, with the advantage that the AI interpretation shows a higher level of detail, particularly on the North-South trending faults (red). The biggest difference between the two interpretation is the time it took to achieve the results.

Using this volume, we can isolate faults with the same orientation. With it, we can clearly identify the different phases of tectonic deformation which the Beagle Sub-basin underwent. These results can also aid the manual fault interpretation (or automatic fault stick extraction), it can be used for regional or detail, reservoir level interpretation. The detailed interpretation produced by the CNN, with a high level of vertical and lateral continuity is an invaluable tool to interrogate the data in terms of possible compartmentalization. Specially if the faults are rendered in 3D, Fig. 6.3C).

The benefits of AI fault interpretation for rapid structural screening at a large scale are clearly illustrated in this case study. In a very short amount of time, the interpreter has an unbiased, high-resolution interpretation with more detail that can be analyzed three dimensionally. With subsequent processes applied to the CNN fault confidence volume the faults belonging to the different tectonic deformation phases could be

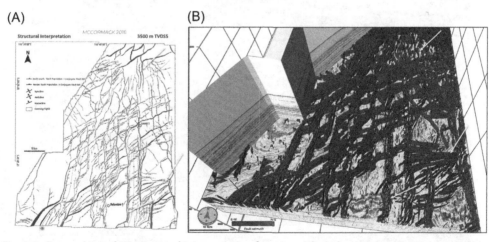

Fig. 6.5 Comparison of (A) structural interpretation from (modified after 16) and (B) AI interpretation. Faults are color coded by azimuth and show the latest Triassic–Early Cretaceous north–south-striking normal faults (*red faults*) and latest Triassic–Early Cretaceous NE–SW-striking normal faults (*blue*).

isolated and analyzed individually for a greater understanding of the tectonic evolution of the basin. All this information was generated without a single slice being interpreted or any other products fed to the network.

6.4.2 East Shetland Basin, UK

The previous case study perfectly illustrated the advantages of applying AI fault interpretation tools to new, unexplored areas at the basin scale. But the application of AI is not limited to those scenarios. In this case study, we go down to the well scale to showcase the advantages of being able to analyze the data spatially, voxel-by-voxel to de-risk a well. In 2013, the Tern development well (210/25a-A49) encountered wellbore stability issues which led to the subsequent sidetracking of the well. Using data provided by the UK Oil & Gas Authority, we put the AI Fault Interpretation tools to the test to see if the CNN could identify the faults in the area and so could have allowed the operator to avoid the drilling issues they encountered. The data has an overall good quality (Fig. 6.6), but there is still enough noise present to merit the application of a noise attenuation workflow.

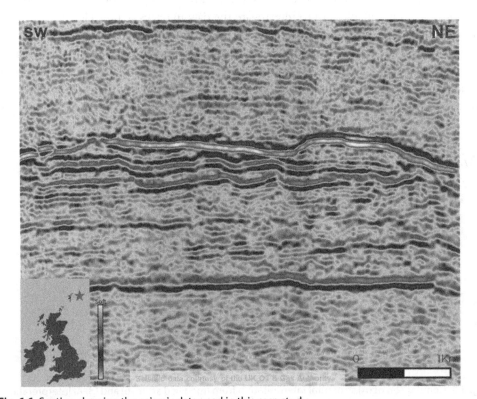

Fig. 6.6 Section showing the seismic data used in this case study.

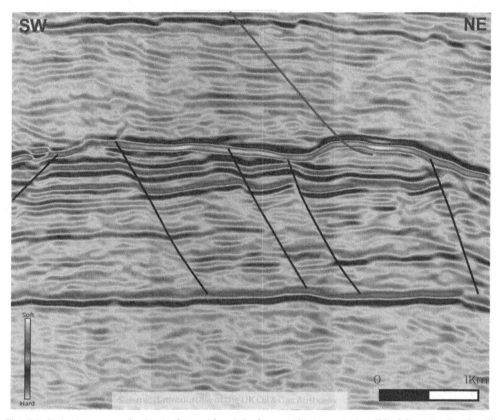

Fig. 6.7 Noise attenuated seismic data with original manual interpretation (*black lineation*) and well path up until the depth the well collapsed (*red line*).

The workflow consists of the application of several structurally oriented, edge preserving filters to attenuate coherent and random noise. All filters are relative amplitude preserving. Fig. 6.7 shows the same seismic section as in Fig. 6.6 after the application of the noise attenuation workflow along with the existing manual interpretation and location of the well when it collapsed.

The seismic volume was processed with the 3D CNN and a fault interpretation was generated. The fault confidence volumes were then used to calculate fault azimuths, Fig. 6.8. As can be seen in Fig. 6.8, the network identified the faults that were manually mapped, (red/yellow faults), these faults have an NNW-SSE orientation, roughly perpendicular to the inline direction. But it also identified faults that were not part of the manual interpretation, most notably at the well's total depth location and where it collapsed. The network not only identified a fault with the same orientation as the manually picked ones but also one along NNE-SSW (pink and purple colors). This family of faults runs almost perpendicular to the inline direction. This highly faulted area creates a

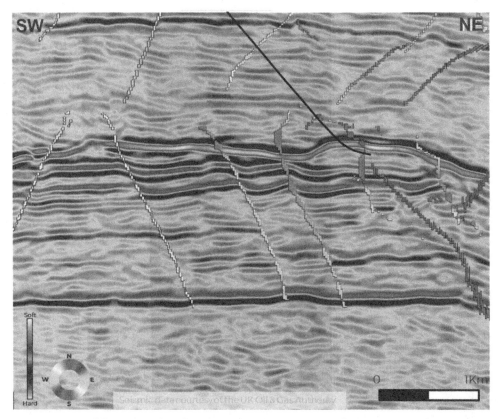

Fig. 6.8 *Results of the CNN overlayed on the seismic.* The color on the faults represents fault azimuth.

very unstable region and would justify why the well collapsed at this location, Zoback.[17] Looking at a 3D rendering of the interpreted faults, it is very easy to identify this area of additional structural complexity, Fig. 6.9.

At the end of well report, the recommendation is for a more detailed seismic and overburden analysis to better assist in future well planning and de-risking.[18] Time constraints and the amount of data that the geoscientist must go through make it very unlikely that faults with this geometry and relative direction in relation to the inline/ crossline would be interpreted. This is where the CNN 3D networks with their voxel-by-voxel analysis of the seismic show their true potential. It only takes a couple of hours to generate the volumes that would highlight the faults in this area, allowing the geoscientist to make the correct risk assessment making this an event free drilling campaign.

6.4.3 Main Pass, Gulf of Mexico, USA

We keep our journey west and now, move to the Gulf of Mexico in the United States of America. The main objective of this case study was to compare the traditional and AI-based tools available for fault imaging: seismic attributes, 2D CNN,

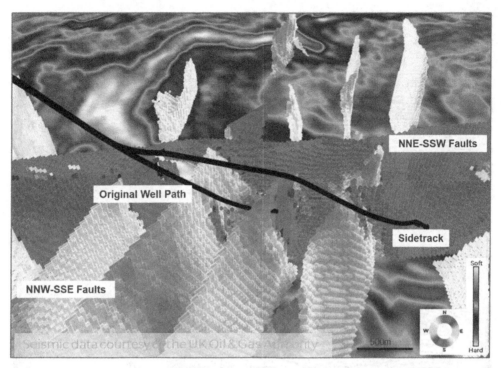

Fig. 6.9 *Seismic time section with 3D rendered AI interpretation.* Interpreted faults are colored by azimuth.

3D CNN and 3D trained network. For this test, we used a vintage data set from the Main Pass area, shallow water just off the mouth of the Mississippi River, Fig. 6.10.

The data set, acquired in 1995 by Western Geco, exhibits several imaging challenges: the quality of the data degrades very rapidly with depth, there is a very high noise content and several stratigraphic features (slumps, channels of different scales, contourite edges, etc.). In terms of noise, we have random noise in the shallow section of the data and in the deeper section there are migration artifacts (migration smiles) and diffraction tails. The stratigraphic elements, present a challenge to the CNN in the same way they do for the edge attributes: they feature an edge (channel edge, carbonate buildup edge, etc.) that can be interpreted as a fault. If the seismic character of the stratigraphic edge is very similar to the seismic character of faults they can be interpreted (by the CNN and/or edge attribute) as a fault. Further labeled examples must be given to the CNN for it to be able to distinguish between the two seismic characters.

The results are shown along two time slices: t = 1.4 seconds – where the data quality is still acceptable and t = 3.4 seconds – where data quality decreases substantially and there are several imaging problems, Fig. 6.10.

To attenuate the noise and reduce the number of false positives a very gentle noise attenuation workflow was applied to the original data. The parameters (filter sizes) and filters (structurally orientated, edge preserving filters) used in the workflow design were

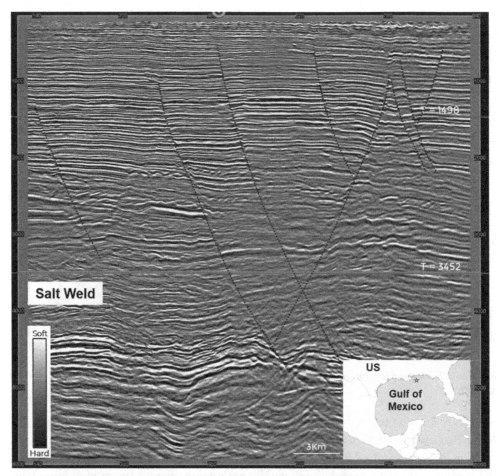

Fig. 6.10 *Vertical section showing the seismic data used in this case study.* Horizontal lines show the positions of the time slices where the results will be shown. Dotted lines show manually interpreted main regional faults.

chosen to guarantee that only noise was being removed. We are conscious that a more aggressive noise attenuation workflow could be applied, one that would remove more noise, but this would be done at the expense of losing some signal. Since we wanted to preserve as much signal as possible and wanted to test the robustness of the CNN to the noise, we designed a gentle noise attenuation workflow. This noise attenuated volume was used as an input for all subsequent edge detection workflows.

The attribute-based workflow that consisted of the combination of three different edge attributes (tensor, amplitude analysis; structurally oriented semblance, phase analysis; dip, geometrical analysis) was the one with the worst performance. In the shallow section (Fig. 6.11, top left) the combination of attributes was able to image the

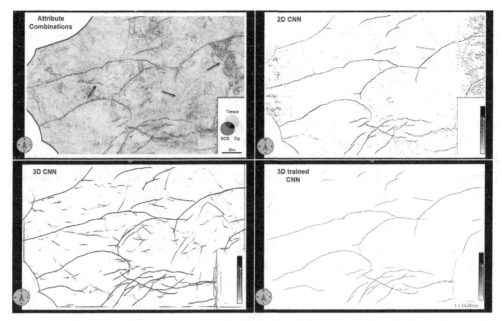

Fig. 6.11 *Composite image showing the results of the four different workflows in the shallow section (time slice t = 1.4 sec):* top left – attribute combination of tensor, SO semblance and dip; top right – 2D CNN interpretation; lower left – 3D CNN interpretation; lower right – 3D trained CNN interpretation.

main regional faults (black lineations) but it also picks, virtually, all stratigraphic edges, The arrows on Fig. 6.11, top left image point to a contourite edge (topmost arrow), channel edges of different scales and a channel complex (right most arrow). These responses to the stratigraphic edges are, in some cases, as strong as the imaged structural edges making it virtually impossible to separate them, even in post imaging. There is also a lot of noise being picked up by the attributes while the strongest responses are from the edges of the survey.

The 2D CNN results are a lot cleaner, in relation to the attributes (Fig. 6.11, top right image). The network picks a lot less noise and it is only picking segments of the channel complex in the right side of the survey. The channel response is weaker in relation to the regional faults making it possible to isolate and eliminate those responses in subsequent processes. This is a steep change in quality, compared with the edge attributes. This volume would provide a much better input for automatic fault stick extraction. The 2D network is also picking the edges of the survey, this case with a response like the faults. Another quality step change happens with the 3D CNN, Fig. 6.11 (bottom left image). The fault responses are much more continuous, and the complexity of certain areas (lower section of the data) is captured correctly. There are still some limitations: the network is more influenced by the noise (compared with the

2D network), in the areas in-between the regional fault it is trying to make sense of the seismic response, possibly interpreting coherent noise as small-scale faults. Similar to the other two methods, it also picks up the edges of the survey.

In the next example, we show the effect of training the network. To train the network, we interpreted a couple of slices and showed the interpretation to the network. The interpretation focused on the regional faults for positive reinforcement and on the stratigraphic features and edges of the survey for negative reinforcement. The results of the trained network can be seen in Fig. 6.11, bottom right image. The improvement in the result quality is in the order of magnitude compared with the other methods (attributes, 2D, and 3D networks). No stratigraphic features are imaged, minimal noise is present, and the edges of the survey are not present in the final result, the volume only contains structural information. The faults are continuous, and the structural complexity is properly imaged. This is a superior structural volume and an ideal input for the automatic fault stick extraction. In Fig. 6.12 we show the results of the different methods 3D rendered to further illustrate their differences.

Although the conventional seismic attributes show a reasonable result, identifying the main regional faults with high vertical and lateral continuity, they are poor in

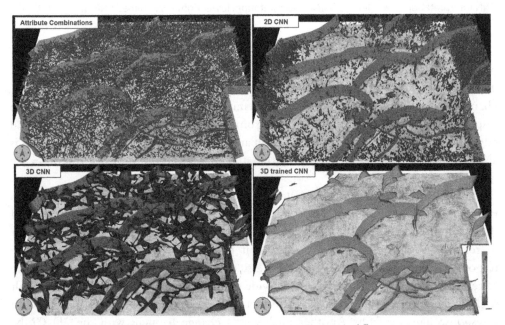

Fig. 6.12 *Composite image showing the results of the four different workflows in the shallow section (time slice t = 1.4 seconds) rendered in 3D:* top left – attribute combination of tensor, SO semblance and dip; top right – 2D CNN interpretation; lower left – 3D CNN interpretation; lower right – 3D trained CNN interpretation.

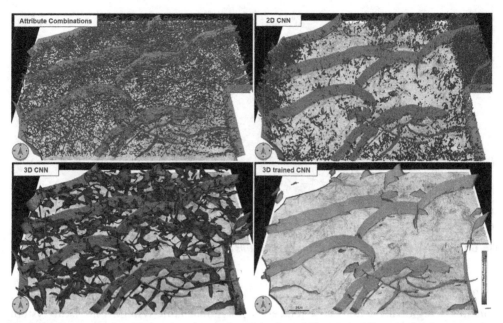

Fig. 6.13 *Composite image showing the results of the four different workflows in the deeper section (time slice t = 3.4 sec) rendered in 3D:* top left – attribute combination of tensor, SO semblance and dip; top right – 2D CNN interpretation; lower left – 3D CNN interpretation; lower right – 3D trained CNN interpretation. The color scale indicates increasing fault probability (towards red).

comparison with the AI-based results, with the 3D trained network outshining them all. The difference between the results is even more staggering in the deeper section, Fig. 6.13.

This case study illustrates the benefits of CNN, in comparison with the traditional attribute-based workflows, especially where it matters the most, in the deeper section of the data, where data quality decreases dramatically. Also evident is the benefit of model training. The interpreter's input into the network is essential to its performance improvement. In the more challenging areas, the interpreter's participation is vital to optimize the network. Of course, the CNN also has its own limitations. Like all other processes that rely on seismic data, the quality of said data still influences the quality of the results. The ability to identify a fault and the quality of that identification is completely dependent on the quality and quantity of training. Same of this shortcoming can be mitigated through further training but again the CNN's performance improvement is dependent of the quality of the interpretation that is used during the training phase.

6.4.4 Cooper-Eromanga Basin, Australia

We now complete our journey around the world and return to Australia. This time we are going to look at an onshore seismic data set.

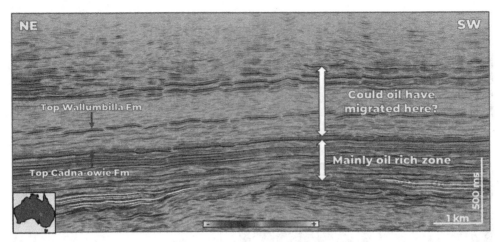

Fig. 6.14 *Section showing the seismic data used in this case study.*

The Cooper–Eromanga Basin is one of the biggest onshore basins in Australia. It contains a regionally extensive Cretaceous polygonal fault system. This fault system contains vast numbers of small faults with no preferred direction of either strike or dip, meaning the pattern of faulting on one horizon will not persist to overlying or underlying horizons. Understanding of this structure would have substantial implications for hydrocarbon migration pathway models and structural and stratigraphic models.

When drilling for deeper targets there were some hydrocarbon shows in the Wallumbilla Formation, the hypothesis was that the oil-rich Cretaceous source rocks in the underlying Cadna-Owie Formation were likely charging the reservoirs above and that faults were acting as conduits for flow, Fig. 6.14.

The objectives of this case study were to interpret the complex polygonal fault system, filter the identified faults by direction to determine what faults could be providing potential migration pathways to the reservoir in the Wallumbilla Formation.

The workflow consisted of running the 3D CNN on the original seismic. The fault confidence volume was then used to extract the faults as single voxel lineations and these lineations were re-embedded in to the seismic. To determine what faults are open and can act as migration pathways we utilized the work of Kulikowski et al., 2017.[24] In this work, the authors claimed that the faults that extended into the Lower Cretaceous oil-rich reservoir, with strikes between 060°N and 140°N and a near-vertical dip angle, had most likely been acting as conduits for the tertiary migration of hydrocarbons from known Lower Cretaceous hydrocarbon reservoirs into shallow Cretaceous sediments. This angle range was used to classify the faults as open and closed. Different colors were assigned to the two different classes of faults, Fig. 6.15. Yellow for the open faults and blue for the closed faults. The ability to analyze the data in a true 3D sense enabled the network to image the faults with the correct geometry, angular position, and continuity with a degree of confidence that allowed for these results.

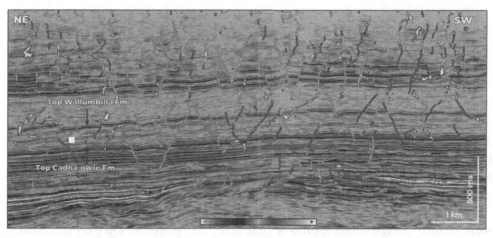

Fig. 6.15 *Section showing the faults interpreted by the 3D CNN.* The color depicts the classification that was made based on the work of Kulikowski et al., 2018.[24] Blue faults – closed faults; yellow faults – open faults.

6.5 Discussions

The four case studies presented show the benefits of applying convolutional neural networks for fault interpretation in a variety of geological settings, objectives, and data types. Compared to the traditional attribute-based workflows, the AI networks improve the results by orders of magnitude, especially the 3D CNN. The networks are extremely robust to noise, giving meaningful results even in the more challenging data (where edge attribute results are uninterpretable). In most cases just the CNN, with its generic training, is enough to greatly improve the results compared with the edge attributes, with the added benefit that it performs an unbiased analysis. The CNN is not without its limitations. The network's performance is only as good at the training it got. This applies, not only to its ability to identify a fault but also its ability to see through the noise in the seismic data. The number of labeled examples also limits the network's ability to identify particular faults.

Low-angle faults, reverse faults, and faults that change angle along the depth are very challenging. In most cases, the seismic character of this types of faults is very similar to actual reflectors making it very hard for the network to interpret them as faults. In the case of these particular faults, even training the network has its limitations since adding these faults, this seismic character to the knowledge of the network will also increase the number of reflectors being interpreted as faults. The interpreter must find the right balance between incorporating this seismic character into the networks knowledge for the network to identify that fault but not giving it too much weight to the point that too many reflectors are also being interpreted as faults.

The 2D networks are less resistant to noise (compared with the 3D networks) and the slice-by-slice, inherent analysis produces less continuous results, (still better than the edge attributes) but can run on a standard workstation/laptop. Some of these shortfalls can be mitigated with training the network.

With training of the 3D CNN models the interpreter, not only we can improve the network's knowledge but also can teach the network the difference between stratigraphy edges, survey edges, steep-dipping reflectors, noise, and structure. The ability of the 3D CNN to acquire new knowledge and analyze the data voxel-by-voxel gives these networks the capability of interpreting structurally complex areas. Task-oriented model training also gives the interpreter the ability to increase or decrease the networks' sensitivity, making the network to interpret the subsurface on a regional or reservoir scale.

By teaching the network that subtle seismic changes are faults will make that network perform a more detailed, small-scale detailed interpretation. On the other hand, if more emphasis is given to the regional faults during training (example Gulf of Mexico case study), the network will undertake a more regional interpretation. There is also the possibility to overtrain the network. If too many examples of the same fault type are given to the network, then this will be the only type of fault the network will know and interpret, even if they are not present in the data.

A key aspect of the CNN, the ability to learn, puts the interpreters in the center of the process. Not only that, but human interpreters are an indispensable part of the workflow. Without the interpreter, the network cannot learn past the initial, generic training. To fully optimize the results and take advantage of all the network capabilities the interpreter needs to be part of the process. Another important aspect is that the AI results cannot be isolated in themselves. To take full advantage of the improved results, the AI interpretation needs to be incorporated in the industry's workflows. This integration also makes the QC of the AI results much easier. In order to trust the AI's results, they must be compared with known, trustworthy, and independent workflows as well as geological ground-truth (image logs, analogs, outcrops, etc.). In the case studies above, for QC, the AI results are always combined, at least, with the frequency decomposition results.

The quality and integration of the AI fault confidence volumes allows for the direct automatic extraction of high-quality fault sticks and/or fault surfaces. This is a fundamental characteristic for a future fully automated workflow that generate fault framework from the seismic data.

These networks have the potential to revolutionize seismic interpretation, raising it from the typical 2D section-based analysis to full 3D fault mapping, taking full advantage of the 3D seismic. All of this with the interpreter at the center of the process.

6.6 Conclusions

Although many still consider the use of machine learning-based algorithms in seismic interpretation as the future, or, at best, a niche workflow, these solutions have inevitably started transforming the way geoscientists interpret and understand the subsurface. The four case studies presented here clearly illustrate what 3D convolutional neural networks bring into fault interpretation, regardless of the size of the data volume, or the scale of the objective: small-scale risk assessment along a planned well path or basin-scale structural reconnaissance equally benefit from the 3D networks. The CNN can quickly deliver highly detailed, accurate fault delineations enabling the interpreters to start their analysis from a very different level than previously. As these networks were built around the interpreter, they are open and ready to learn from the geoscientists – they are digital coworkers augmenting the interpretation workflow.

In the (very) near future, improved deep learning algorithms will be able to analyze a wider range of data and offer a more complete description of the structural framework: drilling information, well logs, image logs, subsurface or outcrop analogues and the physical laws governing rock deformations will also be considered.

All these changes mean that the interpreters also need to adapt and must learn and understand how the different machine learning models make decisions, how to provide accurate information for the algorithms to learn from, and how to control and QC the results of the computer algorithms. These can only be done with value-driven solutions that aim to build on the human knowledge, experience, and intelligence, and offer interactivity instead of a black box approach.

6.7 Acknowledgments

We would like to thank the UK Oil & Gas Authority, the Government of South Australia, Department of Energy and Mining, Geoscience Australia, and the U.S. Bureau of Ocean Energy Management for making the seismic data used in the case studies publicly available.

We would also like to thank Geoteric for supporting us in the writing of this chapter.

References

1. Chaves MU, Oliver F, Kawakami G, Di Marco L. Visualization of geological features using seismic volume rendering, RGB blending and geobody extraction: 12th International Congress of the Brazilian Geophysical Society held in Rio de Janeiro, Brazil, August 15-18, 2011; 2011.
2. Haslina M, Trisakti K, Farisa MZ, M Rahmani H, Nurhakimah M, Luis Gomez M et al. Integrated 3D Quantitative Analysis and Interpretation of Pay Sand in Basin Floor Channel Complexes - Case Study Example in Deep Water Sabah: APGCE 2015 Kuala Lumpur, Malaysia, 12-13 October 2015; 2015.
3. Bahorich MS, Farmer SL. 3-D seismic discontinuity for faults and stratigraphic features: the coherence cube. *The Leading Edge* 1995;**14**:1053–8.
4. Marfurt KJ, Kirlin RL, Farmer SL, Bahorich MS. 3-D seismic attributes using a semblance-based coherency algorithm. *Geophysics* 1998;**63**:1150–1165.
5. Gersztenkorn A, Marfurt KJ. Eigenstructure-based coherence computations as an aid to 3-D structural and stratigraphic mapping. *Geophysics* 1999;**64**:1468–1479.
6. Iacopini D, Butler RWH, Purves S. Seismic imaging of thrust faults and structural damage: a visualization workflow for deepwater thrust belts. *First Break* 2012;**30**:77–84.
7. Grechka V, Li Z, Howell Bo, Garcia H, Wooltorton T. High-resolution microseismic imaging. *The Leading Edge*. the Society of Exploration Geophysicists 2017, pp. 822–888.
8. Wang Z, Di H, Shafiq MA, Alaudah Y, AlRegib G. Successful leveraging of image processing and machine learning in seismic structural interpretation: A review. *The Leading Edge* 2018;**37**:451–461.

9. Wu X, Shi Y, Fomel S, Liang L. Convolutional neural networks for fault interpretation in seismic images, SEG Technical Program Expanded Abstracts, 2018:1946–1950.

10. Guitton A. 3D Convolutional Neural Networks for Fault Interpretation, Conference Proceedings, 80th EAGE Conference and Exhibition 2018, 2018; **2018**:1–5.

11. Huang L, Dong X, Clee TE. A scalable deep learning platform for identifying geologic features from seismic attributes. *The Leading Edge* 2017;**36**:249–256.

12. Pochet A, et al, Seismic fault detection using convolutional neural networks trained on synthetic poststacked amplitude maps, Monografias em Ciência da Computação n° 03/2018, ISSN 0103-9741.

13. Heinonen P, Neuvo Y. FIR-median hybrid filters, IEEE Transactions on Acoustics, Speech, and Signal Processing 351987:832–838. http://doi.org/10.1109/TASSP.1987.1165198.

14. Nieminen A, Heinonen P, Neuvo Y. A new class of detail-preserving filters for image processing. IEEE Transactions on Pattern Analysis and Machine Intelligence, PAMI-91987:74–90. http://doi.org/10.1109/TPAMI.1987.4767873.

15. Perona P, Malik J. Scale space and edge detection using anisotropic diffusion. IEEE Transactions on Pattern Analysis and Machine Intelligence 12;**1990**:629–639. http://doi.org/10.1109/34.56205.

16. McCormack KD, McClay KR. Orthorombic faulting in the Beagle Sub-basin, North West Shelf, Australia. In: K.R. McClay, J.A. Hammerstein (Eds.), Passive Margins: Tectonics, Sedimentation and Magmatism. Special Publications, Geological Society, London 2018, pp. 476.

17. Zoback M. Reservoir Geomechanics. 2007 Cambridge: Cambridge University Press doi:10.1017/CBO9780511586477.

18. Dampier T, Sayers N, Pearce A, Miles P, Bevan J and Atkin T, 2013, End of Well Report, TA07S3/TA07S4, 210/25a-A49 /Slot TA07, 210/25a-A49z / Slot TA07

19. Zhan Y, Mory AJ. Structural interpretation of the northern Canning Basin, Western Australia. The Sedimentary Basins of Western Australia edited by M Keep and SJ Moss: West Australian Basins Symposium, Perth, 18–21 August 2013, 2013. Petroleum Exploration Society of Australia:18.

20. Horstman EL, Purcell PG. The offshore Canning Basin - a review P.G. Purcell, The North West Shelf, Australia: Proceedings of the Petroleum Exploration Society of Australia Symposium Perth 1988; **1988**, pp. 253–257.

21. Han C, Cader A. Interpretational applications of artificial intelligence-based seismic fault delineation. *First Break* 2020;**38**(3):63–71. https://doi.org/10.3997/1365-2397.fb2020020.

22. AAPG Wiki. Seismic Interpretation. https://wiki.aapg.org/Seismic_interpretation. Date accessed: February 17, 2022.

23. SEG Wiki. Structural Interpretation. https://wiki.seg.org/wiki/Structural_interpretation. Date accessed: February 17, 2022.

24. Kulikowski D, Amrouch K, Cooke D, Gray ME, Basement structural architecture and hydrocarbon conduit potential of polygonal faults in the Cooper-Eromanga Basin, Australia. Geophysical Prospecting, January 2018;**66**(2):366–396. https://doi.org/10.1111/1365-2478.12531.

PART 3

Physics-based machine learning approaches

CHAPTER 7

Applying scientific machine learning to improve seismic wave simulation and inversion

Lei Huang, Edward Clee, Nishath Ranasinghe
Department of Computer Science, Prairie View A&M University, Prairie View, TX, United States

Abstract

Accurate simulation of wave motion for the modeling and inversion of seismic wave propagation is a classical high-performance computing (HPC) application using the finite difference, finite element, and spectral element methods to solve the wave equation numerically. The article presents a new method to improve the performance of seismic wave simulation and inversion by integrating the deep learning software platform and deep learning models with the HPC application. The paper has three contributions: (1) instead of using traditional HPC software, the authors implement the numerical solutions for the wave equation employing recently developed tensor processing capabilities widely used in the deep learning software platform of PyTorch. By using PyTorch, the classical HPC application is reformulated as a deep learning recurrent neural network framework; (2) the authors customize the automatic differentiation of PyTorch to integrate the adjoint state method for an efficient gradient calculation; (3) the authors build a deep learning model to reduce the physical model dimensions to improve the accuracy and performance of seismic inversion. The automatic differentiation functionality and a variety of optimizers provided by PyTorch are used to enhance the performance of the classical HPC application. Additionally, methods developed in the paper can be extended into other physics-based scientific computing applications, such as computational fluid dynamics, medical imaging, nondestructive testing, as well as the propagation of electromagnetic waves in the earth.

Keywords

Inverse problem; Scientific machine learning; Seismic imaging; Seismic wave propagation

7.1 Introduction

This chapter explores the use of Scientific Machine Learning (SciML)[1] concepts in addressing the problem of estimating properties of the Earth associated with the propagation of seismic waves. SciML has recently emerged as a new method to solve the scientific computing problems using machine learning models and vice versa. The method leverages the success of traditional scientific computational models and the advances in data-driven machine learning models to augment the efficiency and accuracy of scientific simulation and inversion. Moreover, it facilitates the scientific discovery by modeling both well-known scientific rules and the unknown patterns based on observed data.

Advances in Subsurface Data Analytics
DOI: https://doi.org/10.1016/B978-0-12-822295-9.00011-X

167

In this chapter, SciML is studied and applied to simulate the wave propagation and to enhance the quality of the seismic inversions. In seismic reflection surveys, sound waves from controlled sources are recorded by a large number of surface- or borehole-deployed sensors (receivers) which capture signals both directly transmitted as well as reflected from structures at depth, for the purpose of deducing structural variations in the velocity and density properties of the subsurface rock formations.

7.1.1 Seismic imaging

Advanced seismic imaging is the process of constructing a model (or "image") of the Earth properties through an "inversion" process[2], starting with an initial guess, simulating the wave propagation for the field survey and computing a misfit with respect to the field-recorded data, and iterating through a gradient-based optimization process until an acceptable Earth model is achieved. The particular method of Full-Waveform Inversion (FWI) is typically implemented using the adjoint state method[3] to calculate the gradients.

Fig. 7.1 shows the workflow of seismic inversion. The initial model M_0 is a guess of the true model M that is the unknown "target" of the inversion. These early experiments, comprising several shots of a synthetic seismic reflection survey over a small 2-D Earth model, used for convenience an initial model guess that is a smoothed version of the true model. The seismic traces are either observed via real-world survey recordings or are simulated using the seismic wave forward function in this paper. The residual is obtained by comparing the synthetic data and observed data. The gradient $\frac{\partial u}{\partial M}$ is calculated based on the residual with respect to the initial model. The gradients are used by a gradient-based optimizer to update the initial model to get a step closer

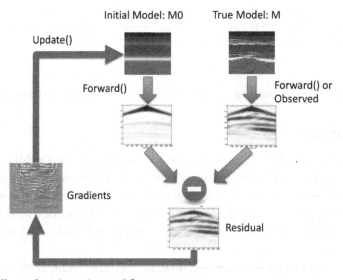

Fig. 7.1 *The full waveform inversion workflow.*

to the real model. The entire process ends when the initial model and the true model are adequately converged or the specified limit of iterations is exceeded. Enhancements to this process using machine learning are treated in a later section, after discussion of additional required concepts.

7.1.2 Computational issues for wave propagation

Physical simulation and inversion are classical scientific computing applications to examine the physical phenomenon and reveal the underlying medium properties. The simulation solves the partial differential equations (PDEs) that govern the physical phenomenon using numerical approximation methods, while the inversion applies the gradient-based optimizations to find the underlying properties by minimizing the misfit between the observed data and the simulated results. The entire process takes significant computing resources to achieve the required accuracy. However, the inverse problem is naturally challenging since it is ill-posed and nonlinear for most cases[4]. In particular, the optimization process in the non-convex system frequently attempts to converge to a local minimum instead of the desired global minimum, due to the effect of "cycle skipping" in which the alignment of seismic events is improperly discerned because of insufficient sampling in time or space. Practical approaches to mitigate these problems are discussed by Virieux and Operto[5], and still need to be applied even when using machine-learning alternative methods.

Another challenge in implementing the seismic inversion process is the need to correctly compute the adjoint-state solution whereby the gradient for updating the physical parameters in the model space must be calculated by backward propagation of the residual data field at each observation point for each source. This calculation is at least as challenging as the original forward propagation through the model, and represents a differential process with stringent accuracy requirements.

7.1.3 Opportunity for data analytics

A potential application of data analytics technology in the analysis of seismic reflection imaging data is to utilize the characteristic patterns exhibited in seismic reflection recordings to learn the full-waveform responses of various Earth structure patterns for different survey geometries by means of a large number of model experiments[6], and train a very deep neural network to recognize such responses in actual field-recorded surveys and discern likely structures that are generating such responses. Inference of structure from such a data-driven approach could then be faster than traditional inversion operations that require many forward and back-propagated simulations in the optimization process. Such a model-driven process can be made efficient by the using a recurrent neural network (RNN) implementation of the wave propagation forward model simulations, which can be transparently implemented on specialized hardware through the newer tensor processing systems, such as TensorFlow and PyTorch.

Ultimately, the objective is to integrate the machine learning models that are data-driven into the scientific computational models that are physics-driven. The differentiable programming method has the potential to smoothly integrate the two models together with a global optimization. It is expected that this study will lead to more interesting research findings in the topic of SciML and to discovering ways to combine the power of these two different methods to facilitate scientific discovery.

7.1.4 Scientific machine learning

A recent push to incorporate domain specific knowledge with machine learn ing has shown great promise and it is emerging as a sub-field under the name of SciML. In a recent workshop organized by the US Department of Energy to identify basic research needs for SciML, a group of leading researchers in the field identified domain-aware, interpretable and robust machine learning as fundamental to the development of SciML[7]. They further highlighted the growing enthusiasm on SciML among the wider scientific community is due to the broad availability of HPC resources; developments in computationally effective data analysis algorithms as well as the availability of large amounts of high-quality discrete and continuous data collected from scientific instruments[7]. Furthermore, SciML could replace meshing associated with classical numerical discretization techniques, such as finite element method, finite difference method with neural networks that could approximate the solutions of physical problems described as PDEs with the use of automatic differentiation[8], and also break free from the curse of dimensionality[9]. The main idea behind obtaining approximate solutions to PDEs in SciML is to constrain the neural network to minimize the residual of the PDE. The SciML research currently includes the physics-informed neural networks (PINNs)[10,11], the universal differential equation (UDE) method[12], the Hamiltonian neural networks (HNNs)[13,14], and the neural ordinary differential equation (NODE)[15] to learn the non-linear dynamics in physics.

The SciML method with integration of data and physics modeling is demonstrated in two ways in this chapter:
1. By expressing wave propagation in the form of a RNN, allowing the gradients calculated by automatic differentiation to be used for the FWI of seismic reflection survey data, and
2. Incorporating existing forward wave propagation and adjoint-state inversion solutions through extensions to the deep learning auto-differentiation tools, thus allowing seismic inversion to be integrated with a deep learning model to create an end-to-end differentiable programming environments.

The contributions of this work are summarized in the following:
1. The PDE solver in the seismic forward model is reformulated using RNN and implemented with the deep learning software package, PyTorch, which allows taking advantage of the tensor processing software and its accelerator implementation.

2. The automatic differentiation feature implemented in PyTorch enables calculating a model gradient to solve the seismic inverse problem to uncover the earth's interior physical properties.
3. Efficiency of the automatic differentiation feature is improved by extending it to create a hybrid back propagation using the adjoint-state method to calculate the gradients.
4. The optimization is incorporated into an autoencoder network to reduce the dimensions of the inverted parameters to augment the convergence process and get more accurate results for the ill-posed problem.

7.2 Related work

7.2.1 Seismic wave simulation and inversion

In this study, the application of a tensor-based machine learning system for solving physical simulation and gradient-based optimization problems is illustrated using seismic wave propagation simulation and full waveform inversion (FWI) as the physical case study.

Both Richardson[16] and Zhu and Xu[17] have mathematically demonstrated the equivalence of wave-propagation inversion by the adjoint-state method with a RNN and its back-propagation gradient computed through automatic differentiation. Richardson uses this equivalence to incorporate an existing implementation of forward modeling and adjoint-state inversion as an extension to the automatic differentiation feature of tensor processing, allowing interventions for explicitly managing memory and computational demands. Zhu and Xu, taking an opposite approach, have formulated the wave propagation problem as an RNN for which the adjoint-state inverse need not be explicitly coded, but instead is implemented automatically using the automatic differentiation capability built into the tensor processing system. This allows both forward and back propagation to be transparently managed on CPU and GPU processors, with memory and computational demands satisfied simply by allocating more resources in parallel as needed.

7.2.2 Surrogate models using machine learning

Longer run times and slower convergence inhibit the rapid integration of scientific models in real time applications and critical decision making processes. Additionally, the "curse of dimensionality" is encountered when the number of samples required to cover the parameter space increases with the number of parameters in the model. The issue of longer run times demands that models be simplified, ignoring physical process and reduction in the numerical accuracy of the models. On the other hand, surrogate models facilitate the execution of complex physical models while preserving numerical accuracy. Surrogate models are also referred to as meta models[18], model emulators[19], reduced order models[20,21], physics based proxy models[22], multifidelity models[23], and response surfaces[24].

7.2.3 Dimensionality reduction

Large realistic seismic inversions in geologically complex settings often utilize global optimization algorithms, such as grid search[25], simulated annealing[26], very fast simulated annealing[27], Monte Carlo optimization[28], particle swarm optimization[29] due to the presence of many local minima in the misfit function between the model and the data. However, global optimization modeling suffers from slow convergence rates as the dimensionality of the model space is increased. In machine learning, autoencoders are widely utilized to reduce dimensionality of the model space. An autoencoder neural network[30] is an unsupervised learning algorithm that implements backpropagation in an effort to generate output values that are equal to the input values. It consists of an encoder and decoder; the encoder operator reduces the dimensionality of the model space and the decoder operator tries to regenerate the original input data from the low-dimensional representation.

Moseley et al.[31] used WaveNet network architecture[32] and a conditional autoencoder to simulate the earth response to the propagation of acoustic waves in horizontally layered mediums and more general faulted 2D mediums, respectively. The encoder network in the autoencoder is composed of 10 convolutional layers, which reduces the spatial dimension of the input velocity model to a latent vector of 1*1 with 1,024 hidden channels. The decoder network utilizes 14 convolutional layers to expand the latent space back into the original dimension. All the hidden layers in the autoencoder network utilizes ReLU activation functions[33] except the final output layer which utilizes an identity activation function. For the both networks, they observed the resulting seismic inversion are 20–500 times faster than finite difference modeling. But the authors noted, extending the deep neural network based seismic inversion into more complex, elastic and 3-D models will be challenging.

Gao et al.[34] used an autoencoder neural network to efficiently conduct a 1-D seismic impedance inversion implementing trace-by-trace inversion with regularization using a fully connected convolutional neural network. They used differential evolution[35], a stochastic, population-based global optimization algorithm, which solves optimization problems based on a population of individuals in which each is a candidate solution. The authors initially trained the autoencoder network to learn to encode the large dimension problem into a reduced order dimension problem and then decode it back to the initial large dimension problem. They solve the large dimension problem by searching the best acceptable solution in the reduced order space and then map the selected solution back to the higher order space to obtain the most acceptable model for the initial problem. They further observed a good initial model supplemented with well-log data is necessary to obtain an accurate inversion. The authors further showed their method converges faster and produces more robust results than common inversion methods for both synthetic and field data.

Another study by Chen and Schuster[36] used skeletonized seismic data, which consists of the low-rank latent-space variables produce by an adequately trained autoencoder network, to obtain a subsurface velocity model by inverting the wave equation. The authors coined the name Newtonian machine learning (NML) for their method as it inverts for the model parameters by utilizing both forward and backward operations of the Newtonian wave propagation while utilizing dimension reduction capabilities of machine learning. They used seismic traces as the input to the autoencoder network, which calculates the perturbation of skeletonized data with regard to velocity perturbations (Fréchet derivative). Then the gradient is calculated by migrating the shifted observed traces weighted by the the skeletonized data residual. Finally, the model which best determines the observed latent-space parameters are chosen as the final velocity model. The authors also noted the their method could reduce the effects of cycle skipping which plagues FWI and it does not require manual picking of important features as the skeletal data are automatically generated by the autoencoder network. They further noted that the velocity model produced from their model has low resolution and it can be used as a starting model for FWI.

A recent FWI study by Sun and Alkhalifah[37] used meta-learning to train a neural network to learn the optimization algorithm instead of using a predesigned optimization algorithm. They used gradient of the misfit function as the input of a RNN and the history information of the gradient was used as the hidden states in the RNN[37]. The authors formulated the loss function for the training as a weighted summation of the L2-norm of the data residuals. Furthermore, they accelerated the training process of the neural network by minimizing randomly generated quadratic functions by locally approximating the optimization as a a linear convex problem. Additionally, they used variational autoencoder[38] methods to project and represent the model in latent space to achieve more accurate and robust velocity model. They obtained faster convergence rates compared to conventional FWI for both Marmousi[39] and the Overthrust[40] models.

Mosser et al.[41] used deep generative models in the context of inverting the acoustic wave equation to obtain subsurface models. They used Wasserstein generative adversarial networks (GAN)[42-44] to create subsurface geological structures and their respective petrophysical properties as a priori model. Then they combined the models with the acoustic wave equation and performed Bayesian inversion using an approximate Metropolis-adjusted Langevin algorithm (MALA)[45] drawing samples from a posterior seismic observations. The authors used the adjoint method to calculate the gradients with respect to the model parameters governing the forward problem and they exploited the differential nature of the deep neural network to calculate the gradient mismatch. Swischuk et al. demonstrated projection-based model reduction.[46]

7.2.4 Differentiable programming

Differentiable programming is a programming paradigm that the whole numerical computation in the program is differentiable via automatic differentiation (AD)[47]. Differentiability is the core to machine learning and inverse problems since the program can be optimized via the gradient-based optimization algorithms. AD is also known as algorithmic differentiation has been widely used in machine learning software to calculate the derivatives of a sequence of operations. AD stores the sequence of operations and applies the chain-rule to calculate the derivative either in the forward or reverse model. The back-propagation process, which is the backbone of machine learning, relies on AD to calculate the accurate gradients for training a model. The development of domain-aware SciML by combining capabilities of AD with the domain knowledge will help to understand the full potential of SciML. AD is natural for scientists to apply since it is automatically implemented, but it comes with a hefty price of efficiency[48]. The computation of AD requires to store all operations, and it needs to follow the chain-rule with a sequential execution fashion. In contrast, many scientific computing applications use the adjoint-state method to calculate derivatives based upon numerical solutions to the associated partial differential equation systems. The numerical solution methods offer opportunity for critical time and memory savings that improve the efficiency while allowing preservation of sufficient accuracy.

AD is the key component for a programming model that can potentially bridge scientific computing and machine learning[49]. Julia[50] is such a programming model built with AD as its core component to provide a high-level and high-performance programming model for numerical computing with differentiable programming. XU et al.[51] developed the AD seismic package based on Julia differential programming and mathematically proved that the adjoint-state method is equivalent to the reverse-mode AD, which can be used very effectively for seismic inverse problems. The authors also learned hidden dynamics using intelligent AD[52] that demonstrates a promising usage of applying the differentiable programming on bridging scientific computing and machine learning domains.

7.3 Wave equations and RNN

7.3.1 Wave equations

The wave motion is governed by physical rules that can be expressed in the following partial differential equation (PDE) (1) along with the boundary conditions (2)(3). The 1D scalar wave equation is used for simplicity purpose in this paper:

$$\frac{1}{c^2(x)}\frac{\partial^2 u(x,t)}{\partial t^2} - \frac{\partial^2 u(x,t)}{\partial x^2} = f(x,t) \tag{1.1}$$

$$\frac{1}{c(0)}\frac{\partial u(0,t)}{\partial t} - \frac{\partial u(0,t)}{\partial x} = 0 \tag{1.2}$$

$$\frac{1}{c(1)}\frac{\partial u(1,t)}{\partial t} - \frac{\partial u(1,t)}{\partial x} = 0 \tag{1.3}$$

where, $c(x)$ is the spatial velocity distribution, $u(x, t)$ is the wave field distribution in space and time, and $f(x, t)$ is the energy source distribution in space and time.

The Eq. (1.1) can be solved numerically using a finite difference approximation:

$$f(x,t) = -\frac{u(x - \Delta x, t) - 2u(x,t), +u(x + \Delta x, t)}{\Delta x^2}$$
$$+ \frac{1}{c^2}\frac{u(x, t - \Delta t) - 2u(x,t) + u(x, t + \Delta t)}{\Delta t^2} \tag{1.4}$$

After factoring, the Eq (1.4) can be expressed as:

$$u(x, t + \Delta t) = f(x,t)c^2\Delta t^2 + (2u(x,t) - u(x, t - \Delta t))$$
$$+ c^2 \frac{\Delta t^2}{\Delta x^2}(u(x - \Delta x, t) - 2u(x,t) + u(x + \Delta x, t)) \tag{1.5}$$

which shows that the next wave field in time $u(x, t + \Delta t)$ can be calculated based on the current and prior wave fields, as well as spatial neighbors in the current wave field. The wave motion simulation follows the time sequence to produce the next state based on the prior ones, which is similar to RNN in deep learning to model a time sequence function.

7.3.2 Recurrent neural network

RNN is used to model the pattern in a sequence of data, mostly in time sequence. In recent years, RNN and its variants have been applied successfully to problems such as speech recognition, machine translation, and text-to-speech rendering. It has an internal cell that repeatedly processes an input, carries a hidden state, and produces an output at each step. The RNN cell can be designed to be simple or complex to model a problem with a forgettable memory mechanism (long short-term memory [LSTM][53]) or/and a gating mechanism (gated recurrent unit [GRU][54]).

Fig. 7.2A shows a typical RNN structure that repeatedly takes an input, updates its hidden state, and produces an output at every step. The RNN model can be unfolded as shown in Fig. 7.2B that learns the recurrence relationship from a sequence of data. The hidden state h_i remembers the prior state of the process and is updated at each step. The hidden state enables RNN to learn the temporal relationships among the inputs since most of the time sequence data do contain temporal patterns. LSTM allows RNN to forget long-term relationships built up in the hidden state and emphasizes the short-term relationships, which can be useful for many cases.

A simple RNN can be expressed in the Eq. (1.6):

$$h_t = \sigma_h(W_h x_t + W_h h_{t-1} + b_h)$$
$$y_h = \sigma_y\left(W_y h_t + b_y\right) \tag{1.6}$$

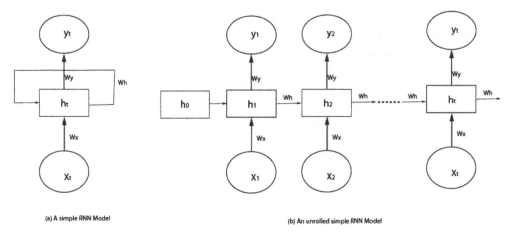

(a) A simple RNN Model   (b) An unrolled simple RNN Model

Fig. 7.2 A simple RNN model (A) with feedback loop and (B) with loop unfolded.

where x_t is the input, h_t is the hidden state, W represents the weights, b is the bias, and σ is the activation function.

The discretized wave Eq. (1.5) can be restructured as an RNN with two hidden states $u(x, t)$ and $u(x, t - \Delta t)$. There is also a spatial stencil relationship of neighboring velocity distribution. The internal operation of the RNN can be defined using a new function F with input of $f(x, t)$, two hidden states $u(x, t)$ and $u(x, t - 1)$, and the constant velocity distribution c:

$$
\begin{aligned}
&F\big(f(x,t),u(x,t),u(x,t-1),c\big) \\
&= f(x,t)c^2\Delta t^2 + \big(2u(x,t)-u(x,t-1)\big) \\
&+ c^2\frac{\Delta t^2}{\Delta x^2}\big(u(x-1,t)-2u(x,t)+u(x+1,t)\big)
\end{aligned}
\tag{1.7}
$$

Then, the Eq. (1.5) can be restructured as an RNN format:

$$
\begin{aligned}
h_{t+1} &= \sigma\big(F\big(f(t),h(t),h(t-1),c\big)\big) \\
y_{t+1} &= P\big(h_{t+1}\big)
\end{aligned}
\tag{1.8}
$$

where, P is the projection function to get the sample of a trace from a receiver.

The Eq. (1.8) is then a nonlearnable, deterministic physical solution represented as the deep learning RNN model. Fig. 7.3 shows the resulting RNN model that solves the wave equation with four inputs $f(x, t)$, $h(t)$, $h(t - 1)$, and c, the velocity distribution, which is constant in the equation. The output y_t is the trace sample of a receiver at each time step.

7.3.3 PyTorch RNN implementation

The wave equation RNN model presented in Fig. 7.3 enables use of the deep learning software platform to solve the wave equations. The benefits of using a deep learning

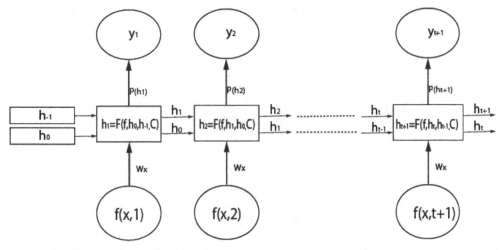

Fig. 7.3 *A RNN model for wave equation.*

model to represent an HPC application include the ability to: (1) leverage the HPC implementation of the deep learning model exploiting the advantages of GPUs/multi-cores and vectorization for better performance, (2) have an automatic gradients calculation using the built-in automatic differentiation package in deep learning, (3) utilize the variety of built-in optimizers to apply the gradients to find the global/local optimums, and (4) use the data- and model-parallelism framework implemented in deep learning package to run the application on a HPC cluster.

The implementation in this work uses PyTorch[1] v1.5 to build the RNN model. PyTorch is an open source machine learning framework developed by Facebook by merging Torch and Caffe2, which supports a variety of hardware platforms including multiple CPUs, GPUs, distributed systems, and mobile devices. Besides the machine learning and deep learning functions, one unique feature of PyTorch is that it contains a just-in-time compiler to optimize the code if it complies with TorchScript, which is a subset of Python. It has a built-in automatic differentiation package for calculating derivatives, as well as a distributed training module to train a model on a HPC cluster. PyTorch has both Python and C++ frontends.

The following Python code snippet illustrates an RNN-similar implementation of wave equation using PyTorch. The two classes derived from torch.nn. Module for RNN cell and RNN driver respectively are called Wave PGNNcell and Wave Propagator. The wave PGNNcell implemented a cell function in RNN that computes the wavefield at a time step. The Wave Propagator iterates over all time steps and takes the seismic source waveform sample as the input at each time step. The hidden state (self.H) contains the previous and current wave fields, which are fed into the cell for the next iteration. The trace is collected by projecting the current wavefield based on the receiver location. The program returns the simulated wavefield and sampled trace at the end.

[1]https://pytorch.org/

```
class Wave_PGNNcell(torch.nn.Module):
   def forward(self, H, src ):
      uC,uP = [ H[0], H[1] ]
      ...

      return [uN,uC]


class Wave_Propagator(torch.nn.Module):
   self.cell = Wave_PGNNcell(C, config)


 def forward(self):
   us = []        # list of output wavefields
   traces = []
   rcv = self.rcvrs
   for it in range(self.nt):
     self.H = self.cell.forward(self.H, self.ws[it])
     us.append( self.H[0].detach().numpy() )
     # Extract wavefield sample at each receiver
     samps = rcv.sample( self.H[0].clone() )
     traces.append( samps )
   trc = torch.stack(traces,dim=1)
   return us, trc
```

7.3.4 Seismic wave simulation

For a simple demonstration of seismic wave simulation, the RNN model is used to simulate the acoustic wave propagation for the scalar wave equation. A "true" synthetic model is used to generate a target "recorded" data set, and an initial model can be constructed as a smoothed version of the true model or some other separately chosen function. A time-symmetric Ricker wavelet is used as a waveform for one or more energy sources (shots), and the wavefield is sampled as traces at an array of receivers. Rock density is assumed constant in these models.

As stated earlier, one benefit of using deep learning technology is to take advantage of its multiple CPUs and GPUs implementation. It is required only to specify which devices the code will operate on and define tensors to these devices. All remaining device-specific implementation and optimizations are done internally by PyTorch; porting of application code using CUDA or OpenACC is not required. Another benefit is to use the data-parallelism implemented in PyTorch. We can parallelize the code by the number of the sources/shots to run the code on multiple GPUs and distributed clusters.

Fig. 7.4 shows a 1D seismic velocity Inversion case applying the physics-ruled RNN implementation. The Fig. 7.4A shows a true synthetic velocity model and an initial model; 4B shows the inverted model comparing with the true model (up) and a slightly smoothed final inverted model (down); 4C shows the comparison of the true traces and the inverted traces; and 4D shows the wave field illustrating how the seismic

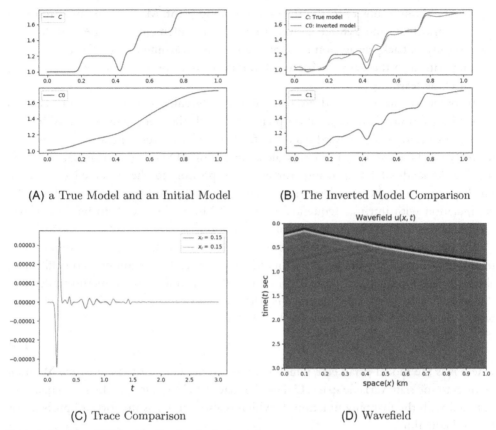

(A) a True Model and an Initial Model   (B) The Inverted Model Comparison

(C) Trace Comparison   (D) Wavefield

Fig. 7.4 *Applying RNN for 1D seismic velocity inversion.*

wave propagates with respect to space and time. The 1D inversion experiment finds a solution close to the true model after 100 iterations, using the Adam optimizer[55] with L2 regularization.

Ongoing efforts include working on 2D cases by revising PySIT package, and performing more test cases to evaluate the performance with both data and model parallelism provided by PyTorch on a CPU cluster and multiple GPUs.

7.4 Differentiable programming

7.4.1 Automatic differentiation and adjoint-state method

The method of automatic differentiation (AD), also called algorithmic differentiation, calculates the derivatives for any arbitrary differentiable program. Unlike using the numerical differentiation of the adjoint state method that is an approximation to calculate the derivatives, the automatic differentiation returns the exact answer of the derivatives, though subject to the intrinsic rounding error. Machine learning software such as TensorFlow and Pytorch all have the built-in implementation of AD as the core functionality of backpropagation to optimize machine learning models. Accurate gradients are critical to the gradient-based optimizations used in both scientific computing and machine learning.

In order to calculate the derivatives of any differentiable programs, AD needs to store all operations on the execution path along with the intermediate results. It then propagates derivatives backward from the final output for every single operation connected with the chain rule. For largescale applications, AD faces the challenge of meeting the demands of fast-growing storage in proportion to the executed operations. Furthermore, the individual derivative function for each operation also slows down the computation with intrinsic sequential execution. More work needs to be done if AD can be directly applied to a real scientific application.

Computationally expensive scientific applications typically use the adjoint state method to calculate the gradient of a function with much better computation efficiency, although it is a numerical approximation. In FWI, the adjoint state method calculates the derivative of a forward function $J(m)$ that depends on the wavefield $u(m)$. The forward function J can be defined using h, as following[3]:

$$J(m) = h(u(m), m) \tag{1.9}$$

where m is the model parameter, which belongs to the model parameter space $\mathbf{M}$ and u belongs to the state variable space, $\mathbf{U}$. The the state variables, u follow the state equations outlined with the mapping function, F, which is also known as the forward problem or forward equation[3]:

$$F(u(m), m) = 0. \tag{1.10}$$

The mapping function F is mapping from $\mathbf{U} * \mathbf{M}$ to $\mathbf{U}$ and is satisfied by the state variable u. If the condition $F(u, m) = 0$ is satisfied, the state variable u becomes a physical realization. Then, the adjoint state equation can be given as following, where λ is the adjoint state variable and $\tilde{u}$ is any element of $\mathbf{U}$[3]:

$$\left[\frac{\delta F(u, m)}{\delta \tilde{u}} \right]^* \lambda = \frac{\delta h(u, m)}{\delta \tilde{u}} \tag{1.11}$$

This adjoint-state gradient calculation involves computing the reverse-time propagated residual wavefield, then combining with the saved forward-propagated wavefield snapshots at specified time intervals to provide adjustments to the medium properties (the gradient) at each spatial mesh point. In summary, the forward propagation computes data observations representing the response of the model, and the residual between the model response and actual observed data is backward propagated and combined with the forward model response to compute adjustments to the current model estimate.

Intervening in the calculation of the gradient in this manner allows for management of the required computational resources by saving the forward wave fields only as often as numerically required, explicitly managing data resources through staging to disk or check-pointing as needed, implementing shot-level parallelism, and other specially tailored techniques.

7.4.2 Extended automatic differentiation

A difficulty with the autodifferentiation (AD) procedure is that memory requirements for the back-propagation graph can become excessive, as noted by Richardson[16]. Applying chain-rule differentiation on elemental network nodes over thousands of RNN time steps for a large mesh of physical parameter values is a reasonably-sized task for 1D problems, but the graph quickly becomes intractable for 2D and 3D models. This issue usually renders impractical the use of pure AD for such model inversion problems.

An alternative solution to the problem, involves extending the AD backward process using the PyTorch AD workflow to integrate the adjoint-state method for the more efficient gradient calculation. In PyTorch, the AD workflow is extended by providing a backward function to calculate the gradients of the corresponding forward function. The required parameters of the forward function, the model parameters and loss function are provided to allow the backward function to pick up these parameters for the adjoint-state calculation.

Control over this auto-differentiation process is available through use of a PyTorch extension to the Autograd feature pictured conceptually in Fig. 7.5, wherein the RNN layer of the network can be replaced by a forward propagation loop and corresponding adjoint back-propagation loop for an equivalent gradient calculation provided

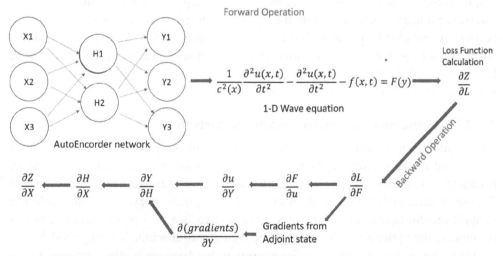

Fig. 7.5 *Adjoint gradient vs automatic differentiation.* Derivatives with respect to model parameters in the backward automatic differentiation are replaced by gradients from adjoint state.

by the user. This alternative gradient calculation can take advantage of well-known techniques in seismic inversion processing, enabling existing performance enhancements to be applied using the extended PyTorch capability for specially designed back-propagation.

In the present case, the physical medium properties to be optimized are provided to the "forward" wave propagation problem implemented using the publicly available PySIT seismic inversion toolkit[56], creating a simulated seismic response. This simulated response is to be compared with the observed seismic trace data from the corresponding actual field data recording (or recordings from a "true" model in our synthetic studies). The corresponding "backward" propagation consists in using the residual wavefield represented by the difference between the simulated data and the observed seismic trace data, and implementing the "adjoint-state" solution to provide the required gradient of the model parameters. Other implementations of wave propagation solutions may also be used in this framework, such as spectral element methods[57] for 2D, 3D, and spherical 3D wave propagation.

The beneficial end result is that traditional adjoint-state solution methods are incorporated into the AD workflow, so that seismic inversion calculations can be integrated within the broader deep learning process with efficient calculation.

7.5 Seismic inversion

7.5.1 Seismic inversion using neural network

These new concepts from machine learning may now be applied to the problem of seismic inversion as presented in Section 1.1. As described in Section 1.3 and Section 1.4, by reconstructing the forward problem using deep learning software, the seismic inversion problem can be solved by the automatic differentiation package, a variety of optimizers provided by PyTorch, and a customized loss function. The automatic differentiation package in PyTorch implements the methodology of automatic differentiation by recording all the forward operations in sequence and performing backward derivative computation based on the chain rule.

7.5.2 Autoencoder for dimensionality reduction

The seismic inversion process needs to uncover the physical properties at every point represented in the geological space, which quickly leads to a large number of model parameters to optimize in the traditional FWI process. The nature of the nonlinear and ill-posed inverse problem often falls into the local minimum traps. It is a sound solution to apply the dimensionality reduction technique to reduce the optimization parameters to improve the optimization accuracy by engaging with machine learning models.

Since the automatic differentiation workflow has been customized by integrating the adjoint state method for the FWI gradients (described in Section 1.4), it is now

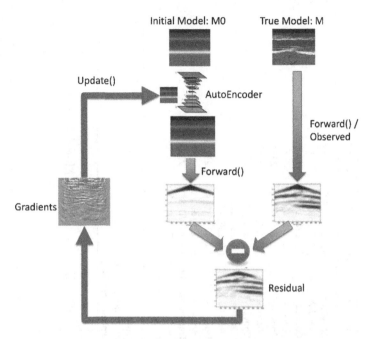

Fig. 7.6 *Full waveform inversion enhanced with autoencoder dimensionality reduction.*

feasible to integrate the machine learning models into the FWI workflow and keep the program differentiable. Since the autoencoder $A(x)$ is differentiable and the forward model $F(x)$ is differentiable, the composition of the $F(A(x))$ is differentiable. The autoencoder is applied as the dimensionality reduction method applied before the forward model as shown in Fig. 7.6.

For the 60×80 2D model in the current example, the autoencoder contains 743,938 parameters as shown in Fig. 7.7A and B. The autoencoder is an unsupervised learning model that compresses the information representation of the input data to a sparse latent variable with less dimensions at the middle of the encoded layer. It then reconstructs the data from the encoded latent variable to the original or enhanced data. The compression process is called encoder and the reconstruction is called decoder. The encoder learns how to compress the input data and describes it with the latent variable, while the decoder learns how to reconstruct the data from the latent variable.

We start the autoencoder training by generating a large number of random seismic velocity models. In this work, we are using some simple and flat velocity layers representing the velocities of different earth interiors including water and rocks. Specifically, these models contain one or more low velocity layers in the middle or bottom of these layers that is challenging for the low velocity inversion. All of these models have the fixed dimensions of 60×80. As indicated in Fig. 7.7A, the autoencoder has two components: a encoder and a decoder. The encoder compresses the input model with

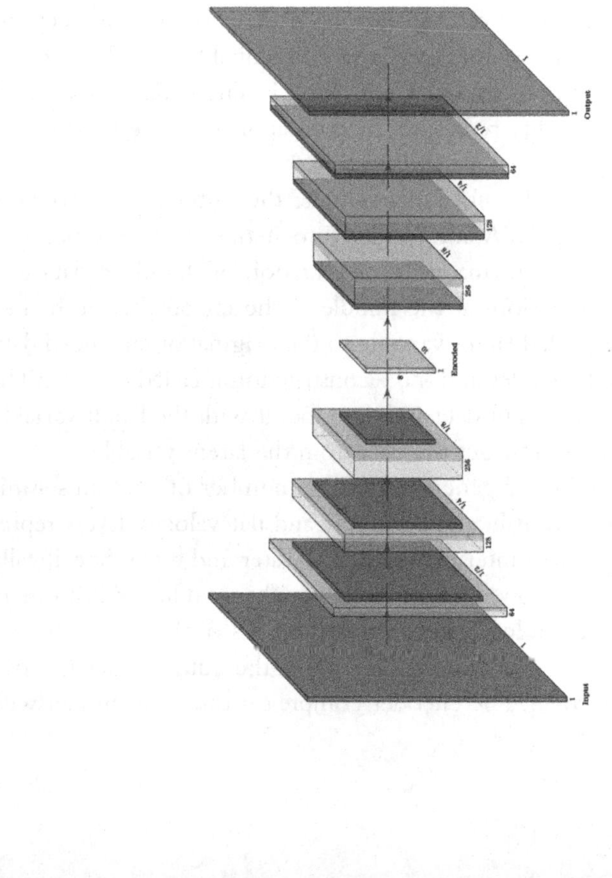

Layer (type)	Output Shape	Param #
Conv2d-1	[-1, 64, 30, 40]	640
ReLU-2	[-1, 64, 30, 40]	0
Conv2d-3	[-1, 128, 15, 20]	73,856
ReLU-4	[-1, 128, 15, 20]	0
Conv2d-5	[-1, 256, 8, 10]	295,168
Tanh-6	[-1, 256, 8, 10]	0
Conv2d-7	[-1, 1, 8, 10]	2,305
ConvTranspose2d-8	[-1, 256, 15, 20]	2,560
ReLU-9	[-1, 256, 15, 20]	0
ConvTranspose2d-10	[-1, 128, 30, 40]	295,040
ReLU-11	[-1, 128, 30, 40]	0
ConvTranspose2d-12	[-1, 64, 60, 80]	73,792
ReLU-13	[-1, 64, 60, 80]	0
ConvTranspose2d-14	[-1, 1, 60, 80]	577

Total params: 743,938
Trainable params: 743,938
Non-trainable params: 0

(B) The AutoEncoder Model Parameters

(A) The AutoEncoder Network Structure

Fig. 7.7 *Traditional seismic velocity inversion.*

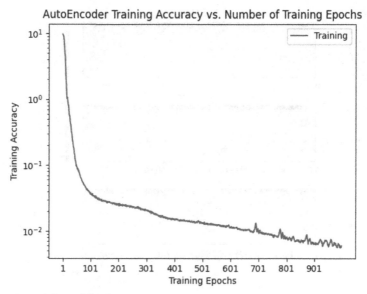

Fig. 7.8 *The autoencoder training loss.*

dimension of 60 × 80 to an encoded latent variable with dimension of 8 × 10, which is 1/60 of the original dimension. The latent variable is then decompressed by the decoder to restore to its original dimension.

The loss function we used to train the autoencoder is the mean-squre-error (MSE) loss and the optimizer is Adam with learning rate of 0.001. The batch size used is 128. The loss values during the training process are shown in Fig. 7.8.

Fig. 7.6 shows the autoencoder-enhanced FWI process, where the autoencoder is inserted before the forward function simulation starts. Note that the encoder is only applied to the first iteration to get the encoded latent variable. For the remaining optimization iterations, the decoder is applied to decompress the encoded latent variable to get a new velocity model with the original dimension. During the gradient-based optimization process, the gradients are calculated with respect to the encoded latent variable, instead of the original model, which reduces the dimensionality of the optimization search space by a factor of 60.

7.5.3 Results

PyTorch has a list of optimizers, including Adam[55], RMSprop[58], stochastic gradient descent (SGD), Adadelta[59], Adagrad[60], LBFGS, and their variants. The learning rate, scheduler and regularizations can be specified to fit different optimization problems. There are also multiple regression and classification loss functions implemented in PyTorch. All of these packages provide a rich environment to solve inverse problems.

The present implementation has demonstrated how to invoke the extended automatic gradient calculation for the velocity model. The Adam optimizer and the MSE

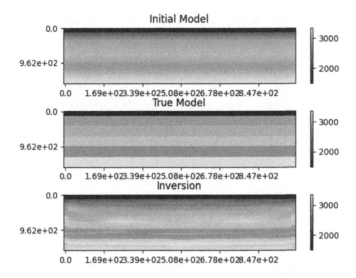

(A) The Initial, True and Inverted Model Comparison

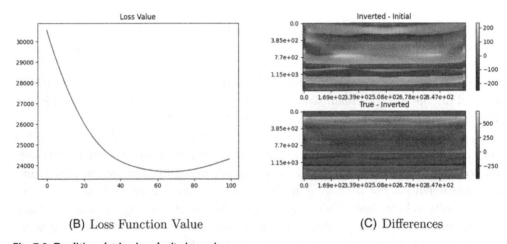

(B) Loss Function Value   (C) Differences

Fig. 7.9 *Traditional seismic velocity inversion.*

loss function are used to compare the misfit of the simulated traces and observed traces after each iteration of the forward model. The partial derivative (the gradient) of the loss function with respect to the initial model and the encoded latent variable is calculated by the automatic differentiation process, which is applied by the optimizer to minimize the misfit. These iterations gradually find an approximation of the true velocity distribution.

Fig. 7.9 and Fig. 7.10 show the differences of the traditional FWI and the auto-encoder enhanced FWI results. Within each figure (1) shows the initial model, the true model (flat layers with a low velocity feature around 1 km depth), and the inverted

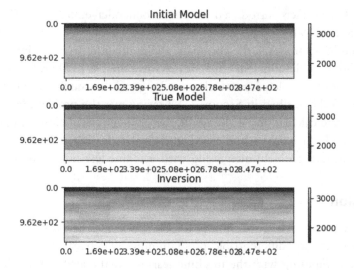

(A) The Initial, True and Inverted Model Comparison

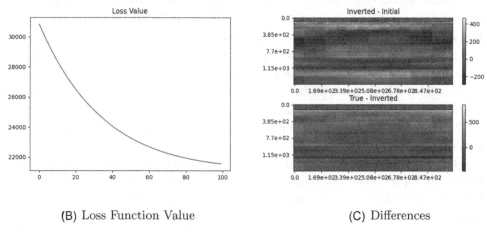

(B) Loss Function Value   (C) Differences

Fig. 7.10 *The autoencoder-enhanced seismic velocity inversion.*

model; the loss graph (B) shows the loss values (at different scales) after each optimiza-tion iteration, and Figure (C) shows the difference between the inverted model and the initial model (top), as well as the difference between the inverted model and the true model. It appears that the traditional FWI does not optimize well in the low velocity layer case after 40 iterations ended with a high loss value, which falls into a local trap. The autoencoder-enhanced FWI discovers the low velocity layer very well and contin-ues to optimize the misfit for all 100 iterations. The difference graphs also confirm that the autoencoder case identifies all layers well showing less structured misfits. Noticeably, there are also less artifacts introduced in the autoencoder-enhanced FWI compared with the traditional FWI.

As described in Section 1.4, the automatic differentiation provided by the PyTorch software does not provide sufficient efficiency to solve the FWI 2D problem. The gradients calculated for the whole program takes too long and too much space to store them. The hybrid method described in Section 1.4.2 overcomes the problem by incorporating the adjoint state calculation. As the result, the gradient calculation using the hybrid approach achieves both accuracy and efficiency, which is feasible for use in a large scale scientific computation problem integrating with machine learning models.

7.6 Discussion

There are a few points that worth noting for the work. The first is that the automatic differentiation is key for differentiable programming, which can bridge the physics-based scientific computing with the machine learning (ML)/artificial intelligence (AI) technologies. ML/AI methods do not have physics principles built-in, which may create an infeasible solution given the fact that most of the scientific inverse problems may be ill-posed. In our prior work[61], the convergence of ML with a scientific application without differentiable programming may not find a generalized solution since optimizations of the two different methods are disconnected.

The second point is that the automatic differentiation needs additional improvements to make it feasible to other applications. In the present method, integrating the adjoint-state method makes it feasible to solve a large case; however, the solution is an approximation. If the automatic differentiation method can be more memory-efficient and parallelizable, it can be much more useful to compute the exact gradients for the large complex problems.

The last point is the deep learning model autoencoder requires a revisit to reduce the loss during decoding. Although it reduces the dimension by compressing the input data into a sparse latent variable, the reconstruction is not lossless. There are some errors introduced during the reconstruction process that may hinder the optimization process. There is a trade-off to take into consideration when designing the convergence of ML/AI with scientific computing.

7.7 Conclusions

Two cases have been demonstrated of restructuring the wave equation using finite difference method in a deep learning RNN model framework and an autoencoder enhanced FWI process. The benefits of the work include fully utilizing the high-performance tensor processing and optimization capabilities implemented in the deep learning package PyTorch, as well as the deep integration of machine learning models with the inverse problem. By integrating an HPC application with a deep learning framework with differentiable programming, we can explore a large number of combinations of machine learning models with physical numerical solutions to achieve better accuracy and efficiency.

Moreover, the work can be extended to more applications in reservoir characterization, identification of sweet spots, and reservoir simulation. With the scientific principles embedded in the machine learning, the method enhances the interpretability and reliability of machine learing models in the engineering fields.

7.8 Acknowledgment

This research work is supported by the US National Science Foundation (NSF) awards ##1649788, #1832034 and by the Office of the Assistant Secretary of Defense for Research and Engineering (OASD(R&E)) under agreement number FA8750-15-2-0119. The U.S. Government is authorized to reproduce and distribute reprints for Governmental purposes notwithstanding any copyright notation thereon. The views and conclusions contained herein are those of the authors and should not be interpreted as necessarily representing the official policies or endorsements, either expressed or implied, of the US NSF, or the Office of the Assistant Secretary of Defense for Research and Engineering (OASD(R&E)) or the U.S. Government. This work used the Extreme Science and Engineering Discovery Environment (XSEDE), which is supported by National Science Foundation grant number ACI-1548562. Specifically, it used the Bridges system, which is supported by NSF award number ACI-1445606, at the Pittsburgh Supercomputing Center (PSC).

References

1. Baker N, Alexander F, Bremer T, Hagberg A, Kevrekidis Y, Najm H, et al., Workshop report on basic research needs for scientific machine learning: core technologies for artificial intelligence. http://doi.org/10.2172/1478744. https://www.osti.gov/biblio/1478744.
2. Schuster G, Seismic Inversion, Society of Exploration Geophysicists, 2017. doi: http://doi.org/10.1190/1.9781560803423.
3. Plessix R-E. A review of the adjoint-state method for computing the gradient of a functional with geophysical applications. *Geophys J Int* 2006;**167**(2):495–503. http://doi.org/10.1111/j.1365-246X.2006.02978.x.
4. Adler J, Oktem O, Solving ill-posed inverse problems using iterative deep neural networks. https://arxiv.org/pdf/1704.04058.pdf. Accessed Feb. 15, 2021.
5. Virieux J, Operto S. An overview of full-waveform inversion in exploration geophysics. *Geophysics* 2009;**74**(6):WCC127–52. http://doi.org/10.1190/1.3238367.
6. Moseley B, Markham A, Nissen-Meyer T, Fast approximate simulation of seismic waves with deep learning. https://arxiv.org/pdf/1807.06873.pdf. Accessed Feb. 15, 2021.
7. Baker N, Alexander F, Bremer T, Hagberg A, Kevrekidis Y, Najm H, et al., Workshop report on basic research needs for scientific machine learning: core technologies for artificial intelligence. http://doi.org/10.2172/1478744.
8. Baydin AG, Pearlmutter BA, Radul AA, Siskind JM. Automatic differentiation in machine learning: a survey. *J. Mach. Learn. Res.* 2017;**18**(1):5595–637.
9. Poggio T, Mhaskar H, Rosasco L, Miranda B, Liao Q. Why and when can deep-but not shallow-networks avoid the curse of dimensionality: a review. *Int J Autom Comput* 2017;**14**(5):503–19. http://doi.org/10.1007/s11633-017-1054-2.
10. Raissi M, Perdikaris P, Karniadakis G. Physics-informed neural networks: a deep learning framework for solving forward and inverse problems involving nonlinear partial differential equations. *J Comput Phys* 2019;**378**:686–707. https://doi.org/10.1016/j.jcp.2018.10.045.

11. Kharazmi E, Zhang Z, Karniadakis GE, Variational physics-informed neural networks for solving partial differential equations (2019). arXiv: 1912.00873.
12. Rackauckas C, Ma Y, Martensen J, Warner C, Zubov K, Supekar R, Skinner D, et al., Universal differential equations for scientific machine learning, arXiv preprint arXiv:2001.04385.
13. Greydanus S, Dzamba M, Yosinski J, Hamiltonian neural networks (2019). arXiv:1906.01563.
14. Mattheakis M, Protopapas P, Sondak D, Giovanni MD, Kaxiras E, Physical symmetries embedded in neural networks (2020). arXiv:1904.08991.
15. Chen RTQ, Rubanova Y, Bettencourt J, Duvenaud DK. Neural ordinary differential equations-Bengio S, Wallach H, Larochelle H, Grauman K, CesaBianchi N, Garnett R. Advances in Neural Information Processing Systems, Curran Associates, Inc. 2018:6571–83. https://proceedings.neurips.cc/paper/2018/file/69386f6bb1dfed68692a24c8686939b9-Paper.pdf.
16. Richardson A, Seismic full-waveform inversion using deep learning tools and techniques. https://arxiv.org/pdf/1801.07232v2.pdf. Accessed Feb. 15, 2021.
17. Zhu W, Xu K, Darve E, Beroza1 GC, A general approach to seismic inversion with automatic differentiation. https://arxiv.org/pdf/2003.06027.pdf. Accessed Feb. 15, 2021.
18. Blanning RW. The construction and implementation of metamodels. *Simulation* 1975;**24**(6):177–84. http://doi.org/10.1177/003754977502400606.
19. O'Hagan A, Bayesian analysis of computer code outputs: a tutorial, the Fourth International Conference on Sensitivity Analysis of Model Output (SAMO 2004), Reliab Eng Syst Saf. 2006;**91**(10)1290–1300. https://doi.org/10.1016/j.ress.2005.11.025. URL http://www.sciencedirect.com/science/article/pii/S0951832005002383.
20. Willcox K, Peraire J. Balanced model reduction via the proper orthogonal decomposition. *AIAA J* 2002;**40**(11):2323–30. http://doi.org/10.2514/2.1570.
21. Wilson KC, Durlofsky LJ. Computational optimization of shale resource development using reduced-physics surrogate models, the SPE Western Regional Meeting. March 2012. Bakersfield, California, USA: Society of Petroleum Engineers. https://doi.org/10.2118/152946-MS.
22. Sethian JA. A fast marching level set method for monotonically advancing fronts. *Proc Natl Acad Sci* 1996;**93**(4):1591–5. http://doi.org/10.1073/pnas.93.4.1591.
23. Robinson TD, Eldred MS, Willcox KE, Haimes R. Surrogate-based optimization using multifidelity models with variable parameterization and corrected space mapping. *AIAA J* 2008;**46**(11):2814–22.
24. Regis RG, Shoemaker C. Constrained global optimization of expensive black box functions using radial basis functions. *J Glob Optim* 2005;**31**:153–71.
25. Sandvol E, Seber D, Calvert A, Barazangi M. Grid search modeling of receiver functions: implications for crustal structure in the middle east and north africa. *J Geophys Res Solid Earth* 1998;**103**(B11):26899–917. http://doi.org/10.1029/98JB02238.
26. Sen MK, Stoffa PL. Nonlinear one-dimensional seismic waveform inversion using simulated annealing. *Geophysics* 1991;**56**(10):1624–38.
27. Zhao L-S, Sen MK, Stoffa P, Frohlich C. Application of very fast simulated annealing to the determination of the crustal structure beneath Tibet. *Geophys J Int* 1996;**125**(2):355–70.
28. Press F. Earth models obtained by monte carlo inversion. *Journal of Geophysical Research (1896-1977)* 1968;**73**(16):5223–34.
29. Fernández Martínez JL, García Gonzalo E, Fernández Álvarez JP, Kuzma HA, Menéndez Pérez CO. Pso: a powerful algorithm to solve geophysical inverse problems: application to a 1d-dc resistivity case. *J Appl Geophys* 2010;**71**(1):13–25. https://doi.org/10.1016/j.jappgeo.2010.02.001.
30. Kramer MA. Nonlinear principal component analysis using autoassociative neural networks. *AIChE J* 1991;**37**(2):233–43.
31. Moseley B, Nissen-Meyer T, Markham A. Deep learning for fast simula-tion of seismic waves in complex media. *Solid Earth* 2020;**11**(4):1527–49. http://doi.org/10.5194/se-11-1527-2020.
32. van den Oord A, Dieleman S, Zen H, Simonyan K, Vinyals O, Graves A, Kalchbrenner N, et al., WaveNet: a generative model for raw audio, arXiv e-prints (2016) arXiv: 1609.03499arXiv:1609.03499.
33. Nair V, Hinton GE. Rectified linear units improve restricted boltzmann machines. In: Proceedings of the 27th International Conference on International Conference on Machine Learning, ICML'10, Omnipress, Madison, WI, USA 2010:807–14.

34. Gao Z, Li C, Liu N, Pan Z, Gao J, Xu Z. Large-dimensional seismic inversion using global optimization with autoencoder-based model dimensionality reduction. *IEEE Trans Geosci Remote Sens* 2020:1–5. https://doi.org/10.1109/TGRS.2020.2998035.

35. Storn R, Price K. Differential evolution – a simple and efficient heuristic for global optimization over continuous spaces. *J. Glob Optim* 1997;**11**(4):341–59. http://doi.org/10.1023/A:1008202821328.

36. Chen Y, Schuster GT. Seismic inversion by newtonian machine learning. *Geophysics* 2020;**85**(4):WA185–200.

37. Sun B, Alkhalifah T, Ml-descent: an optimization algorithm for FWI using machine learning, GEOPHYSICS **85**:R477-R492. arXiv: https://doi.org/10.1190/geo2019-0641.1.

38. Kingma DP, Welling M, Auto-encoding variational bayes, CoRR abs/1312.6114.

39. Versteeg R. The marmousi experience: velocity model determination on a synthetic complex data set. *The Leading Edge* 1994;**13**(9):927–36.

40. Appendix B. The SEG-EAGE salt data set, 2012, pp. 217–8. arXiv: https://library.seg.org/doi/pdf/10.1190/1.9781560801689.appendixb.

41. Mosser L, Dubrule O, Blunt MJ, Stochastic seismic waveform inversion using generative adversarial networks as a geological prior (2018). https://arxiv.org/abs/1806.03720. Accessed Feb 15, 2021.

42. Arjovsky M, Chintala S, Bottou L, Wasserstein GAN (2017). https://arxiv.org/abs/1701.07875. Accessed Feb. 15, 2021

43. Gulrajani I, Ahmed F, Arjovsky M, Dumoulin V, Courville A, Improved training of wasserstein gans (2017). https://arxiv.org/abs/1704.00028. Accessed Feb. 15, 2021.

44. Tolstikhin I, Bousquet O, Gelly S, Scholkopf B, Wasserstein autoencoders. https://arxiv.org/pdf/1711.01558.pdf. Accessed Feb. 15, 2021.

45. Roberts GO, Rosenthal JS. Optimal scaling of discrete approximations to langevin diffusions. *J R Stat Soc: Series B StatMethodol* 1998;**60**(1):255–68.

46. Swischuk R, Mainini L, Peherstorfer B, Willcox K. Projection-based model reduction: Formulations for physics-based machine learning. *Comput Fluids* 2019;**179**:704–17. http://doi.org/10.1016/j.compfluid.2018.07.021.

47. Baydin AG, Pearlmutter BA, Radul AA, Automatic differentiation in machine learning: a survey, CoRR abs/1502.05767. arXiv:1502.05767. http://arxiv.org/abs/1502.05767. Accessed Feb. 15, 2021.

48. Margossian CC, A review of automatic differentiation and its efficient implementation, CoRR abs/1811.05031. arXiv:1811.05031. http://arxiv.org/abs/1811.05031. Accessed Feb. 15, 2021.

49. Innes M, Edelman A, Fischer K, Rackauckas C, Saba E, Shah VB, et al., A differentiable programming system to bridge machine learning and scientific computing, CoRR abs/1907.07587. arXiv:1907.07587. http://arxiv.org/abs/1907.07587. Accessed Feb. 15, 2021.

50. Bezanson J, Edelman A, Karpinski S, Shah VB. Julia: a fresh approach to numerical computing. *SIAM Rev* 2017;**59**(1):65–98. https://doi.org/10.1137/141000671.

51. Zhu W, Xu K, Darve EF, Beroza G, A general approach to seismic inversion with automatic differentiation, arXiv: Computational Physics. https://arxiv.org/abs/2003.06027. Accessed Feb. 15, 2021.

52. Xu K, Li D, Darve EF, Harris JM, Learning hidden dynamics using intelligent automatic differentiation, ArXiv abs/1912.07547. https://arxiv.org/abs/1912.07547. Accessed Feb. 15, 2021.

53. Hochreiter S, Schmidhuber J. Long short-term memory. *Neural Comput* 1997;**9**(8):1735–80.

54. Chung J, Gulcehre C, Cho K, Bengio Y, Empirical evaluation of gated recurrent neural networks on sequence modeling (2014). arXiv:1412.3555. https://arxiv.org/abs/1412.3555. Accessed Feb. 15, 2021.

55. Kingma DP, Ba J, Adam: A method for stochastic optimization (2014). arXiv:1412.6980. https://arxiv.org/abs/1412.6980. Accessed Feb. 15, 2021.

56. Hewett RJ, Demanet L, Team TP, PySIT: python seismic imaging toolbox (2020). http://doi.org/10.5281/zenodo.3603367. Accessed Feb. 15, 2021.

57. Tromp J, Komatitsch D, Liu Q. Spectral-element and adjoint methods in seismology. *Commun Comput Phys* 2008;**3**(1):1–32.

58. Ruder S, An overview of gradient descent optimization algorithms (2016). arXiv:1609.04747. https://arxiv.org/abs/1609.04747. Accessed Feb. 15, 2021.

59. Zeiler MD, Adadelta: An adaptive learning rate method (2012). arXiv: 1212.5701. https://arxiv.org/abs/1212.5701. Accessed Feb. 15, 2021.
60. Duchi J, Hazan E, Singer Y. Adaptive subgradient methods for online learning and stochastic optimization. *J. Mach. Learn. Res.* 2011;**12**:2121–59.
61. Huang L, Polanco M, Clee TE. Initial experiments on improving seismic data inversion with deep learning. *New York Scientific Data Summit (NYSDS)* 2018;**2018**:1–3. http://doi.org/10.1109/NYSDS.2018.8538956.

CHAPTER 8

Prediction of acoustic velocities using machine learning and rock physics

Lian Jiang[a], John P. Castagna[a], Pablo Guillen[b]
[a]Department of Earth and Atmospheric Sciences, University of Houston, Houston, TX, United States
[b]Department of Computer Science, University of Houston, Houston, TX, United States

Abstract

Synthetic sonic logs are generated with a theoretical rock physics model (RPM) and are used to train three different machine learning algorithms for velocity prediction: support vector regression, random forest, and multi-layer perceptron. Random forest model best emulates the theoretical modeling of acoustic P-wave and S-wave velocities among the three models. For measured well logs from the Volve oil field in the North Sea, machine learning methods achieve significantly smaller prediction error than does the representative, but highly simplified, RPM that we used, with the best average coefficient of determination (R^2 score) increasing by 8.0% for P-wave velocity prediction, and 63.9% for S-wave velocity prediction relative to the RPM. Finally, a proposed hybrid approach, combining machine learning and RPM, is found to improve prediction robustness, with the average R^2 score increasing by 13.3% for P-wave velocity prediction relative to the RPM.

Keywords

Acoustic velocities; Hybrid machine learning; Machine learning; Rock physics modeling; S-wave velocity

8.1 Introduction

Rock physics modeling (RPM) can be used to estimate the elastic parameters of rocks with known mineral and fluid composition, porosity, and pore geometry, etc. (e.g.,[1–6]). In general, we can group these rock physics models into three classes: theoretical, heuristic, and empirical.[7] Many of the rock physics models achieve good results in the estimation of elastic parameters such as *P*- and *S*-wave velocity, but they all contain many underlying assumptions. For example, many inclusion theoretical models assume the inclusions have regular but usually unrealistic shapes, such as spheres or ellipsoids.[2,4,8–10] The heuristic models (e.g.,[11,12]) make fewer assumptions than theoretical models, but still we may not be able to capture the underlying complexity of many rocks, or may match observations under restricted circumstances. Empirical rock physics models (e.g.,[5,6,13,14]) make few assumptions but may be limited outside the range of calibrated data.

Alternatively, machine learning has been used for velocity prediction (e.g.,[15–17]). Example inputs are the P-wave velocity, gamma-ray (GR), deep/shallow resistivity (RD/RS), and bulk density (RHOB) logs. The mapping methods that have been used

Advances in Subsurface Data Analytics
DOI: https://doi.org/10.1016/B978-0-12-822295-9.00003-0

in velocity prediction include the neuro-fuzzy method, the multi-layer perceptron (MLP), and long short-term memory (LSTM) networks (e.g.,[15,16,18] Rajabi et al.[16] used the neuro-fuzzy method to predict P- and S-wave velocities using well logs, achieving good prediction error. Mehrgini et al.[18] proposed a method to predict S-wave velocity from well logs using the Elman network. The results from their work showed that the Elman network is more accurate and robust than the results predicted using three empirical relations, including Carroll,[19] Castagna et al.,[20] and Brocher,[21,22] as well as two artificial neural networks, including the RPM trained with two different optimization methods. Yang et al.[17] used the LSTM network to predict S-wave velocity from well logs with good success.

Various machine learning methods can be used to predict the velocities from well logs with high accuracy given sufficient calibration data and reliable inputs. In any specific application, however, several open questions requiring further study and analysis remain: What is the best method and architecture for the application at hand? How much, and with what distribution of, training data is required for a robust and accurate prediction? What happens when geologic occurrences that are not captured in the training dataset occur? How sensitive is the prediction to errors in the training data and in the input logs? How general is the resulting network and how far can it be extended?

In this work, we hypothesize that, despite the known inadequacies of theoretical and empirical rock physics models, incorporating them in the training and prediction process may aid in velocity prediction and in investigation of the above questions for any particular case. No RPM can be applicable to all data under all circumstances, and we make no attempt to construct an all-inclusive model. Rather, we use a representative theoretical RPM and investigate the extent to which it (1) can be replicated for velocity prediction using various machine learning algorithms given a limited local training dataset, and (2) can improve prediction robustness when combined with machine learning methods.

The RPM establishes the relationships between mineral composition, porosity, fluid properties, water saturation, temperature, and pressure with density and the P-wave and S-wave velocities. We use the RPM to create various training datasets and investigate the ability of three machine learning algorithms to learn to replace the RPM for velocity prediction. The three algorithms are: (1) support vector regression (SVR),[23,24] (2) random forest (RF),[25] and (3) multilayer perceptron (e.g.,[26,27]). We apply and compare the RPM and the machine learning algorithms to a field dataset individually and in a hybrid combined prediction method.

8.2 Conventional rock physics modeling

The purpose of a RPM is to predict the properties of the composite rock from constituent properties, volume fractions, and environmental conditions. We know the RPM to be oversimplified and, in our application, we do not require the RPM to be able to predict

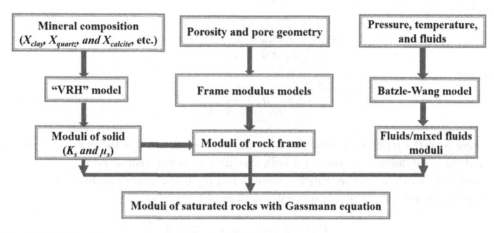

Fig. 8.1 *Workflow for conventional theoretical RPM.*

velocities quantitatively for field data given the limited information that usually is available and the consequently limited mathematical parameterization that can be employed. For example, pore shape distributions are important factors affecting velocities (e.g.,[2]) but are largely unknown given only well log measurements. Given the complexity of real rocks and our simplified assumptions about them, we know the RPM could be wrong, but we hope that it is useful in capturing the types of behavior that we can expect from the ground-truth for the purpose of evaluation of machine learning methods.

Fig. 8.1 is the workflow of the theoretical RPM we employ in this work. We use the Voigt-Reuss-Hill (VRH) average equation[28] to calculate the bulk and shear moduli of the zero-porosity mineral composite and the Batzle-Wang (BW) equations[29] to calculate the fluid bulk moduli.

To model rock frame moduli, we choose the modified Hashin-Shtrikman bound[30–32] and Walton's model[33] as appropriate approaches for granular rocks. The moduli of dry frame rocks at porosity ϕ are calculated using the modified Hashin-Shtrikman bound from the two end-points: one is the bulk and shear modulus of solid matrix calculated using the VRH average equation, and the other is the bulk and shear modulus of the dry frame at the critical porosity ϕ_c calculated utilizing Walton's model. Walton's model assumes that: (1) the rock is composed of a random packing of identical spheres, (2) spheres are either infinitely rough or perfectly smooth, (3) the intergranular porosity is below or close to the critical porosity, and (4) the relative displacement of the contacting particles' center is computed from the average strain.[31,33]

The modified Hashin-Shtrikman bound equations are:

$$K_{fr} = \left(\frac{\phi/\phi_c}{K_W + 4/3\mu_s} + \frac{1-\phi/\phi_c}{K_s + 4/3\mu_s} \right)^{-1} - \frac{4}{3}\mu_s, \tag{8.1}$$

and

$$\mu_{fr} = \left(\frac{\phi/\phi_c}{\mu_W + \xi} + \frac{1-\phi/\phi_c}{\mu_s + \xi} \right)^{-1} - \xi, \tag{8.2}$$

where

$$\xi = \frac{1}{6} \mu_s \frac{9K_s + 8\mu_s}{K_s + 2\mu_s}, \tag{8.3}$$

where K_{fr} is the frame bulk modulus, μ_{fr} is the shear modulus, ϕ is the porosity, μ_s is the solid shear modulus, K_s is the solid bulk modulus, ϕ_c is the critical porosity, K_W is the frame bulk modulus at the critical porosity and μ_W is the corresponding frame shear modulus, which are calculated using Walton's model:

$$K_W = \frac{C(1-\phi_c)\mu_s}{3\pi(1-\sigma_s)} \left[\frac{3\pi}{2} \frac{(1-\sigma_s)}{(1-\phi_c)C} \frac{P_d}{\mu_s} \right]^{1/3}, \tag{8.4}$$

and

$$\mu_W = \frac{3K_{HM}}{5} \frac{\left[2-\sigma_s + 3\gamma(1-\sigma_s) \right]}{(2-\phi_c)}, \tag{8.5}$$

where σ_s is the solid Poisson's ratio, P_d is the differential pressure, C is the average number of contacts per grain or coordination number, and γ is the friction coefficient characterizing the friction between two grains. The average coordination number for a random dense pack of identical spheres is about 9, and the higher the porosity, the lower the coordination number.[34] $\gamma = 0$ indicates that the surface is smooth, and $\gamma = 1$ suggests that the surface is infinitely rough.[31] The solid properties are determined from logged lithology, the nominal mineral composition for each lithology, rock physics handbook-based values of mineral elasticity, and VRH averaging.

After we obtain the bulk and shear moduli of the rock's solid matrix, rock frame, and fluids, we then calculate its moduli saturated with fluids using the Gassmann equation as follows[1]:

$$K_{sat} = K_{fr} + \frac{\left(1 - \frac{K_{fr}}{K_s} \right)^2}{\frac{\phi}{K_f} + \frac{1-\phi}{K_s} - \frac{K_{fr}}{K_s^2}}, \tag{8.6}$$

where K_{sat} is the bulk modulus of saturated rocks, K_m is the bulk modulus of the solid matrix, K_{fr} is the bulk modulus of the dry frame, and K_f is the bulk modulus of the fluids in pores, ϕ is porosity. Finally, the P- and S-wave velocity can be calculated by

$$V_P = \sqrt{\frac{K_{sat} + \frac{4}{3}\mu_{sat}}{\rho}}, \tag{8.7}$$

$$V_S = \sqrt{\frac{\mu_{sat}}{\rho}},\tag{8.8}$$

where ρ is the density, and μ_{sat} is the shear modulus of the rocks saturated with fluids, equal to the shear modulus of the rock's dry frame.

8.3 Machine learning methods

Machine learning uses observations to establish relationships and make predictions. Prediction success depends on a variety of factors, including the amount and quality of training data, the machine learning algorithm employed, and its suitability for a particular situation. It is often unknown in advance which algorithm would perform the best, and the performance of models strongly depends on the features and quality of the provided data set. To reach an acceptable performance, it is necessary to train a variety of algorithms multiple times with different hyper-parameter options to achieve the best prediction accuracy in validation with an acceptable stability. In this study, we apply three different machine learning algorithms: (1) SVR, (2) RF, and (3) MLP. To better illustrate and compare the pros and cons of each algorithm for our specific problems, we introduce their fundamental principles one by one in the following sections.

8.3.1 Support vector regression

The SVR is a supervised machine learning algorithm specializing in handling the regression problems with continuous values.[23,24] Given a training set $\{(x_1, y_1), (x_2, y_2), ..., (x_m, y_m)\}$ with m training instances, where $x_i \in R^n$ is an n-dimensional feature vector, and $y_i \in R$ is its corresponding label value. The primal optimization problem for SVR is described as follows:[24]

$$\min_{w, \xi, \xi^*} \left[\frac{1}{2} w^T w + \lambda \left(\sum_{i=1}^m \xi_i + \sum_{i=1}^m \xi_i^* \right) \right],$$

$$\text{subject to } w^T \phi(x_i) + b - y_i \leq \varepsilon + \xi_i,\tag{8.9}$$

$$y_i - w^T \phi(x_i) - b \leq \varepsilon + \xi_i^*,$$

$$\xi_i \geq 0, \xi_i^* \geq 0, i = 1, ..., m \quad \text{and} \quad \lambda > 0, \varepsilon > 0,$$

where $\phi(x_i)$ maps x_i to a higher-dimensional space, λ is the regularization parameter, w is the weight matrix, b is the bias; ε is a parameter that defines a tube so that: 1) $|w^T\phi(x_i) + b - y_i - \alpha| = 0$ if the predicted value is within the tube (i.e. $|w^T\phi(x_i) + b - y_i|$ $\leq \varepsilon$), 2) ξ_i is the positive difference between the predicted and observed values and ε if the observed point is above the tube (i.e. $y_i - w^T\phi(x_i) + b > \varepsilon$), and 3) ξ_i^* is the positive difference between the observed and predicted values and ε if the observed point is below the tube (i.e. $w^T\phi(x_i) + b - y_i > \varepsilon$).

The objective function of the primal problem is not differentiable; thus, we need to solve the following dual problem,

$$
\min_{\alpha,\alpha^*} \left[\frac{1}{2}(\alpha - \alpha^*)^T Q(\alpha - \alpha^*) + \varepsilon \sum_{i=1}^{m}(\alpha_i + \alpha_i^*) + \sum_{i=1}^{m} \gamma_i \left(\alpha_i - \alpha_i^* \right) \right],
$$

$$
\text{subject to } e^T \left(\alpha - \alpha^* \right) = 0 \quad \text{and} \quad 0 \le \alpha_i, \alpha_i^* \le \lambda, i = 1, ..., m,
$$

(8.10)

where $e = [1, ..., 1]T$, Q is a m by m positive semidefinite matrix and $Q_{i,j} \equiv K(x_i, x_j)$ is the kernel function (e.g., polynomial and Gaussian radial basis function). If the sample point is inside the tube, ξ_i or ξ_i^* is zero, α_i and α_i^* will be zero. If the observed point is "above" the tube, α_i is nonzero. Similarly, α_i^* is nonzero if the observed point is below the tube.

After solving the problem, the prediction function is, if we have a new vector $x^{(p)}$:

$$
y^{(p)} = \sum_{i=1}^{m} \left(\alpha_i^* - \alpha_i \right) K \left(x_i, x^{(p)} \right) + b.
$$

(8.11)

There also are two other algorithms to implement the SVR, called ν-SVR and linear-SVR.[35] Since ν-SVR is based on a similar concept as SVR and linear-SVR only can handle a linear relationship, we only test SVR in this work. SVR uses kernel functions to project the information into spaces of a greater dimensionality and then transform them to fit all the training data points with a certain error threshold by learning a hyperplane to process non-linear relationships. The selection of an appropriate kernel function is essential since the kernel function defines the feature space in which the training set will be used to build the predictive model. The kernel functions we tested in this work include polynomial, radial basis (RBF), and sigmoid functions. We choose the best kernel function based on the cross-validation performance of the training data.

The SVR algorithm is a learning machine based on the concept of support vector, which could make it perform better in a high dimensional space, since SVR optimization is not dependent on the input space's dimensionality. The training of this algorithm involves optimizing a convex cost function, whereas the testing predicts out-of-sample observations using the decision functions produced in the training process. SVR is adopted in this work due to its high accuracy relative to other machine learning methods and its ability to deal with high-dimensional data and complex non-linear relationships.

8.3.2 Random forest

Random forest (RF)[25] combines the predictors generated from many individual decision trees to improve prediction accuracy for various tasks, such as regression and classification. The "random" here means a random selection of samples and features from the training set for each tree predictor. After the training on each tree predictor is finished, the final prediction result for an RF predictor is simply the average values of all the trees.

Given a training set $\{(x_1, y_1), (x_2, y_2), \ldots, (x_m, y_m)\}$ with m training instances, $x_i \in R^n$ is an n dimensional feature vector and $y_i \in R$ is its corresponding label value. Starting from the root node, a decision tree recursively splits the data space into different new nodes with a rule that the data points with the same labels are grouped together. Denoting the data at node m as Q, for each candidate, $\theta = (j, t_m)$ is set, where t_m is the threshold for feature j. The data is split into $Q_{left}(\theta)$ and $Q_{right}(\theta)$ subsets:

$$Q_{left}(\theta) = (x, y) \mid x_j <= t_m, \tag{8.12}$$

$$Q_{right}(\theta) = Q \backslash Q_{left}(\theta). \tag{8.13}$$

We use an impurity or loss function H to evaluate the quality of a split at node m:

$$G(Q, \theta) = \frac{n_{left}}{N_m} H\left(Q_{left}(\theta)\right) + \frac{n_{right}}{N_m} H\left(Q_{right}(\theta)\right), \tag{8.14}$$

where n_{left} is the number of samples split on the left and n_{right} is the number of samples split on the right, N_m is the total samples number at node m.

Parameters are chosen that minimize the impurity,

$$\theta^* = argmin_\theta G(Q, \theta). \tag{8.15}$$

The process is repeated for the remaining subsets $Q_{left}(\theta^*)$ and $Q_{right}(\theta^*)$ until reaching the maximum allowable depth (e.g. $N_m < min_{samples}$ or $N_m = 1$).

Since the targets in our work are continuous values, we represent a region R_m with N_m observations at node m. The criterion we use to minimize the objective function for future splits is mean square error (MSE), which minimizes the L_2 error using mean values at terminal nodes, and mean absolute error (MAE), which minimize the L_1 error using median values at terminal nodes:

For MSE:

$$\bar{y}_m = \frac{1}{N_m} \sum_{i \in Q} y_i, \tag{8.16}$$

and

$$H(Q) = \frac{1}{N_m} \sum_{i \in Q} (y_i - \bar{y}_m)^2. \tag{8.17}$$

For MAE:

$$median(y)_m = \underset{y \in Q}{median}(y), \tag{8.18}$$

and

$$H(Q) = \frac{1}{N_m} \sum_{i \in Q} |y_i - median(y)_m|. \tag{8.19}$$

Compared with a single decision tree, RF is less prone to overfitting by training many single tree predictors.[25] This is because each tree is trained using a random drawing of samples with replacement from the training set and split from either all input features or a random subset with a predefined size. The tree predictors trained in this process could produce somewhat decoupled prediction errors, and some of the errors could cancel out from each other when averaged. Therefore, the randomness from these two sources helps reduce the variance of the forest estimator and could help improve the model generalization overall, although the bias could be slightly increased.

8.3.3 Multilayer perceptron

A MLP is a deep artificial neural network (DNN) composed of more than one perceptron (e.g.,[26,27]). It consists of an input layer to receive the signal, an output layer that makes a decision or prediction about the input, and in between those two, an arbitrary number of hidden layers that are the true computational engine of the MLP. Through proper training, the network can learn how to optimally represent inputs as features at different scales or resolutions, and combine them into higher-order feature representations relating these representations to output variables, and therefore learning how to make predictions. To achieve a high accuracy in the prediction, there are two dependent problems that have to be addressed when solving problems with DNNs: network architecture and training routine.[36] Defining network architectures involves setting fine-grained details such as activation functions and the types of layers as well as the overall architecture of the network. Defining training routines consists of setting the learning rules, the loss functions, regularization techniques, and hyper-parameter optimization. The hyper-parameters here include learning rate, the size of each hidden layer, epoch number, and batch size.

Fig. 8.2 is the workflow of MLP we used in this work. First, the training set is randomly split into the different batch sizes of sub-training sets, and a batch size of sub-training sets are fed into the network for training. Second, the hyper-parameters, such as learning rate and the number of nodes for each hidden layer, will be initialized. Third, we design the network's architecture, such as how to connect the nodes and how many hidden layers, etc. Fourth, the weights and the threshold of the network are initialized with a random normal distribution of numbers. With these settings and initialization, the output of the network is then computed. The weights and threshold for each layer will be updated by minimizing the cost function (e.g., mean square error) with a gradient descent algorithm and backpropagation rule. We use the prediction error of the network calculated using the validation set to determine if the training is overfitting. Finally, we repeat the previous steps with a different batch size of sub-training sets until the decrease in the training and validation errors becomes stable. Cross-validation and random search can be used to find the best combination of hyper-parameters and optimal models. The dropout technique can be used to alleviate the overfitting of the network, if necessary.

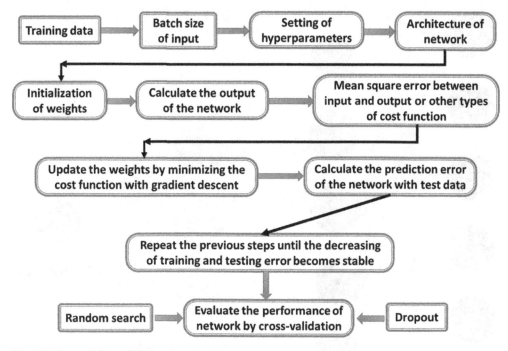

Fig. 8.2 *The workflow of MLP.*

Fig. 8.3 is the MLP architecture we used in this study. The first layer is the input layer, the second and third layers are hidden layers, and the last layer is the output layer. For each hidden layer, we use the ReLU activation function. The performance of the network is the mean square error between the output layer and target values.

8.3.4 Metrics for model performance evaluation

We used two metrics to evaluate the performance of each result. One is the prediction accuracy, and the other one is the coefficient of determination (R^2 score).

Given a data set $y \in R^m$ with m values $\{y_1, y_2, \ldots, y_m\}$ and the corresponding predicted values are $\{y_1^*, y_2^*, \ldots, y_m^*\}$, the prediction accuracy τ is defined by:

$$\tau = \frac{1}{m} \sum_{i=1}^{m} \left(1 - \frac{|y_i - y_i^*|}{y_i} \right). \tag{8.20}$$

The prediction accuracy describes average percentage of the true values that are correct, whereas the mean square error (MSE) describes the mean squared error between the true and the predicted values. Both metrics require the calculation of the difference between the true and predicted values; however, the prediction accuracy provides a more popular way (yet with caveats) to understand the model performance. Therefore, we chose prediction accuracy as a metric in this work.

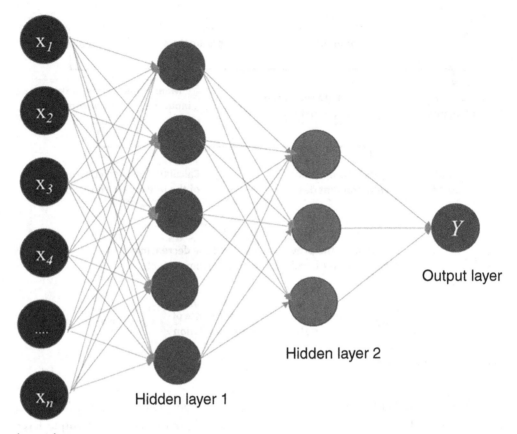

Fig. 8.3 *Architecture of MLP.*

The R^2 score is a metric used to measure the goodness of the prediction and the original data. It is defined as follows:

$$R^2 = 1 - \frac{SS_{res}}{SS_{tot}},$$ (8.21)

where

$$SS_{res} = \sum_{i=1}^{m}(y_i - \bar{y})^2 \text{ and } SS_{tot} = \sum_{i=1}^{m}(y_i - y_i^\star)^2,$$ (8.22)

where $\bar{y}$ is the mean of the observed data y. When the predicted values exactly match the measured data y, the R^2 score is equal to 1.0; when the predicted values are the baseline model (i.e., always predicts $\bar{y}$), the R^2 score is equal to 0.0; when the predicted values are worse than the baseline model, the R^2 score is less than 0.0. the R^2 score describes the fraction of variance that is unexplained; therefore, we can apply it to evaluate the goodness of a predictive model.

We use the Python Scikit-learn, Keras, and Tensorflow libraries to perform the model-building process for all the machine learning algorithms.

8.4 Synthetic data test

In order to test the efficacy of the machine learning algorithms at emulating a RPM, we first created synthetic wells by modifying measured and calculated logs such that each realization has different lithology, porosity, water saturation, differential pressure, etc. The procedure used was:

I.   Produce random variations in sand (V_{sand}), shale (V_{sh}), and carbonate (V_{carb}) volume fraction for each layer summing to 1.0;

II.   Produce random variations to the differential pressure with perturbations from –5.0 MPa to 5.0 MPa from the mean pressure;

III.   Produce the random variation of the porosity for each layer with a range of 0.0% to 35.0%;

IV.   Produce the random variation on the water saturation for layers with V_{sand} > 50.0%;

V.   Produce random variation of thickness for all layers;

VI.   Fill each layer with the randomly produced rock properties using steps I. through IV;

VII.   Calculate P-wave velocity (V_p), S-wave velocity (V_s), and density (ρ) for each synthetic well using the RPM (described further in the next section).

Fig. 8.4 is an example of a pseudo well produced using the RPM proposed in this work. As one can see in this figure, there are many variations for the rock elastic properties for each layer between the pseudo and original well logs.

The input data for each machine learning method are 10 synthetic wells for training purposes, two wells for validation, and one well for blind best. We used the random

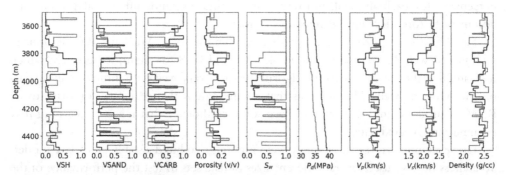

Fig. 8.4 *An example of a synthetic well created for this study.* The logs from the left to right are the volume percentage of shale, sand, and carbonates, porosity, water saturation, differential pressure, P-wave velocity, S-wave velocity, and density, respectively. The black curves are the measured log data, and the red curves are the new synthetic well logs produced by perturbing the original rock properties and using the RPM.

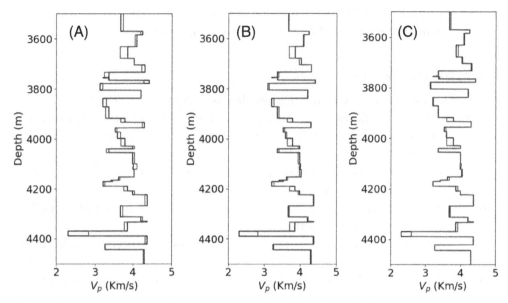

Fig. 8.5 The P-wave velocity predicted using the (A) SVR, (B) MLP, and (C) RF algorithms for a blind test well. The R^2 score is 0.94, 0.95, and 0.98 for them, respectively.

search method to find the best hyper-parameters for each machine learning method. We also normalized all the data before training the machine learning models. Each well has 2,000 training instances and 6 features (V_{sand}, V_{sh}, V_{carb}, ϕ, S_w, P_d) for each training instance; thus, the size of the training set is 20,000 instances, and the size of the validation set is 4,000. The rocks are composed of various minerals, pores, and fluids, and they are also subject to the differential pressure in the subsurface; therefore, the volumetric percentage for each mineral, porosity, water saturation, and differential pressure are dominant factors affecting the velocities of rocks.

Fig. 8.5 shows P-wave velocity predicted using the SVR, MLP, and RF methods in a blind test well. The R^2 scores are 0.94, 0.95, and 0.98 for them, respectively, which indicates that the machine learning algorithms work very well with limited training data on the synthetic wells. This suggests that, to the extent that real data conforms to an RPM of similar complexity, we can have some confidence that machine learning may be applicable to field data.

However, one may ask: how many wells is enough to build a stable predictive model? To address this, we used several different sizes of data sets to test the performance of the predictive models created using RF. Fig. 8.6 shows P-wave velocity prediction using the RF based on 3, 4, 5, and 7 training wells, and its corresponding R^2 scores are 0.72, 0.86, 0.92, and 0.97 for them, respectively. Not surprisingly, we can generalize that the greater the number of wells in training, the higher the prediction accuracy.

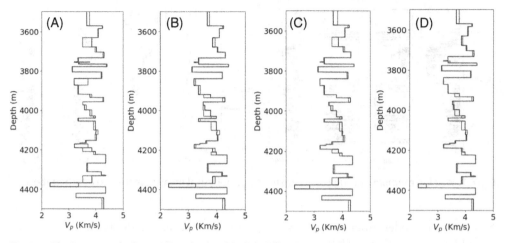

Fig. 8.6 *The P-wave velocity predicted using RF with different sizes of training sets (3, 4, 5, and 7) wells. The R^2 scores are 0.72, 0.86, 0.92, and 0.97 for them (from left to right), respectively.*

8.5 Real data test

The well data we used is from the Volve oil field, located in the central part of the North Sea.[37] The target interval is from the upper Permian and Jurassic to Cretaceous, with an average depth of about 3000.0 m. The lithology includes sandstone, shale, and carbonates with various grain sizes and sorting. The porosity varies from 0 to 30%. Four wells in the field have complete well logging suites. We edited the logs before using them in our calculation since any noise in the logs could introduce a bias in the training of a machine learning model.

To use Eqs. (8.1–8.5) to estimate the bulk and shear moduli of the rock frame, we need to know the model parameters (C, γ, and ϕ_c). Although we do not know their exact values, we know their possible range. For instance, the critical porosity can be assumed to be between 35.0% and 45.0%. The values of the model parameters for field data used for training are estimated using the following procedure:

1. Calculate K_{dry} and μ_{dry} with well log data (including S_w), using the Gassmann equations. This requires reliably measured V_p, V_s, and density, and known fluid and solid grain properties.
2. Set reasonable ranges for the three model parameters according to their physical meaning: C, γ, and ϕ_c, with discrete values within each range.
3. For each combination of model parameters in step 2), predict K_{dry} and μ_{dry} at each depth using Eqs. (8.1–8.5).
4. Separate measured and predicted K_{dry} and μ_{dry} from steps 1) and 3) into shale, sand, and carbonates based on their respective volume percentage (from logs).
5. For each lithology, calculate the combined R^2 score of the calculated versus predicted K_{dry} and μ_{dry}.

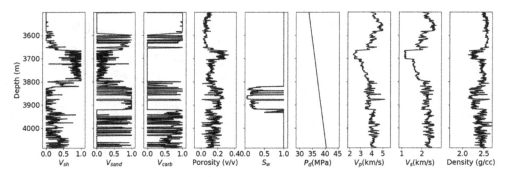

Fig. 8.7 *Well log example.* The log curves are from petrophysical measurement and interpretation.

6. Search all model parameter combinations for the optimal model parameters set that achieve the highest total R^2 score for each lithology.

We will take a single well as an example to illustrate how the above method works. The well logs are shown in Fig. 8.7, in which V_{sh}, V_{sand}, and V_{carb} are the volumetric percentage of shale, sand, and carbonates, respectively; S_w is water saturation, P_d is the differential pressure, V_p is the P-wave velocity, and V_s is the S-wave velocity.

Then we tried to estimate the bulk and shear modulus of the solid matrix using an empirical relationship between total porosity and bulk and shear moduli. The reason why we do not want to use the commonly available rock physics handbook-based values for minerals is that in real cases, it is difficult to estimate the mineral composition accurately. For each specific lithology (shale, sand, and carbonate), we generated the crossplot between bulk modulus and total porosity and the crossplot between shear modulus and total porosity. Fig. 8.8 is the result for the estimation of shale moduli, with $K_{shale} = 27.11$ GPa and $\mu_{shale} = 10.67$ GPa being values extrapolated to zero total porosity. These values are simi-

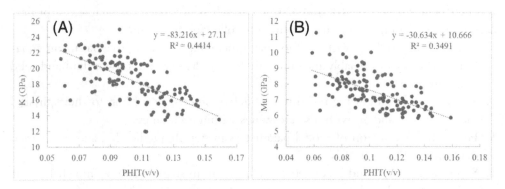

Fig. 8.8 *Estimation of the moduli of solid matrix.* (A) Crossplot of bulk modulus and total porosity and (B) crossplot of shear modulus and total porosity. $K_{shale} = 27.11$ GPa, $\mu_{shale} = 10.67$ GPa at zero total porosity.

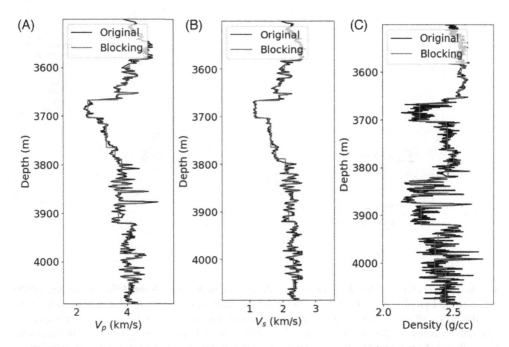

Fig. 8.9 *Blocking of well log curves in the real data.* (A) P-wave velocity, (B) S-wave velocity, and (C) density. The black curves are the measured logs and the red curves are the blocked logs.

lar to reported laboratory measurements.[34] Assuming that the scatter is mostly due to compositional variation and experimental error, we assume that the trend line provides a modulus estimate for the average shale composition at any given porosity.

We also blocked all the log curves to make it easier to estimate the model parameters and create synthetic well log data (Fig. 8.9).

We estimated the differential pressure using the following empirical equation:

$$P_d = 23.0 \star TVD - P_f, \tag{8.23}$$

where P_d is the differential pressure (MPa), TVD is the calculated true vertical depth, and P_f is the formation pressure. Since we do not have a complete density log for this well, we used an empirical overburden pressure gradient, which is 23.0 MPa/km.

To better evaluate the RPM created using this workflow, we calculated the bulk and shear frame moduli by changing one free parameter but fixing the other two free parameters rotationally. For instance, in Fig. 8.10, we calculated the bulk and shear frame moduli by using different friction coefficients by fixing the critical porosity and coordination number. As one can observe in this figure, the friction coefficient does not have any influence on the calculation of bulk modulus and the boundary value K_W, but has a significant effect on the calculation of shear frame modulus and μ_W, which increase with increased friction coefficient.

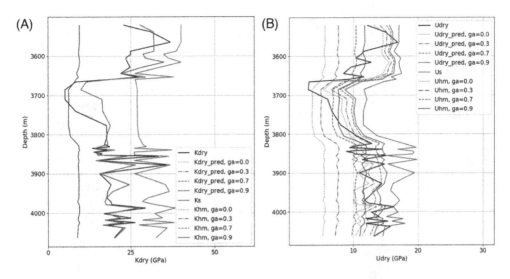

Fig. 8.10 *Predictions of the bulk and shear moduli of rock frame from different friction coefficients using Eqs. (8.1–8.5).* (A) Bulk frame modulus and (B) shear frame modulus. The black curves are the bulk and shear frame moduli calculated using well log data and the Gassmann equation. The green curves are the moduli of the solid matrix. The red and violet solid and dash curves are the bulk and shear frame moduli calculated using the optimized model parameters by varying the friction coefficients.

Fig. 8.11 shows that the optimized model log curves (red) match the bulk and shear moduli (black) calculated using the Gassmann equation reasonably well (see the detailed optimal model parameters in Table 8.1), with a R^2 score of 0.52 for the estimation of frame bulk modulus, and 0.20 for the estimation of frame shear modulus.

To make a prediction on one well, we use the remaining three wells to train the machine learning model and then use the trained model to predict this well. We repeat this process until we finished the prediction on all four wells. We first predicted P-wave velocity using the RPM built with the workflow illustrated in Fig. 8.1 with the optimized model parameters. Fig. 8.12 displays the P-wave velocity predictions for the four different wells. We calculated the average prediction accuracy and R^2 score using these four wells. The average prediction accuracy is computed by averaging the prediction accuracy for all the wells. The average prediction accuracy is approximately 94.4%, but the average R^2 score is only about 0.75. We use both the prediction accuracy and R^2 score to evaluate the goodness of our results more objectively.

We also predicted P-wave velocity using the three machine learning algorithms introduced previously. Fig. 8.13 shows the prediction results from the MLP. The prediction accuracy is better than that of RPM. with an average prediction accuracy of 95.6% and an average R^2 score of 0.81.

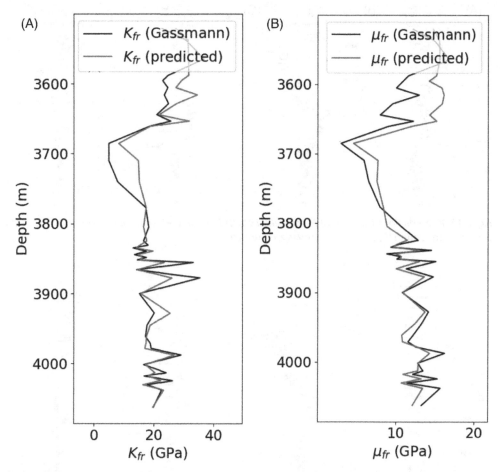

Fig. 8.11 *Predictions of the bulk and shear moduli of rock frame.* The black curves are the bulk and shear frame moduli calculated using the Gassmann equation. The red curves are the bulk and shear frame moduli computed using the Eqs. (8.1–8.5) with optimized model parameters. (A) Bulk frame modulus and (B) shear frame modulus.

Table 8.1 The optimized model parameters for each lithology.

	ϕ_c	C	γ
Shale	0.35	2.0	0.01
Sand	0.40	4.1	0.03
Carbonates	0.35	3.7	0.65

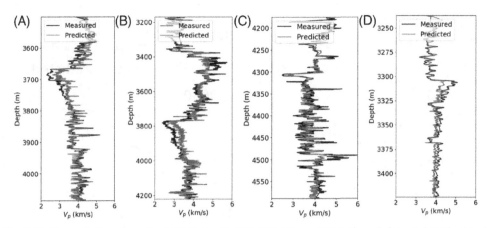

Fig. 8.12 V_p *predictions from the RPM for wells 19A, BT2, SR, and F1-B (from left to right).* The black curves are measured V_p, and the red curves are predicted V_p.

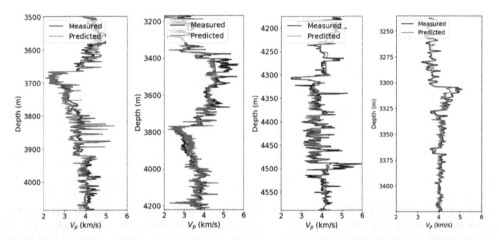

Fig. 8.13 *The V_p predictions using the MLP for wells 19A, BT2, SR, and F1-B (from left to right).* The black curves are measured V_p, and the red curves are the predicted V_p.

Fig. 8.14 shows the comparison of the S-wave velocity predictions between the MLP and conventional RPM for two different wells. We observe an obvious improvement from the conventional RPM to the MLP results, with an average R^2 score increasing from 0.46 to 0.75, though the predictions are not as good as the V_p predictions.

To combine the advantages of both conventional RPM and machine learning, we investigate a hybrid method (HM) to improve the prediction accuracy. The steps for this workflow are as follows:

1. Calculate the prediction error between the measured velocity and RPM-predicted velocity;
2. Train a machine learning model on the residual errors;

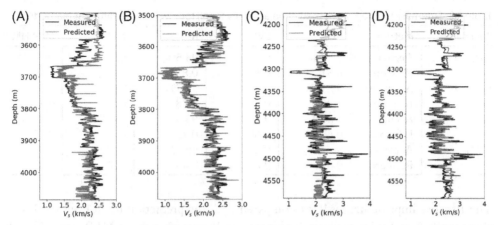

Fig. 8.14 *The V_s predictions from the conventional RPM and MLP.* (A) The prediction result from well 19A using the RPM, with accuracy = 90.6% and R^2 score = 0.71. (B) The prediction result from well 19A using the MLP, with accuracy = 93.4% and R^2 score = 0.86. (C) The prediction result from well SR using the conventional RPM, with accuracy = 89.3% and R^2 score = 0.29. (D) The prediction result from well SR using the MLP, with accuracy = 91.6% and R^2 score = 0.66. The black curves are measured V_s, and the red curves are predicted Vs.

3. Use machine learning to predict the residual for a blind test well using the trained model;
4. Add the predicted residual back to the predicted velocity from the RPM.

We predicted the P-wave velocities for all four wells using the HM and displayed two of them in Fig. 8.15. The R^2 score is 0.89 for well 19A, and 0.87 for well F1-B.

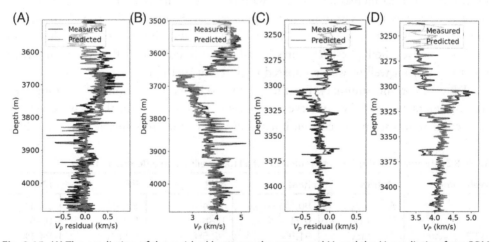

Fig. 8.15 (A) The prediction of the residual between the measured V_p and the V_p prediction from RPM for 19A well, with R^2 score of 0.39. (B) The V_p predicted using the proposed hybrid method for 19A, with R^2 score of 0.89. (C) The prediction of the residual between the measured V_p and the V_p prediction from RPM for F1-B well, with R^2 score of 0.48. (D) The V_p predicted using the proposed hybrid method for F1-B, with R^2 score of 0.87. The black curve is measured V_p or V_p residual, and the red curve is the predicted one.

Table 8.2 Summary for the *P*-wave prediction accuracy for all four wells using different methods.

Well name	Metrics	RPM	SVR	RF	MLP	MLP-HM
19A	Accuracy	93.4%	95.4%	94.5%	95.1%	95.7%
	R^2 score	0.81	0.88	0.83	0.86	0.89
BT2	Accuracy	93.1%	94.2%	92.7%	94.4%	95.1%
	R^2 score	0.85	0.86	0.78	0.85	0.89
SR	Accuracy	94.7%	93.2%	95.5%	95.3%	95.1%
	R^2 score	0.67	0.60	0.73	0.71	0.75
F1-B	Accuracy	96.3%	93.2%	95.3%	97.5%	97.8%
	R^2 score	0.67	0.07	0.45	0.82	0.87

The R^2 score improvement is 9.9% compared with the prediction result from the RPM alone for 19A and 3.5% compared with the prediction result from MLP alone. The R^2 score improvement is 29.9% compared with the prediction result from RPM for F1-B, and 6.1% compared with the prediction result from MLP. Combining the RPM with machine learning using the hybrid method improves the prediction accuracy relative to using either prediction approach alone.

Table 8.2 is the summary for the *P*-wave predictions for all four wells using different methods. In general, we can observe a noticeable improvement of the prediction results from the RPM to the machine learning methods. Different machine learning algorithms behave differently for different wells, though the difference between them is not large. The difference between the average prediction accuracy for all the wells from method to method is not noticeable, but the average R^2 score of the predictions from RPM to HM increases by 13.3%, with the lowest increase of 4.7% to the highest increase of 29.9%. The hybrid method, in general, has the best accuracy, with the exception of well SR, where the random forest (RF) is slightly better. However, the hybrid method achieves the highest R^2 scores for all the wells combined, though the improvement varies from well to well compared to different methods.

Table 8.3 is the summary for the *S*-wave velocity predictions of all four wells from different methods. We observe a noticeable improvement in the *S*-wave velocity prediction results from the conventional RPM to various machine learning methods.

Table 8.3 Summary for the *S*-wave predictions of all four wells using different methods.

Well name	Metrics	RPM	SVR	RF	MLP
19A	Accuracy	90.6%	94.4%	92.4%	93.4%
	R^2 score	0.71	0.88	0.79	0.86
BT2	Accuracy	91.9%	92.5%	90.6%	94.4%
	R^2 score	0.78	0.76	0.69	0.83
SR	Accuracy	89.3%	89.6%	91.4%	91.6%
	R^2 score	0.29	0.50	0.54	0.66
F1-B	Accuracy	94.9%	91.3%	91.8%	97.1%
	R^2 score	0.05	-0.82	-1.0	0.65

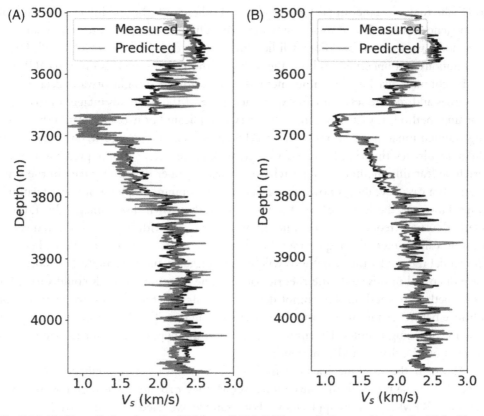

Fig. 8.16 *The V_s predictions from the MLP for well 19A.* (A) The prediction result using the MLP without using V_p as an input, with accuracy = 93.4% and R^2 score = 0.86. (B) The prediction result using the MLP after adding V_p as an input, with accuracy = 96.4% and R^2 score = 0.95.

The average prediction accuracy for all the wells from RPM to MLP increases by 2.7%, and the average R^2 score of the predictions increases by 63.9%, with the lowest increase of 6.4% to the highest increase of 127.6%.

We also tried to predict the S-wave velocity by adding P-wave velocity as an input in the training of the neural network. The results shown in Fig. 8.16 reveal that we can improve the prediction accuracy of S-wave velocity by using P-wave velocity as a constraint, with the R^2 score improved from 0.86 to 0.95.

The hybrid method did not improve V_S prediction over standard machine learning, presumably because the RPM used was so far off in the initial prediction.

8.6 Discussion

As we have shown in this study, the whole RPM process is quite complex, involving many different steps, which demand relatively more experience and domain knowledge to perform good work. Many assumptions have been made in this process, such as (1) the rock is a random packing of identical spheres, (2) the solid matrix is made of homogenous minerals and rock fragments from different data points to points, and (3) the model parameters for different lithologies do not change from well to well. These assumptions and uncertainties limit the utility of the velocity prediction using RPM.

In contrast, machine learning methods are relatively straightforward data–driven processes and require less domain experience. One of the main advantages of machine learning methods is that they can model very complicated non–linear relationships in a high–dimensional space. In our case, the RPM process is very complex, involving many different physics theories; however, we can make a relatively robust prediction using machine learning methods using a relatively small number of training data in the synthetic data case, revealing a powerful capacity for mapping complex nonlinear relationships. However, the learning-based methods have their own disadvantages. First, these methods usually require a certain amount of training data with high quality and diversity. If we cannot meet this requirement, we have to resort to conventional RPM because the model trained by these methods will be very unstable and inaccurate. Second, as we have discussed in this study, the theories of machine learning methods could vary a lot from method to method. We cannot determine which method is best for us to use in advance before testing them using the corresponding data. Third, the training process of machine learning is not a deterministic process as the hyper-parameter tuning process is based on the dataset and experience.

Although the machine learning methods can achieve better prediction results in general than the results from conventional RPM, they may not be as robust as the conventional RPM in various applications. For example, we tried to quantitatively analyze the effect of water saturation changes on P- and S-impedance changes, one of the most common applications in 4-D seismic inversion studies. We found that there is a distortion for the predictions compared to the data points calculated from the theoretical RPM, though the trends are similar. In addition, the conventional RPM can handle anisotropy relatively easily, whereas it is not easy to train an anisotropic machine learning model with conventional well logs.

The hybrid method proposed in this study proved to be useful to integrate the advantages of the RPM and machine learning methods, though this method still suffers the issues illustrated previously in both RPM and machine learning. In our future work, we would like to explore how to use this hybrid method to better predict the S-wave velocity using various kinds of geophysical data. In addition, testing how to build a generalized RPM that has a broader application from a large amount of data using machine learning is another important research area.

8.7 Conclusions

In this study, we tested how accurate and robust various machine learning methods can be in predicting velocities from measured and computed well logs. Using synthetic well logs, created by using a representative RPM to perturb an initial set of log measurements, we found that the model performance was very dependent on the machine learning algorithm employed as well as the size and quality of the training set. The synthetic data tests reveal that the machine learning algorithms can predict P-wave velocity, with an R^2 score as high as 0.98 for RF. In this synthetic test, the predictive model trained from machine learning algorithms reached what might be acceptable prediction accuracy (R^2 score = 0.90) by using a relatively small number of training wells (i.e., 5 wells with 10,000 training instances).

We then used the modified Hashin-Shtrikman bound and Walton's model to calculate the moduli of dry frame rocks by optimizing the model parameters with a lithology constraint. Synthetic trials demonstrated that we could model the bulk and shear moduli of dry rock frame, with a R^2 score of 0.52 for the estimation of frame bulk modulus and 0.20 for the estimation of frame shear modulus.

Finally, we proposed a hybrid approach to predict the P- and S-wave velocities that combined the RPM with the machine learning algorithms to predict velocities. Compared to conventional RPM, we found that:

1. The machine learning methods proposed in this work perform better than the RPM method in general, with the best average R^2 score increasing by 8.0% for P-wave velocity prediction relative to the RPM and 63.9% for P-wave velocity prediction relative to the machine learning results.
2. The proposed hybrid method achieves a better prediction result than the RPM or machine learning methods alone, with the average R^2 score increasing by 13.3% for P-wave velocity prediction relative to the RPM and 5.0% for S-wave velocity prediction relative to the machine learning prediction results.
3. Different machine learning methods behave differently for different wells, with the RF achieving the best performance for the synthetic data set and the MLP for the real data set.
4. The S-wave velocity prediction accuracy is improved by using P-wave velocity as an input to the machine learning training process.

This physics-informed machine learning approach shows its strength regarding uncertainty quantification and its robustness on how one could augment machine learning models with physics-based domain knowledge.

8.8 Acknowledgments

The authors would like to thank the research computing data core center from the University of Houston for providing computing resources. We thank Equinor AS, the former Volve license partners ExxonMobil Exploration and Production Norway AS and Bayerngas (now Spirit Energy) for permission to use the Volve dataset, and to the many people who have contributed to the work here. We also want to thank Dr. Yi Wang and Dr. Tom Smith for their very useful suggestions and discussion. At last, we want to thank Dr. Leon Thomsen for his very useful and insightful suggestions and help in this work.

References

1. Gassmann F. Uber die Elastizitat poroser Medien. *Vierteljahrsschrift der Naturforschenden Gesellschaft in Zurich* 1951;**96**:1–23.
2. Kuster GT, Toksoz MN. Velocity and attenuation of seismic waves in two-phase media: part 1. Theoretical formulations. *Geophysics* 1974;**39**:587–606.
3. Brown RJS, Korringa J. On the dependence of the elastic properties of a porous rock on the compressibility of the pore fluid. *Geophysics* 1975;**40**:608–16.
4. Berryman JG. Long-wavelength propagation in composite elastic media. *J Acoust Soc Am* 1980;**68**:1809–31.
5. Han D. Effects of porosity and clay content on acoustic properties of sandstones and unconsolidated sediments: Ph.D. dissertation, Stanford University, 1986.
6. Greenberg ML, Castagna JP. Shear-wave estimation in porous rocks: theoretical formulation, preliminary verification and applications. *Geophys Prospect* 1992;**40**:195–209.
7. Avseth P, Mukerji T, Mavko G. Quantitative Seismic Interpretation: Applying Rock Physics Tools to Reduce Interpretation Risk: Cambridge, United Kingdom: Cambridge University Press, 2005.
8. Eshelby JD. The determination of the elastic field of an ellipsoidal inclusion, and related problems. *Proc R Soc Lond* 1957;**A241**:376–96.
9. Walsh JB. The effect of cracks on the compressibility of rock. *J Geophys Res* 1965;**70**:381–9.
10. Jakobsen M, Hudson J, Johansen T. T-matrix approach to shale acoustics. *Geophys J Int* 2003;**154**:533–58.
11. Wyllie MRJ, Gregory AR, Gardner LW. Elastic wave velocities in heterogeneous and porous media. *Geophysics* 1956;**21**:41–70.
12. Amos MN, Mavko G, Dvorkin J, Gal D. Critical porosity: the key to relating physical properties to porosity in rocks. *SEG Technical Program Expanded Abstracts* 1995:878–81.
13. Gardner GHF, Gardner LW, A. R. Formation velocity and density - the diagnostic basics for stratigraphic traps. *Geophysics* 1974;**39**:770–80.
14. Castagna JP, Batzle ML, Eastwood RL. Relationships between compressional wave and shear-wave velocities in clastic silicate rocks. *Geophysics* 1985;**50**:571–81.
15. Rezaee MR, Ilkhchi AK, Barabadi A. Prediction of shear wave velocity from petrophysical data utilizing intelligent systems: an example from a sandstone reservoir of Carnarvon Basin, Australia. *J Pet Sci Eng* 2007;**55**(3-4):201–12.
16. Rajabi M, Bohloli B, Ahangar EG. Intelligent approaches for prediction of compressional, shear and Stoneley wave velocities from conventional well log data: a case study from the Sarvak carbonate reservoir in the Abadan Plain (Southwestern Iran). *Comput Geosci* 2010;**36**(5):647–64.
17. Yang LX, Sun S, Ji LL. S-wave velocity prediction for complex reservoirs using a deep learning method. *SEG Technical Program Expanded Abstracts* 2019:2574–8.
18. Mehrgini B, Izadi H, Memarian H. Shear wave velocity prediction using Elman artificial neural network. *Carbonates Evaporites* 2017;**34**(4):1281–91.
19. Carroll RD. The determination of acoustic parameters of volcanic rocks from compressional velocity measurements. *Int J Rock Mech Min Sci* 1969;**6**:557–79.
20. Castagna JP, Batzle ML, Kan TK, Backus MM. Rock physics - the link between rock properties and AVO response. In: Offset - dependent reflectivity - theory and practice of AVO analysis. *Soc Expl Geophys* 1993:124–57.
21. Brocher TM. Empirical relations between elastic wave speeds and density in the Earth's crust. *Bull Seismol Soc Am* 2005;**95**(6):2081–92.
22. Brocher TM. Key elements of regional seismic velocity models for long period ground motion simulations. *J Seismolog* 2008;**12**(2):217–21.
23. Cortes C, Vapnik V. Support-vector networks. Machine learning 1995;**20**:273–297.
24. Drucker H, Burges LK, Smola A, Vapnik V. Support vector regression machines. Advances in neural information processing systems 1996;**9**:155–161.
25. Breiman L, Random forests. Machine learning, 2001;**45**(1):5–32.
26. Hecht-Nielsen R. Kolmogorov's mapping neural network existence theorem. In IEEE First International Conference on Neural Networks; 1987.

27. Hornik K, Stinchcombe M, H W. Multilayer feedforward networks are universal approximators. *Neural Netw* 1989;**2**(5):359–66.
28. Hill R. The elastic behaviour of a crystalline aggregate. *Proc Phys Soc* 1952;**A65**:349.
29. Batzle M, Wang Z. Seismic properties of pore fluids. *Geophysics* 1992;**57**(11):1396–408.
30. Gal D, Dvorkin J, Nur A. A physical model for porosity reduction in sandstones. *Geophysics* 1998;**63**(2):454–9.
31. Ruiz FJ. Porous grain model and equivalent elastic medium approach for predicting effective elastic properties of sedimentary rocks: Ph.D. dissertation, Stanford University, 2009.
32. Grana D. Bayesian linearized rock-physics inversion. *Geophysics* 2016;**81**:625–41.
33. Walton K. The effective elastic moduli of a random packing of spheres. *J Mech Phys Solids.* 1987;**35**(2):213–26.
34. Mavko G, Mukerji T, Dvokin J. Tools for Seismic Analysis in Porous Media. Cambridge, United Kingdom: Cambridge University Press, 2009.
35. Scholkopf B, Smola A, Williamson RC, Barlett PL. New support vector algorithms. *Neural Comput* 2000;**12**:1207–45.
36. Goodfellow I, Bengio Y, Courville A. Deep Learning: Cambridge, Massachusetts, USA: MIT Press, 2016.
37. Malm OA et al. Discovery evaluation report: Statoil, 1993.

CHAPTER 9

Regularized elastic full-waveform inversion using deep learning

Zhendong Zhang[a], Tariq Alkhalifah[b]
[a]The Earth Resources Laboratory, Massachusetts Institute of Technology, MA, United States
[b]Department of Physical Science and Engineering, King Abdullah University of Science and Technology, Thuwal, Saudi Arabia

Abstract

Elastic full-waveform inversion, which aims to match the waveforms of prestack seismic data, potentially provides more accurate high-resolution reservoir characterization from seismic data. However, full-waveform inversion can easily fail to characterize deep-buried reservoirs due to illumination limitations. We present a deep learning-aided elastic full-waveform inversion strategy using observed seismic data and available well logs in the target area. Seismic facies interpreted from well logs are linked to the inverted P- and S-wave velocities using trained neural networks, corresponding to the subsurface facies distribution. The desired reservoir-related parameters such as velocities and anisotropy parameters are evaluated using a weighted summation given by the neural network classification distribution of facies. Finally, we update these estimated parameters by matching the resulting simulated wave fields to the observed seismic data, which corresponds to another round of elastic full-waveform inversion aided by the prior knowledge gained from ML predictions. A modified Marmousi synthetic example is used to prove the concept of the proposed inversion method. A North Sea field data example, the volve oil field data set, shows that the use of facies as prior helps resolve the deep-buried reservoir target better than the use of only seismic data.

Keywords

Artificial intelligence; full waveform inversion; reservoir characterization

9.1 Introduction

Artificial intelligence (AI) or machine learning (ML) has been used in many disciplines and geoscience is not an exception[1–12]. While not a replacement for the physical laws, ML can be a powerful technique in solving geophysical problems. Not all the geophysical problems lend themselves to the virtue of ML, which is fundamentally dependent on pattern recognition and matching. Actual physical insight is valuable and necessary for solving geophysical problems. In the past years, elastic full-waveform inversion (FWI) is still the most feasible way to image the Earth with high-resolution[13–15]. However, not all the physical parameters are well constrained by the often limited coverage of seismic data. Additional measurements that may illuminate the Earth from a different perspective and resolution can provide considerable value. The integration of different

Advances in Subsurface Data Analytics
DOI: https://doi.org/10.1016/B978-0-12-822295-9.00009-1

measurements for model estimation is challenging due to their different scales and sensitivities. As a result, physics-informed neural networks that respect any given physical laws described by partial differential equations (PDEs) are introduced by Raissi et al.[16]. Recent applications of physics-informed neural networks show promising results in wavefield-based processing[17-19]. Alternatively, we solve the elastic wave equations governing the wave propagation using the traditional staggered-grid finite-difference schemes[20]. Neural networks are sometimes used to find statistically accurate data mapping between different measurements, where closed-form relations are not available[21,22].

Elastic FWI for subsurface imaging was introduced in the late 1970s. Later, Lailly and Bednar[23] and Tarantola[24] addressed the inverse problem by adapting the adjoint-state method. Not many promising FWI applications were reported until 2007, when a blind test was successfully implemented by Brenders and Pratt[25]. Since then, FWI has been developed and applied to a wide range of applications, for example, ultrasound imaging[26,27], exploration geophysics[28,14] and global tomography[29,30]. The ultimate goal of FWI is to use every wiggle in seismic records to determine the structure of Earth's interior with constraints from the physics of seismic wave propagation[13,15]. Despite its elegant theorem, the practical use of FWI faces many challenges. One of them is the risk of converging to a local minimum. In the past decades, lots of effort have been put into finding better objective functions that are immune from cycle skipping[31-37]. The conventional wiggle-to-wiggle subtraction-based measurement fails FWI when the predicted and observed data exceed the half-cycle limit[13]. Choi and Alkhalifah[38] proposed a normalized global crosscorrelation-based objective function for FWI, which reduces the dependency on the seismic amplitudes. It is more sensitive to phase differences in the data, and thus, it is more immune to ambient noise in the field data[39]. However, the global crosscorrelation objective function still suffers from high nonlinearity and the danger of converging to a local minimum when the initial model is far from the actual one. One intuitive remedy to the problem is to initially select parts of the data free from cycle-skipping in the inversion, which is referred to as multiscale inversion[40-43]. A selection of frequencies from low to high is a widely used strategy in multiscale inversion. An alternative choice of multiscale inversion is selecting data with certain offsets, which are free of cycle-skipping. The crucial step of these approaches is the scheme used in selecting data for inversion in a progressive manner. For example, the first arrivals are selected for inversion in the early stages[44,45]. Other strategies based on the half-wavelength criterion are also investigated[41,42]. On the other hand, a better description of the physics of the Earth needs multiple parameters. Estimating elasticity and anisotropy parameters is an ongoing topic of interest in the seismic exploration community. In current practice, due to an inherent trade-off between parameters, for example, P-wave velocity and density, the density model is not usually updated at all to reduce the nonlinearity (null space) of the inversion[46]. Many investigations in vertical transverse isotropy (VTI) multiparameter inversion have shown that not all the anisotropy parameters are recoverable from surface collected data[47-50]. However, to get a better understanding of the subsurface, an accurate multiparameter inversion is necessary.

Multiparameter estimation increases the size of the model space and the ill-posedness of the inverse problem. Regularization, such as total-variation, on the other hand, can effectively constrain the model space, and thus, reduces the nonuniqueness. Besides, total-variation can extend the range of model wavenumbers in updates[51]. As an alternative, utilizing different geophysical data can also help resolve the multi parameters even better[52,53,54,87]. The interpretation of seismic data on their own will provide incomplete information due to the limited extent and illumination provided by seismic data and their limited spatial resolution. However, additional measurements that may illuminate the reservoir with alternative coverage and resolution can provide considerable value[53]. A facies-constrained elastic full-waveform inversion strategy can effectively reduce the crosstalk between different parameters by incorporating known facies[21,55–58]. Facies, defined as groups of seismic properties and conformity layers that share a particular relationship with geological and lithological properties, can be obtained from wells, geological models or maps or other investigations[52]. Estimated models from surface seismic data and those extracted facies from other geophysical surveys, like well logs, are often provided at very different scales and there are no explicit formulas to relate these scales accurately. Previously, a Bayesian inversion was used to connect the additional information in a statistical manner[55]. However, recently emerging ML algorithms can do a better job in finding statistical relationships between different types of data[11,12,59–63]. ML methods, in general, can be categorized as either supervised or unsupervised learning. Their structures can vary quite a lot depending on the purpose. Bergen et al.[64] provided a comprehensive review on ML applications in Geoscience. In this chapter, as an example, we train deep neural networks (DNNs) using labeled facies. In this way, after training, the network's task is to map a list of facies onto a 2D/3D inverted model, which is also known as the facies distribution. The facies distribution can include high-resolution properties from the well, like the elastic properties of waves, or even porosity and permeability.

This chapter will show how can trained neural networks fill the gap between the well log data and the seismic data in reservoir characterization. We first introduce the basics of elastic FWI, deep neural networks and facies constraints in methodology. Then we use synthetic and real data examples to show how the ML-assisted algorithm works. The example training code in the Appendix can help understand the workflow. Finally, we discuss issues that we may face in practice.

9.2 Methodology

9.2.1 Correlation elastic FWI

Crosscorrelation-type objective functions emphasize matching the phases of the predicted and observed data. Thus, it is insensitive to amplitude issues arising from the simplified physics we use to represent the medium[38]. The zero-lag crosscorrelation objective function is given by

$$J_d\left(m^i\right) = -\sum_s \sum_r \hat{u} \cdot \hat{d}, \tag{9.1}$$

where $\hat{u} = \dfrac{u}{\|u\|}$ and $\hat{d} = \dfrac{d}{\|d\|}$ are normalized predicted and observed data, respectively. J_d is the corresponding data misfit for the model $\mathbf{m}^i$, and i is an index of the model that belongs to a model space guided by the way we parametrize the subsurface of interest. The indexes s and r correspond to the source and receiver locations, respectively.

The inverse problem is constrained by a first-order elastic wave equation[14], which is given by

$$\begin{pmatrix} \rho I_3 & 0 \\ 0 & C^{-1} \end{pmatrix} \frac{\partial \Psi(x,t)}{\partial t} - \begin{pmatrix} 0 & E^T \\ E & 0 \end{pmatrix} \Psi(x,t) - f(x_s,t) = 0, \qquad (9.2)$$

where $\Psi(\mathbf{x},\,t) = (v_1,\,v_2,\,v_3,\,\sigma_1,\,\sigma_2,\,\sigma_3,\,\sigma_4,\,\sigma_5,\,\sigma_6)^T$ is a vector containing three particle velocities and six stresses, $\mathbf{I}_3$ is a 3×3 identity matrix. $\mathbf{C}$ is the stiffness matrix, E denotes spatial differentiation, and $\mathbf{f}(\mathbf{x}_s,\,t)$ is the source, located at $\mathbf{x}_s$.

To obtain the gradient function of the proposed objective function, we take its derivative with respect to the model parameters as follows

$$\frac{\partial J_d}{\partial m} = \sum_s \sum_r \frac{\partial u}{\partial m} \cdot \left(\frac{1}{\|u\|} \left(\hat{u}\left(\hat{u} \cdot \hat{d}\right) - \hat{d}\right)\right). \qquad (9.3)$$

For the parameterization of C_{ij}, the Fréchet derivative, $\dfrac{\partial u(C_{ij},x,t)}{\partial C_{ij}}$, is given by Vigh et al.[14]:

$$\frac{\partial u(C_{ij},s,x,t)}{\partial C_{ij}} = \frac{\partial C}{\partial C_{ij}} C^{-1} \left(\frac{\partial \sigma}{\partial t} - f \right)_{i=1,\ldots 6;\, j=i,\ldots,6} \quad \text{and} \quad \left(\frac{\partial C}{\partial C_{ij}} \right)_{pq} = \begin{cases} 1, p=i, q=j \\ 1, p=j, q=i, \\ 0, otherwise \end{cases} \qquad (9.4)$$

where σ denotes the stress component of the forward-propagated wavefield $\dfrac{\partial C}{\partial C_{ij}}$ is a six-by-six matrix with elements defined in Eq. 9.4. Here we show the gradient for the parameterization of C_{ij}, but the gradients for other parameters such as V_P and V_S can be derived using the chain rule. The model is updated iteratively using the Limited-memory Broyden–Fletcher–Goldfarb–Shanno (BFGS) algorithm[65], which is written as

$$\mathbf{m} = \mathbf{m}_0 - \alpha \mathbf{H}^{-1}\mathbf{g}, \qquad (9.5)$$

where α is the step length calculated by satisfying the Wolfe condition[66], and $\mathbf{H}$ is the approximated Hessian matrix. $\mathbf{m}_0$ and $\mathbf{g}$ are vectors of current model and gradient, respectively.

9.2.2 Deep neural networks

A deep neural network has multiple hidden layers between the input and output layers, as indicated by its name. It is nothing but a nonlinear system of equations that maps the input onto the output. With the input layer denoted as x, the kth hidden layer can be expressed as $a_k = \varphi_k\{W_k(\ldots \varphi_1[W_1 x + b_1]) \ldots + b_k\}$, and the output layer is written as $y = W_{l+1}a_l + b_{l+1}$, where l denotes the last hidden layer. The input, x, can be raw data or derived data features (e.g., V_S/V_P). The output, **y**, depends on the application. For example, it can be integers for classification problems. The data moves in one direction, from the input layer, through the hidden layers and to the output layer in a feedforward fashion. In each hidden layer, φ denotes the activation function, which defines the output of that node with a fed input. The activation function can be the sigmoid, rectified linear unit (ReLu), or other functions. Its purpose is to add nonlinearity to the mathematical operations involved in computing outputs given inputs to the network. The training process updates **W** and **b** for each layer through an optimization problem. The objective of the problem is to find a more accurate mathematical manipulation capable of mapping the input to the output using a loss function of sparse softmax cross-entropy[67]. Our designed DNNs use three features, V_P, V_S and V_S/V_P, and labeled facies as inputs for the training, which are from the conventionally inverted models and the well logs, respectively. Four hidden layers with 256 nodes in each layer are deployed as shown in Fig. 9.1. A ReLu activation function is used[68]. For each layer, we use a random dropout of 10% to avoid overfitting[69]. Besides, random data augmentation is applied to balance the proportion of different facies in training the data[70]. The number of labeled data pairs is limited by the available well logs, for example, about 1500 in our examples. We duplicate the labeled data to balance the number of data points in each category. There is no clear criterion for data augmentation, but we compared the data loss (training) and test accuracy curves to decide whether the data augmentation is sufficient or not. A batch

Neural Networks architecture

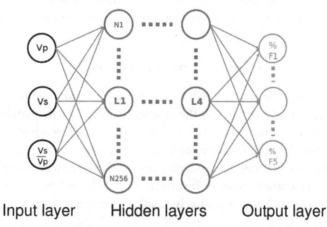

Input layer   Hidden layers   Output layer

Fig. 9.1 *The neural network architecture.* Three features are used in the input layer. Four hidden layers with 256 nodes are fully connected neural networks with a dropout rate of 10%. The output layer provides probabilities of being certain facies for the current input.

size of 128 and 10000 total training steps helped us estimate the network parameters. The training takes less than 5 minutes on an NVIDIA K40 GPU card. The Adam optimizer is used to update the weighting matrix of neural networks. In our application, we output the probabilities for all facies instead of one specific kind. After obtaining the probability distribution of facies, we can calculate the subsurface model parameters (like V_H, V_S, ϵ and η) by a weighted summation over n_f facies, $m^p = \sum_{i=1}^{n_f} p_i m_i \times m^p$ denotes averaged P-, S-wave velocities or anisotropy parameters, which is equivalent to the posterior expectation in Zhang et al.[55]. p_i and m_i are probabilities estimated by the trained DNNs for the corresponding facies, respectively. Such a weighted summation avoids potential bias caused by a particular kind of facies when the DNNs fail. Besides, it can be used to interpolate between different facies. In practice, we can never know all the subsurface facies, and we do not need to know all of them in this ML-assisted method. The probabilities act as interpolation weights for the known facies. Suppose the corresponding facies for certain pairs of V_p and V_S are not available as prior knowledge. In that case, the averaged parameters still have a chance of being (or close to) the correct ones through interpolation. We add an example script to the Appendix to help you understand the procedure.

9.2.3 Facies constraints

Seismic facies are defined abstractly by clustering seismic properties and spatial coherence that will generally have some relationship with geological and lithological properties. Facies can, therefore, provide rock physics relationships, which can be utilized as constraints in the inversion. The space-varying relation is a crucial feature (i.e., rock physics constraints per facies[52,71]) as opposed to assuming only one relationship over the entire area[72]. In a previous study[55], the facies distribution is converted to desired parameters used in inversion, which are then used as model constraints. The process can be expressed as

$$J_{reg}(\mathbf{m}^i) = J_d(\mathbf{m}^i) + \beta \left\| \mathbf{m}^i - \mathbf{m}^p \right\|^2, \tag{9.6}$$

where $J_d(\mathbf{m}^i)$ measures the data mismatch (Eq. 9.1), $\mathbf{m}^i$ and $\mathbf{m}^p$ denote the inverted and prior models, respectively. β balances the amount of seismic data matching and utilizing the known facies in this case. The choice of β is case dependent. Mathematically, it can be determined by the L-curve method[73]. However, in practice, seismic data can be very noisy for the real case, and we may use a relatively large β, which means the extracted facies are trusted more in inversion. For high-quality seismic data, such as marine data, we might choose a relatively small β to assign a larger weight to seismic-data matching.

The workflow of the ML-assisted method is summarized in Fig. 9.2. We first obtain the initial estimates for V_p and V_S using elastic FWI and a list of known facies in the

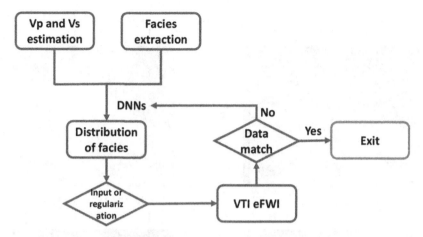

Fig. 9.2 *The workflow for the ML-assisted inversion method.* The facies distribution and the regularly inverted models can be updated, iteratively.

target area. Then we train a deep neural network to map the known facies to the initially estimated V_p and V_S and obtain the distribution of any desired models (V_p, V_S, ϵ and η in VTI media).

Finally, we use the smoothed version of such models as an input model for another round of elastic FWI. In this way, we avoid estimating β in Eq. 9.6. We can update the facies distribution and models iteratively by applying multiple nested inversions.

9.3 Numerical examples

In the following, synthetic and real data examples are shown to demonstrate the ability of the machine-learned prior in helping elastic and anisotropic FWI admit high-resolution models.

9.3.1 A synethetic Marmousi example

The Marmousi model was created based on a profile through the North Quenguela Trough in the Cuanza Basin[74]. The model includes strong vertical and lateral velocity variations, which pose a serious challenge for depth-migration and velocity tomography methods. The actual S-wave velocity is generated for this Marmousi model by setting $v_s = v_p / \sqrt{3} + 0.1(v_p - 2.4)$. The actual and initial velocities are shown in Fig. 9.3. Initial models are 1D linear-gradient models, which are far from the actual ones $\left(v_s = v_p / \sqrt{3}\right)$. 220 sources and 330 receivers are evenly deployed on the surface of the model and the recorded data are two-component particle velocities. The maximum offset is 6.6 km. A staggered finite-difference scheme is implemented to solve the elastic wave equation[20]. The source wavelet is a Ricker wavelet (f_p = 5 Hz) without frequencies below 5 Hz, in which case the inversion fails to converge to the global minimum

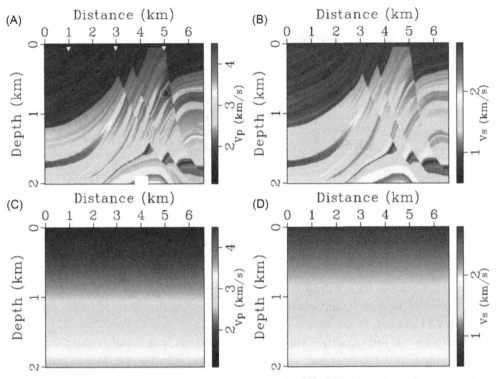

Fig. 9.3 *Velocity models.* Actual V_P (A) and V_S (B). √Initial V_P (C) and V_S (D). There are two low velocity zones in actual V_P; actual $V_S = V_P / \sqrt{3} + 0.1(V_P - 2.4)$. Solid triangles in (A) indicate locations of pseudo√wells used in the training. The initial models are constant gradient models and $Vs = V_P / \sqrt{3}$.

without prior information as shown in Fig. 9.4. We extract ten facies from pseudo wells at X = 1, 3, and 5 km as shown in Table 9.1. There is generally no need to extract all the existing facies from the well, and in our case, we focused on the dominant ones. The weighted summation using probabilities as weights effectively interpolate between facies for the outputs (e.g., V_p and V_S). Then we use the estimated V_P, V_S (as shown in

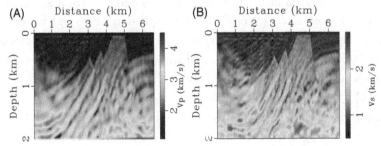

Fig. 9.4 *Estimated velocities without facies constraints.* (A) V_p (B) V_s. The inversion has apparently converged to one of the local minima.

Table 9.1 Ten facies in the Marmousi model.

Facies number	1	2	3	4	5	6	7	8	9	10
V_P (km/s)	1.5	1.7	2.2	2.5	2.65	3.2	3.5	3.8	4.0	4.5
V_S (km/s)	0.78	0.92	1.24	1.45	1.55	1.94	2.15	2.32	2.46	2.78

Fig. 9.4) and their ratio, V_S/V_P, at the same location as data features. The interpreted facies (Table 9.1) from the pseudo wells are labels of the training data set. After the DNNs are well trained, the full dimension of inverted velocities and their ratios are used as input data to generate a possible distribution of facies over the model space. Fig. 9.5 shows the normalized data loss versus iteration at every 100 steps. A total of 70% data loss for the training data set and a 55.6% test accuracy are achieved. K-fold cross-validation can be used to aid the design of neural networks[75]. We did not apply the K-fold cross-validation in this example since the neural networks used can generate acceptable initial models for elastic FWI. Although the test accuracy is relatively low, the converted parameters still have a chance to be close to the actual values thanks to the weighted summation. Besides, the following elastic FWI can improve the accuracy by matching the observed seismic data. The distribution of facies is converted to V_p using a weighted summation $\bar{v} = \sum_{i=1}^{n_f} p_i v_i$, v are given in Table 9.1) as shown in Fig. 9.6A. It has a similar structure as the actual V_p but with some loss in detail. The largest probabilities of area's large values indicate that the trained DNNs can classify the inverted velocities to a particular facies with high confidence (one large p_i and the rest are smaller ones). However, the smaller values in the deep part indicate that the trained DNNs are slightly puzzled in the classification and return similar probabilities to nearby facies (a list of small p_i). The variances of the probabilities ($var = mean((output\ output.mean)^2)$) as shown in Fig. 9.6C indicate a similar conclusion. In this case, the variance can indicate the uncertainties in the inverted velocities from this elastic FWI. In the definition above, a large variance indicates that the estimation matches the known facies well. In contrast, a small variance

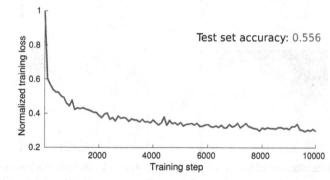

Fig. 9.5 *Normalized training loss at every 100 steps.* A total of 70% training loss is achieved with a random dropout of 10% for each layer. The test set accuracy is 55.6%.

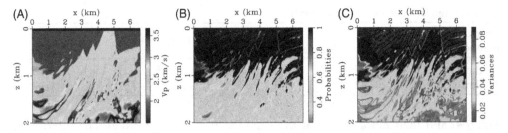

Fig. 9.6 *Classified facies.* (A) Converted to V using a weighted summation $\bar{v} = \sum_{l=1}^{n_f} p_i v_i, v_i$ are given in Table 9.1, (B) The maximum probabilities (softmax) of the classification and (C) the variances of the estimated probabilities. V_s is not shown here since it shares the same probability as V_p.

value indicates less confidence in mapping to a particular facies (i.e., we cannot pick a model from the output probabilities). It is also possible that the classification is biased by particular facies and has a big variance. However, from the data loss (Fig. 9.5) and the converted V_p (Fig. 9.6A), this does not happen in this example. A smoothed version of the estimated distribution of facies (e.g., Fig. 9.6A) shown in Figs. 9.7A and B is used as the initial model for an L_2 norm-based elastic FWI. The final inverted velocities after adding prior information are shown in Fig. 9.7C and D. The inverted model is close to the actual one except for the areas near boundaries. For a better comparison, we also compare profiles of velocities of actual, initial, inverted without regularizations and inverted with regularizations in Fig. 9.8. Estimated S-wave velocities without constraints

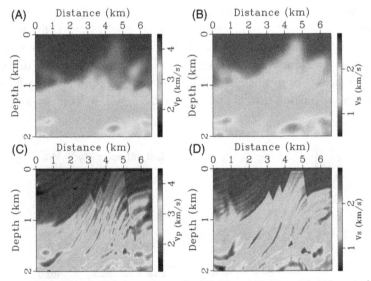

Fig. 9.7 Estimated distribution of facies converted to V_p (A) and V_s (B) and final inverted V_p (C) and V_s (D). (A) and (B) are smoothed versions of the original estimates (e.g., V_p in Fig. 9.6A) and used as initials for obtaining (C) and (D). (C) and (D) are inverted using an L_2 norm based elastic FWI.

x=2.21km

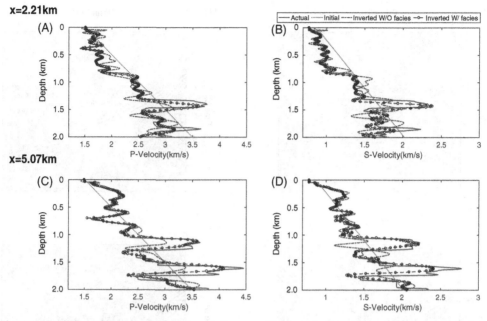

x=5.07km

Fig. 9.8 *Vertical profiles across the low-velocity zones.* The inverted velocities are far from the actual ones without using facies as constraints and V_s suffers from a severe cycle-skipping problem, as the pink arrow indicates in (D). Facies constraints can eliminate artifacts caused by cycle skips.

are trapped in one of the local minima, as the arrow indicates. Data comparison in Fig. 9.9 indicates a similar conclusion. The predicted data using the ML–assisted algorithm (Fig. 9.9D) are much closer to the observed one (Fig. 9.9A) than the one without regularization (Fig. 9.9C). A similarity measurement $\left(\dfrac{u \cdot d}{\sqrt{u \cdot u} \sqrt{d \cdot d}} \right)$ shown in Fig. 9.10. The measured value should be equal to 1 when the predicted data are the same as the observed ones. It shows that the adaptive data-selection objective function alone fails in the far-offset. The ML–assisted approach that utilizes facies can match the observed data in the far-offset.

9.3.2 The North Sea field data example

9.3.2.1 Facies extraction

The Volve field is a small oil field with a dome-shaped structure located in the southern part of the Viking Graben. It was formed by the collapse of adjacent salt ridges during the Jurassic period[76]. The primary seismic imaging goal is to identify the chalk layers and reservoirs below the base Cretaceous unconformity[77]. Seismic facies of the field can be obtained from different sources such as well logs, core analysis and sedimentation history (e.g., Fig. 9.2 in Szydlik et al.[76]). Here, we extract a list of facies from the P- and

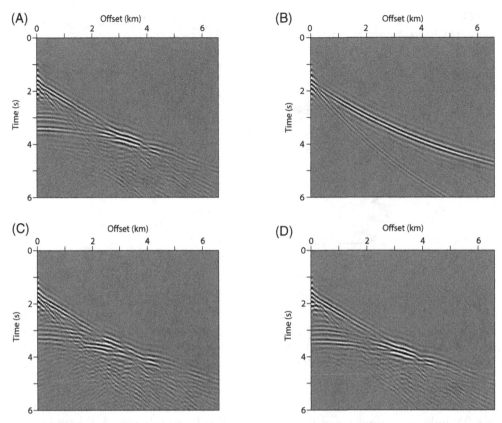

Fig. 9.9 *Data comparison (vertical component).* (A) Observed, (B) initial, (C) inverted without facies constraints, and (D) inverted with facies constraints. The shot gathers are plotted at the same scale so that a direct comparison can be made.

S-wave velocity well log as shown in Fig. 9.12. The reservoir is located at 2.75–3.12 km depth, with an overlying seal rock. The well log covers the depth around the reservoir layer and the well slightly deviates. We calibrate the depth of the top and bottom of the dominant layers using the check shot (red line). Five facies are extracted from the reservoir section by manually grouping the velocities. We only extract the dominant facies in the target area. More experienced interpreters can utilize more advanced classifications of facies and include more desirable model parameters, i.e., porosity and fluid identifier. The interpreted facies are used as labels in supervised learning. We then calculate the anisotropy parameters ϵ and η using Backus averaging[78] as shown in Fig. 9.13 to upscale the sonic well velocities. The delineated facies have different combinations in terms of V_p, V_s, ϵ and η as listed in Table 9.2. The listed parameter values are the averaged values within the facies. The seal rock (f2) has a strong anisotropy, while the reservoir layer (f3) is almost isotropic. Fig. 9.11 shows the cross-plot between the seismic velocity and the anisotropy parameters listed in Table 9.2. Only the velocities are used as input data features in the training and prediction.

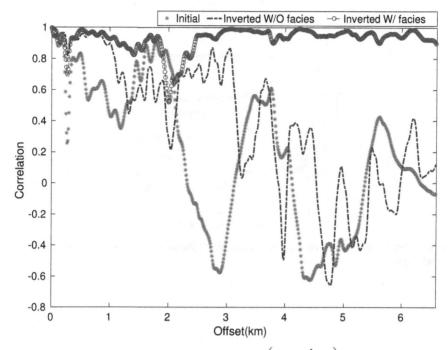

Fig. 9.10 *Correlation of the predicted and observed data* $\left(\dfrac{\boldsymbol{u}\cdot\boldsymbol{d}}{\sqrt{\boldsymbol{u}\cdot\boldsymbol{u}}\sqrt{\boldsymbol{d}\cdot\boldsymbol{d}}}\right)$. The initial model can-not provide accurate prediction in the far-offsets. The adaptive-selection objective function fails when the predicted and observed data are far from each other. The inverted model of the ML-assisted inversion approach can provide accurate prediction at the far-offsets.

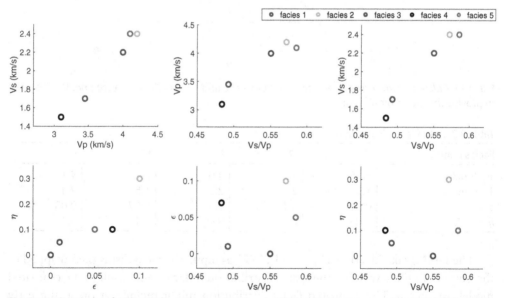

Fig. 9.11 *Cross-plots between the seismic velocities and the anisotropy parameters listed in Table 9.2.* The facies are separable from the combination of V_p, V_s, and V_s/V_p.

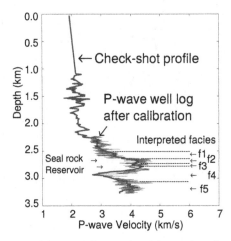

Fig. 9.12 A depth-calibrated well log and the extracted facies. Five dominant facies are extracted at the target depth.

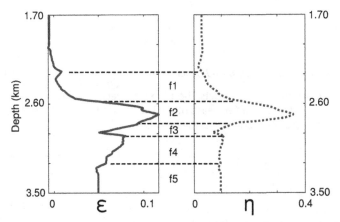

Fig. 9.13 Calculated anisotropy parameters in terms of ϵ and η using Backus averaging. They're used to provide the facies information.

Table 9.2 List of facies in the target area.

Facies number	f1	f2	f3	f4	f5
V_P (km/s)	3.45	4.2	4.0	3.1	4.1
V_S (km/s)	1.7	2.4	2.2	1.5	2.4
ϵ	0.01	0.1	0.0	0.07	0.05
η	0.05	0.3	0.0	0.1	0.1

The DNNs take inverted V_P, V_S and V_S/V_P as input, the facies list is then mapped to the output of DNN, which is the spatial distribution of facies, and result in the estimated model parameters. The estimated facies distribution might include errors at the early stages of the inversion and can be improved by matching the observed seismic data in another scenario of FWI.

9.3.2.2 Inversion results

We apply the ML–assisted inversion strategy to a 2D line of the Volve data set. The seal layer and the reservoir, located at 2.75-3.12 km depth, respectively, are the main imaging goals. We use the raw data set with limited processing applied, including polarity correction, instrumental deconvolution, and data quality control. For the inversion, we use 240 shots and 240 two-component receivers distributed evenly at 50 m and 25 m, respectively. The ocean-bottom cable (OBC) length is 6 km, and the sources are evenly distributed along a 12 km line just below the sea surface. A modified free-surface boundary condition, which can suppress strong surface waves, is used in the simulation[79]. We convolve the observed data with the half-order differentiation of the known wavelet, and thus, we avoid source estimation while correcting the phase discrepancy between the 3D acquisition and the 2D simulation[80,81]. The initial model is a 1D smoothed version of the data owners' model and shown in Fig. 9.14. Only one frequency band (2-12 Hz) is used for the inversion. We first conduct an isotropic elastic FWI using the primary arrivals to improve the 1D initial model (not shown here). Then we use the full data to refine the inverted isotropic model as shown in Fig. 9.15. We apply a hierarchical vertical transverse isotropic (VTI) inversion[82], in which we use the parameterization V_H, V_S, ϵ and η as shown in Fig. 9.16. The diving waves are dominant in the observed data, and thus, the updates focus on shallow depths. The updates of ϵ and η are mainly constrained by reflected and diving waves, respectively. Thus, the updated ϵ and η have different wavenumber features. The high-velocity seal and the relatively low-velocity layer, constrained by deep reflections, are not sufficiently resolved by the relatively weak reflections. Finally, we train a deep neural network to build the connection between the estimated V_p and V_S (Fig. 9.15) and the extracted facies (Table 9.2). As mentioned above, we use a four-layer deep neural network and each layer has 256 nodes with a 10% random dropout. The input features are V_p, V_S and V_S/V_p and the outputs are probabilities of being one of the known facies. We use three vertical profiles from the estimated V_p and V_S (Fig. 9.15) to generate the training data set. The three lines are located at X = 4.5, 6.0 and 7.5 km to cover possible illumination variations. The selection of such a training set should consider the diversity of inversion patterns

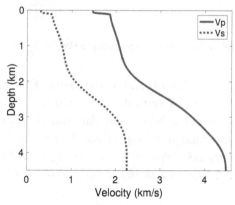

Fig. 9.14 *The initial 1D velocity models.*

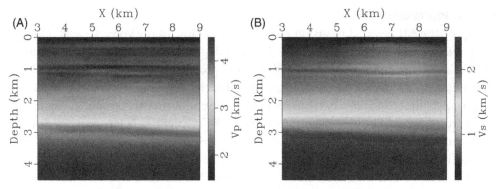

Fig. 9.15 *The inverted models using an isotropic elastic FWI.* (A) V_p and (B) V_s.

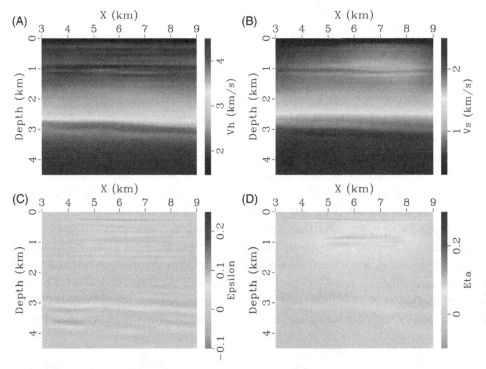

Fig. 9.16 *The inverted models using the anisotropic elastic FWI.* (A) V_h, (B) V_s, (C) ϵ, and (D) η.

(structure and illumination variations) and include as many as possible unique patterns (i.e., facies) for efficient training. A vertical line located at X = 6.75 km is used to generate the test data set. A total of 2158 training data and 704 validation data are used in the network's training and validation, respectively. The training loss and the validation accuracy history after every 100 steps are plotted in Fig. 9.17. The training set accuracy is about 0.9, while the validation set accuracy stays around 0.7. This often indicates that

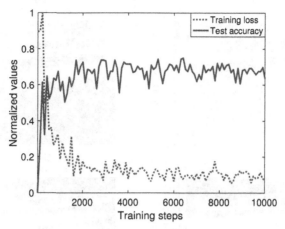

Fig. 9.17 *Normalized training loss at every 100 steps.* A total of 90% training loss is achieved with a random dropout of 10% for each layer, while the validation accuracy stays around 70%.

the training data are not over-fitted. After training, we use all the model pixels to estimate the subsurface facies distribution. The distribution is further converted to V_H, V_S, ϵ and η using the weighted summation, shown in Fig. 9.18. The model parameters show more facies catalogs than the prior information, thanks to the weighted summation. The chalk layer and the reservoir are identifiable by the P- and S-wave velocity drops. The magnitude of these parameters is extracted from well logs, while seismic data mainly constrain the lateral variations. We slightly smooth the converted model (below 2.5 km) and merge it into the VTI inversion model (above 2.5 km) and generate a new initial model for another VTI inversion. The updated model is shown in Fig. 9.19. The high-velocity seal rock with a strong anisotropy above a low-velocity zone is improved. Our ML-assisted inversion managed to obtain a high-resolution ϵ and a lower-resolution η at the reservoir depth, which was guided by the data. Usually, η in the deeply buried seal rocks is not recoverable from the seismic data with limited offsets since it requires a relatively large offset/depth ratio[83]. The interleaved predicted and observed vertical- and horizontal-component data plot (Fig. 9.20) of the initial model indicates that the initial model can provide reasonably good prediction in the near offsets. The inversion considering isotropic elasticity helps to match the data in the far offsets and recovers some dominant reflections as shown in Fig. 9.21. The interleaved data comparisons indicate that adding anisotropy effects can help us obtain simulated data that match the observed data even better (Figs. 9.21 and 9.22). The deep learning aided approach can help improve the matching of deep reflections (Figs. 9.22 and 9.23). A zoomed-in view of the marked area is shown in Fig. 9.24. We can find the improvements in phase matching marked by the arrows. We plot the data-matching history for the different inversion scenarios as shown in Fig. 9.25.

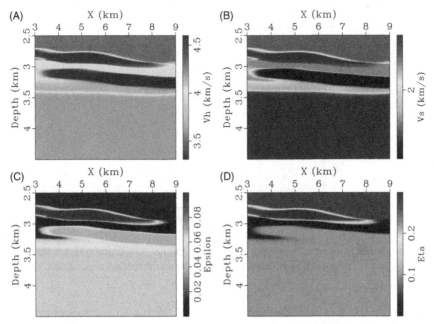

Fig. 9.18 *The predicted facies distribution after training.* They are converted to the parameters of (A) V_H, (b) V_s, (C) ϵ, and (D) η. The number of colors is not equal to the number of facies since the weighted summation can generate more parameter values in between and also different facies can have the same value for certain parameters.

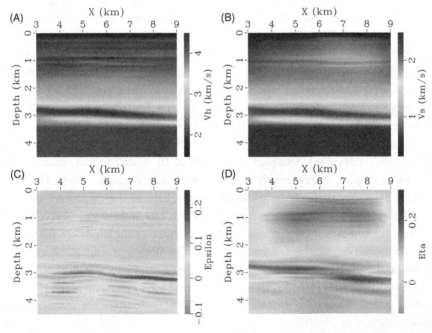

Fig. 9.19 *The inverted models using the anisotropic elastic FWI with facies constraints.* (A) V_H, (B) V_s, (C) ϵ, and (D) η.

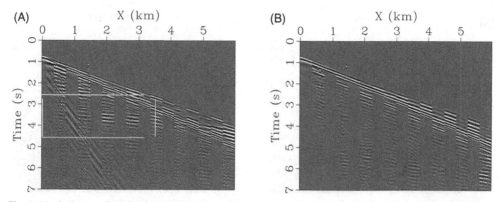

Fig. 9.20 *A shot gather displaying interleaved predicted and observed data using the initial V_p and V_s as shown in Fig. 9.14.* (A) Vertical and (B) horizontal components. The dashed rectangular indicates the zoomed-in view plotted in Fig. 9.24.

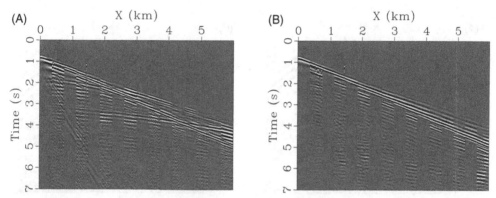

Fig. 9.21 *A shot gather displaying interleaved predicted and observed data using the estimated V_p and V_s from isotropic inversion as shown in Fig. 9.15.* (A) Vertical and (B) horizontal components.

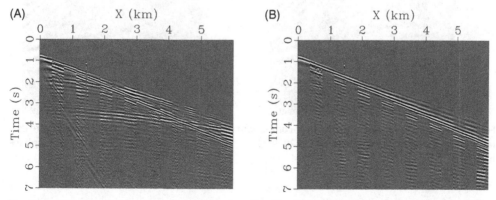

Fig. 9.22 *A shot gather displaying interleaved predicted and observed data using the estimated V_H, V_g, ϵ and η from hierarchic VTI inversion as shown in Fig. 9.16.* (A) Vertical and (b) horizontal components.

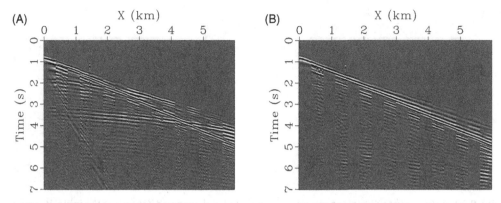

Fig. 9.23 *A shot gather displaying interleaved predicted and observed data using the deep learning aided VTI inversion as shown in Fig. 9.19.* The matching of reflections of the vertical component is improved compared to the hierarchical VTI inversion (Fig. 9.22). (A) Vertical and (B) horizontal components.

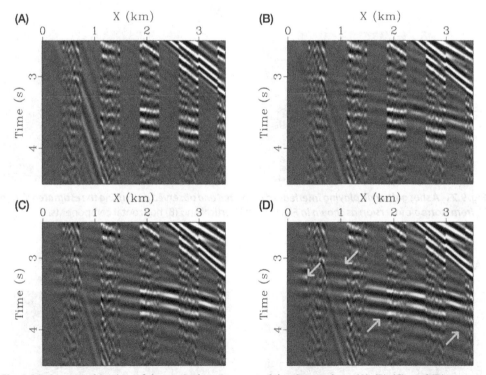

Fig. 9.24 *A zoomed-in view of the vertical-component of the shot gathers.* (A), (B), (C), and (D), are corresponding to the marked areas in Figs. 9.20–9.23, respectively. The arrows mark the improved phase matching.

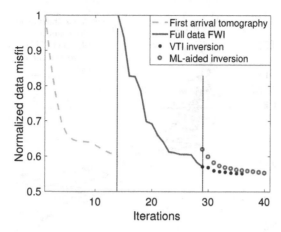

Fig. 9.25 *Data matching history. There are three sequential inversions.* (1) isotropic first arrival FWI, (2) isotropic full data FWI, and (3) VTI inversions. The VTI inversions are parallel inversions. One is the hierarchical VTI inversion and another one is the machine-learning aided inversion. The isotropic inversion reduces the data misfit by 40% and the follow-up deep learning aided inversion can further reduce the data misfit by about an additional 5%.

It shows that the isotropic inversion reduces the data misfit by 40% and the follow-up deep learning aided inversion can further reduce the data misfit by about an additional 5%. The data mismatch suddenly increases after adding facies constraints, but it reduces to the same level as the hierarchical VTI inversion after few iterations. The inversion stops when the updates cannot satisfy the Wolfe line-search condition[66]. The facies distribution was estimated once in this example. We also compare the inverted vertical P-wave velocities with the one from the check shot nearby in Fig. 9.26. The isotropic elastic FWI seems to be able to improve the initial model but with a low resolution. After considering the anisotropic effects, there are some further improvements, but the gains are still mild unless the facies information is used as constraints. The estimated vertical P-wave velocity using the ML-assisted inversion approach is close to the one from the check shot. Remarkably, we did not use well logs or check shots as direct model constraints in the ML-assisted inversion. Also, we were given an inverted VTI model as shown in Fig. 9.27, which was obtained using a layer-stripping tomography approach. To plot our inverted and their reference models using the same color scale, we clip the high values of V_H and V_S in Fig. 9.27. A vertical-profile comparison of the parameters in terms of V_H, V_S, ϵ and η between the ML-assisted inversion and the provided reference model is shown in Fig. 9.28. The facies constraints are only applied around the reservoir section (below 2.3 km depth), where well logs are available. The horizontal velocities (V_H) have a reasonable match with those predicted by the data providers. There is a depth mismatch due to the lack of background ϵ information. The inverted V_S does not match the reference model well without the facies constraints (e.g.,

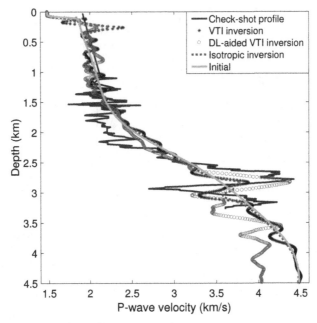

Fig. 9.26 *Vertical P-wave velocity profiles at X = 5.25 km.*

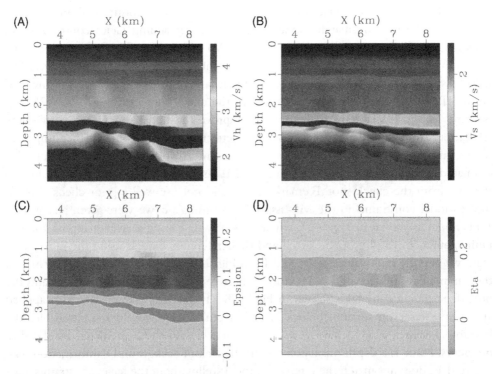

Fig. 9.27 *The reference VTI model.* (A) V_H, (B) V_S, (C) ϵ, and (D) η, obtained using the layer-stripping tomography technique.

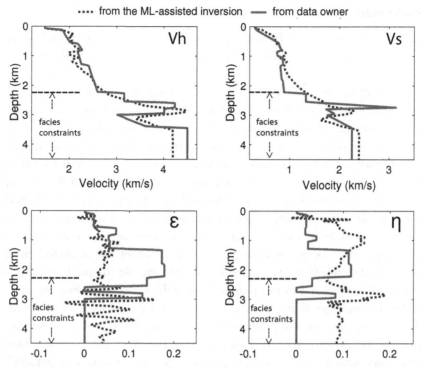

Fig. 9.28 *A vertical-profile comparison of the parameters in terms of V_H, V_S, ϵ, and η between the ML-assisted inversion and the provided reference model at location $X = 5.25$ km.* The facies constraint is only applied below 2.3 km due to the availability of well log data.

the shallow area), which might be caused by the ignorance of S-wave anisotropy. For the deep area with facies constraints, the inverted V_S is closer to the reference model. Otherwise, the inverted V_S might have crosstalks with V_P due to the limited scattering angles. The inverted ϵ has high resolution and it often acts as an absorber of phenomena not addressed in our approximation, like density variations[77]. The inverted anisotropy parameters are closer to those derived using Backus averaging (Fig. 9.13), which has a discrepancy with the reference model. We use the provided tomographic-based and well-assisted model to provide an opportunity to assess our results. We do not expect, what we refer to them as, a reference model to exactly represent the Earth, as they are vulnerable to their limited assumptions of the Earth. In our inverted results, we obtain large η in the shallow part. Although they are not part of the region covered by the well information, the η estimation, considering the large offset to depth ratio in the shallow part, might be accurate. The matching of the moveout at the far offset also seems to support that.

9.4 Discussion

Different geophysical surveys have their advantages in imaging the subsurface. For example, seismic surveys have a larger illumination area than well logs, but with lower resolution.

Often, not all the parameters, such as when using the anisotropy assumption, are resolvable from surface collected seismic data. However, well logs can provide such anisotropy parameters usually corresponding to only a limited area around the well. Full-waveform inversion, in most cases, aims to fit the observed seismic data. Despite its elegant theorem, FWI faces many problems in practice. The actual Earth has too complex physics to be fully represented by numerical simulations, and thus, a wiggle-to-wiggle matching is impractical, especially for real data. In this case, perfectly fitting the seismic data results in an overfitted estimates, which are often incorrect. Incorporating well logs as regularization to the inverse problem can add physical constraints and hopefully reduce the risk of overfitting[84]. The utilization of the predicted distribution of facies is case-dependent. In the synthetic example, we use the predicted model from DNNs as initial models and improve them by matching the noise-free seismic data. In our real data example, we think that the well log is reliable, at least for the region near the well. Thus, we also use the predicted model from DNNs as a solid regularization for removing the negative influence of matching the noisy part of the seismic data. Meanwhile, our seismic inversion can produce generallycoherent structures[85]. With detailed 1D velocities from the well and a reliable trend from FWI, we can image the 2D or 3D subsurface better. A robust FWI algorithm is needed to generate such general coherent structures. Utilizing other geophysical surveys is an approach to stabilize FWI along with modifying the objective function.

The physical laws (e.g., wave equation) and statistical relations (e.g., ML) are more complementary than competitive in solving geophysical problems. The wave-equation-based inversion can compensate for an insufficient network due to the lack of training samples. ML can compensate for the simplified physical processes used by the wave equations by finding a statistically accurate data mapping. The DNN is expected to learn some aspects of the inverted model's varying illumination and amplitude characteristics, specifically near the well locations. It may, however, suffer in areas that are not represented by illumination features covered by the well locations, like at the edges of the model. In terms of illumination distribution, a well-behaved elastic or even anisotropic inversion is important to the application of this DNN. DNN learns from what is used as input and output in the training set. DNN will find it hard to predict those facies if we are missing features in the training set. For example, we might need more wells to obtain the needed facies information in an area with strong lateral variations. The weighted summation might help interpolate the inter-medium values for the missing facies. Also, the extracted list of facies (Table 9.2) uses the mean values for each model parameter, which has an implicit Gaussian-distribution assumption to the measurements within one facies[55]. For non-Gaussian distributions, we may need to modify the weighted summation to consider the actual distribution. Otherwise, we can use finer layers (more labels) to maintain the Gaussian assumption. In future work, we can integrate multiple physical measurements using trained neural networks, which provide better constraints for wave velocities and other reservoir parameters such as pore pressure and temperature.

9.5 Conclusions

Elastic FWI with facies constraints can mitigate the cycle-skipping caused by poor initial models. Facies are usually interpreted from different geophysical observations such as well logs. The ML-assisted inversion algorithm aims to fit not only seismic data but also the well logs. The well logs are not used as a direct constraint in inversion. Dedicatedly designed deep neural networks are trained to find the correct mathematical relation that can connect the well log and seismic measurements. Training the networks is a data-driven process and thus, avoids considering complex physical processes. Although the well logs have limited lateral illumination, the trained DNNs can map them to 2D or 3D models with a structure-guided interpolation. The estimated distribution of facies can be used as a physical constraint for conventional elastic FWI. It can be imperfect at the beginning and can be updated iteratively in elastic FWI. In our synthetic example, both seismic data and well logs are reliable, and thus, the final model can fit both of them. We applied this DNN-assisted elastic full-waveform inversion on OBC data from the North Sea and obtained a reasonable inversion of the reservoir region. The comparison with the well information, as well as the reference model provided with the data, reveal the ability of the approach in mapping the well information to the rest of the model space. One weakness of the problem is that successful training needs many data samples, which might not be available. However, with the data augmentation technique, artificially created data samples can help to train the neural networks.

9.6 Acknowledgments

The original Marmousi model is available from https://www.geoazur.fr/WIND/bin/view/Main/Data/Marmousi. We want to thank Equinor and the former Volve license partners ExxonMobil E&P Norway AS and Bayerngas Norge AS, for the release of the Volve data (https://data.equinor.com/). The views expressed in this paper are the views of the authors and do not necessarily reflect the views of Equinor and the former Volve field license partners. For computer time, this research used the resources of the Supercomputing Laboratory at King Abdullah University of Science & Technology (KAUST) in Thuwal, Saudi Arabia.

9.7 Appendix example training code

We provide an example training script based on Tensorflow 1.x[86]. Note that functions in Tensorflow have evolved rapidly in the past two years. The functions used in the example might be deprecated in the latest version of Tensorflow. However, the readers can build their only classifiers based on the example script.

```
1  """An Example of a DNNClassifier for the facies classification."""
2  #Import modules required by the classifier
3  from __future__ import absolute_import
4  from __future__ import division
5  from __future__ import print_function
6  import argparse
7  import tensorflow as tf
8  import numpy as np
9  import model_data #customized I/O for networks
10 from binaryrw import bina_read,bina_write #customized I/O for binary data
11
12 # Dimension of the model
13 nz   = 720
14 nx   = 1921
15 nn   = nz*nx
16 nfac = 5 # Number of predefined facies
17 nstps=10000 # Number of training steps
18
19 parser = argparse.ArgumentParser()
20 parser.add_argument('--batch_size', default=128, type=int, help='batch
     size')
21 parser.add_argument('--train_steps', default=nstps, type=int,
22                     help='number of training steps')
23
24 def main(argv):
25     args = parser.parse_args(argv[1:])
26     # Fetch the data
27     (train_x, train_y), (test_x, test_y) = model_data.load_data()
28
29     # Feature columns describe how to use the input.
30     my_feature_columns = []
31     for key in train_x.keys():
32         my_feature_columns.append(tf.feature_column.numeric_column(key=key
     ))
33     # Build 4 hidden layer DNN with 256, 256, 256, 256 units respectively.
34     classifier = tf.estimator.DNNClassifier(
35         feature_columns=my_feature_columns,
36         hidden_units=[256,256,256,256],
37         dropout=0.1,
38         n_classes=nfac,
39         config=tf.estimator.RunConfig(
40             save_checkpoints_steps=100,
41             keep_checkpoint_max=200,
42             model_dir="./fckpt/"
43         )
44     )
```

```
45   # Training
46   classifier.train(
47       input_fn=lambda:model_data.train_input_fn(train_x, train_y,
48       args.batch_size), steps=args.train_steps)
49
50   # Evaluate the model.
51   evall_result = classifier.evaluate(
52       input_fn=lambda:model_data.eval_input_fn(test_x, test_y,
53       args.batch_size))
54
55   print('\nTest set accuracy: {accuracy:0.3f}\n'.format(**evall_result))
56
57   ckpts=[i for i in range(1,nstps+2,100)]
58   ckpts[len(ckpts)-1] = nstps
59   print('ckpts=\n',ckpts)
60   acca=ckpts
61   cont=-1
62   for epoch,ckpt in enumerate(ckpts):
63       cont =cont+1
64       eval_result = classifier.evaluate(
65       input_fn=lambda:model_data.eval_input_fn(test_x, test_y,
66       args.batch_size), checkpoint_path="./fckpt/model.ckpt-"+str(ckpt))
67       acca[cont] = eval_result['accuracy']
68
69   # Read input velocities.
70   vpinv  = bina_read('./model_folder/vpinv_iso.bin',nn)
71   vsinv  = bina_read('./model_folder/vsinv_iso.bin',nn)
72   vspinv = vpinv/vsinv
73   # Prepare data features.
74   pred_x = {
75       'Vp':vpinv,
76       'Vs':vsinv,
77       'Vsp':vspinv,
78   }
79   # Priors extracted from the well log data.
80   fac_vp= [ 3.45,      4.2, 4.0,    3.1,   4.0]
81   fac_vs= [ 1.7,       2.4, 2.2,    1.5,   2.4]
82   fac_delt=[ -0.0364, -0.125, 0.0, -0.025, -0.0417]
83   fac_eta= [ 0.05,       0.3, 0.0,    0.1,   0.1 ]
84   # Prediction
85   pred_weights = classifier.predict(
86       input_fn=lambda:model_data.eval_input_fn(pred_x, labels=None,
87                                 batch_size=args.batch_size))
88   # Initialize vectors
89   vp_reg   = np.array(vpinv)
90   vs_reg   = np.array(vpinv)
91   delt_reg = np.array(vpinv)
92   eta_reg  = np.array(vpinv)
93   for i in range(0,nn):
94       pred_dict = next(pred_weights,None)
95       # print(pred_dict['probabilities'])
96       vptemp   = 0.0
97       vstemp   = 0.0
98       delttemp = 0.0
```

```
 99        etatemp   = 0.0
100        # Summation weighted by the predicted probabilities.
101        for j in range(0, nfac):
102            vptemp    = vptemp   +pred_dict['probabilities'][j]*fac_vp[j]
103            vstemp    = vstemp   +pred_dict['probabilities'][j]*fac_vs[j]
104            delttemp  = delttemp +pred_dict['probabilities'][j]*fac_delt[j]
105            etatemp   = etatemp  +pred_dict['probabilities'][j]*fac_eta[j]
106
107        vp_reg[i]   = vptemp
108        vs_reg[i]   = vstemp
109        delt_reg[i] = delttemp
110        eta_reg[i]  = etatemp
111    # Write the distribution of facies in binary format
112    wrtvp   =bina_write('./model_folder/vpreg.bin',vp_reg)
113    wrtvs   =bina_write('./model_folder/vsreg.bin',vs_reg)
114    wrtdelt =bina_write('./model_folder/deltreg.bin',delt_reg)
115    wrteta  =bina_write('./model_folder/etareg.bin',eta_reg)
116
117 if __name__ == '__main__':
118    tf.logging.set_verbosity(tf.logging.INFO)
119    tf.app.run(main)
```

References

1. Zhao X, Mendel JM. Minimum-variance deconvolution using artificial neural networks In: SEG Technical Program Expanded Abstracts 1988. Society of Exploration Geophysicists 1988:738–41.
2. Dowla FU, Taylor SR, Anderson RW. Seismic discrimination with artificial neural networks: preliminary results with regional spectral data. *Bull Seismol Soc Am* 1990;**80**(5):1346–73.
3. McCormack MD. Neural computing in geophysics. *The Leading Edge* 1991;**10**(1):11–5.
4. Tiira T. Detecting teleseismic events using artificial neural networks. *Comput Geosci* 1999;**25**(8):929–38.
5. Esposito AM, Giudicepietro F, Scarpetta S, D'Auria L, Marinaro M, Martini M. Automatic discrimination among landslide, explosion-quake, and microtremor seismic signals at stromboli volcano using neural networks. *Bull Seismol Soc Am* 2006;**96**(4A):1230–40.
6. Lary DJ, Alavi AH, Gandomi AH, Walker AL. Machine learning in geosciences and remote sensing. *Geosci Front* 2016;**7**(1):3–10.
7. Bhattacharya S, Mishra S. Applications of machine learning for facies and fracture prediction using bayesian network theory and random forest: case studies from the appalachian basin, usa. *J Pet Sci Eng* 2018;**170**:1005–17.
8. Wu X, Liang L, Shi Y, Fomel S. Faultseg3d: Using synthetic data sets to train an end-to-end convolutional neural network for 3d seismic fault segmentation. *Geophysics* 2019;**84**(3):IM35–45.
9. Liu M, Jervis M, Li W, Nivlet P. Seismic facies classification using supervised convolutional neural networks and semisupervised generative adversarial networks. *Geophysics* 2020;**85**(4):O47–58.
10. Alkhalifah T, Almomin A, Naamani A. Machine-driven earth exploration: artificial intelligence in oil and gas. *The Leading Edge* 2021;**40**(4):298–301.
11. Feng S, Passone L, Schuster GT. Superpixel-based convolutional neural network for georeferencing the drone images. *IEEE J Sel Top Appl Earth Obs Remote Sens* 2021;**14**:3361–72.
12. Di H, Li C, Smith S, Li Z, Abubakar A. Imposing interpretational constraints on a seismic interpretation CNN. *Geophysics* 2021;**86**(3):1–36.
13. Virieux J, Operto St. An overview of full-waveform inversion in exploration geophysics. *Geophysics* 2009;**74**(6):WCC1–26.
14. Vigh D, Jiao K, Watts D, Sun D. Elastic full-waveform inversion application using multicomponent measurements of seismic data collection. *Geophysics* 2014;**79**(2):R63–77.
15. Tromp J. Seismic wavefield imaging of earth's interior across scales. *Nat Rev Earth Environ* 2020;**1**(1):40–53.

16. Raissi M, Perdikaris P, Karniadakis GE. Physics-informed neural networks: a deep learning framework for solving forward and inverse problems involving nonlinear partial differential equations. *J Comput Phys* 2019;**378**:686–707.

17. Xu Y, Li J, Chen X. Physics informed neural networks for velocity inversion. In SEG Technical Program Expanded Abstracts 2019. Society of Exploration Geophysicists 2019:2584–8.

18. Song C, Alkhalifah T, Waheed UB. Solving the frequency-domain acoustic VTI wave equation using physics-informed neural networks. *Geophys J Int* 2021;**225**(2):846–59.

19. Waheed UB, Haghighat E, Alkhalifah T, Song C, Hao Qi. PINNeik: Eikonal solution using physics-informed neural networks. *Comput Geosci* 2021;**155**:104833.

20. Virieux J. P-SV wave propagation in heterogeneous media: Velocity-stress finite-difference method. *Geophysics* 1986;**51**(4):889–901.

21. Zhang Z-D, Alkhalifah T. Regularized elastic full-waveform inversion using deep learning. *Geophysics* 2019;**84**(5):R741–51.

22. Zhang Z-, Alkhalifah T. High-resolution reservoir characterization using deep learning-aided elastic full-waveform inversion: The North Sea field data example. *Geophysics* 2020;**85**(4):WA137–46.

23. Lailly P, Bednar J. The seismic inverse problem as a sequence of before stack migrations. In Conference on inverse scattering: theory and application, SIAM, Philadelphia, PA 1983:206–20.

24. Tarantola A. Inversion of seismic reflection data in the acoustic approximation. *Geophysics* 1984;**49**(8):1259–66.

25. Brenders AJ, Pratt RG. Full waveform tomography for lithospheric imaging: results from a blind test in a realistic crustal model. *Geophys J Int* 2007;**168**(1):133–51.

26. Guasch L, Agudo OC, Tang M-X, Nachev P, Warner M. Full-waveform inversion imaging of the human brain. *NPJ Digit Med* 2020;**3**(1):1–2.

27. Bachmann E, Tromp J, Davies G, Steingart D. Quantitative ultrasound imaging based on seismic full waveform inversion, March 18 2021. US Patent App. 16/643,321.

28. Brossier R, Operto St, Virieux J. Seismic imaging of complex onshore structures by 2D elastic frequency-domain full-waveform inversion. *Geophysics* 2009;**74**(6):WCC105–18.

29. French SW, Romanowicz BA. Whole-mantle radially anisotropic shear velocity structure from spectral-element waveform tomography. *Geophys J Int* 2014;**199**(3):1303–27.

30. Lei W, Ruan Y, Bozdăg E, Peter D, Lefebvre M, Komatitsch D et al. Global adjoint tomography—model glad-m25. *Geophys J Int* 2020;**223**(1):1–21.

31. Van Leeuwen T, Mulder WA. A correlation-based misfit criterion for wave-equation traveltime tomography. *Geophys J Int* 2010;**182**(3):1383–94.

32. Engquist B, Froese BD, Yang Y. Optimal transport for seismic full waveform inversion. *Commun Math Sci* 2016;**14**(8):2309–30.

33. Métivier L, Brossier R, Mérigot Q, Oudet E, Virieux J. Measuring the misfit between seismograms using an optimal transport distance: Application to full waveform inversion. *Geophys Suppl Mon Not R Astron Soc* 2016;**205**(1):345–77.

34. Liu Y, He B, Lu H, Zhang Z, Xie X-Bi, Zheng Y. Full-intensity waveform inversion. *Geophysics* 2018;**83**(6):R649–58.

35. Sun B, Alkhalifah T. Robust full-waveform inversion with radon-domain matching filter. *Geophysics* 2019;**84**(5):R707–24.

36. Zhang Z-, Alkhalifah T. Local-crosscorrelation elastic full-waveform inversion. *Geophysics* 2019;**84**(6):R897–908.

37. He B, Liu Y, Lu H, Zhang Z. Correlative full-intensity waveform inversion. *IEEE Trans Geosci Remote Sens* 2020;**58**(10):6983–94.

38. Choi Y, Alkhalifah T. Application of multi-source waveform inversion to marine streamer data using the global correlation norm. *Geophys Prospect* 2012;**60**(4):748–58.

39. Chi B, Dong L, Liu Y. Correlation-based reflection full-waveform inversion. *Geophysics* 2015;**80**(4):R189–202.

40. Bunks C, Saleck FM, Zaleski S, Chavent G. Multiscale seismic waveform inversion. *Geophysics* 1995;**60**(5):1457–73.

41. Martínez-Sansigre A, Ratcliffe A. A probabilistic QC for cycle-skipping in full waveform inversionSEG Technical Program Expanded Abstracts 2014. Houston, TX: Society of Exploration Geophysicists; 2014. p. 1105–9.

42. Bi H, Lin T. Impact of adaptive data selection on full waveform inversionSEG Technical Program Expanded Abstracts 2014. Society of Exploration Geophysicists 2014:1094–8.

43. Zhang Z-, Alkhalifah T. Adaptive data-selection elastic full-waveform inversionSEG Technical Program Expanded Abstracts 2018. Society of Exploration Geophysicists 2018:5163–7.

44. Alkhalifah T, Choi Y. From tomography to full-waveform inversion with a single objective function. *Geophysics* 2014;**79**(2):R55–61.

45. Choi Y. Full waveform inversion of exponentially damped wavefield using the global-correlation norm. *Pure Appl Geophys* 2020;**177**(12):5819–32.

46. K¨ohn D, De Nil D, Kurzmann A, Przebindowska A, Bohlen T. On the influence of model parametrization in elastic full waveform tomography. *Geophys J Int* 2012;**191**(1):325–45.

47. Plessix R-E, Cao Q. A parametrization study for surface seismic full waveform inversion in an acoustic vertical transversely isotropic medium. *Geophys J Int* 2011;**185**(1):539–56.

48. Wu Z, Alkhalifah T. Waveform inversion for acoustic VTI media in frequency domainSEG Technical Program Expanded Abstracts 2016. Society of Exploration Geophysicists 2016:1184–9.

49. Zhang Z-D, Alkhalifah T. Full waveform inversion using oriented time-domain imaging method for vertical transverse isotropic media. *Geophys Prospect* 2017;**65**(S1):166–80.

50. Sun B, Alkhalifah T. Automatic wave equation migration velocity analysis by focusing subsurface virtual sources. *Geophysics* 2017;**83**(2):1–40.

51. Alkhalifah T, Sun BB, Wu Z. Full model wavenumber inversion: Identifying sources of information for the elusive middle model wavenumbers. *Geophysics* 2018;**83**(6):R597–610.

52. Kemper M, Gunning J. Joint impedance and facies inversion–seismic inversion redefined. *First Break* 2014;**32**(9):89–95.

53. Hu W, Abubakar A, Habashy TM. Joint electromagnetic and seismic inver- sion using structural constraints. *Geophysics* 2009;**74**(6):R99–109.

54. Bhattacharya S, Verma S. Seismic attribute and petrophysics-assisted interpretation of the Nanushuk and Torok Formations on the North Slope, Alaska. *Interpretation* 2020;**8**(2):SJ17–34.

55. Zhang Z-, Alkhalifah T, Naeini EZ, Sun B. Multiparameter elastic full waveform inversion with facies-based constraints. *Geophys J Int* 2018;**213**(3):2112–27.

56. Aragao O, Sava P. Elastic full-waveform inversion with probabilistic petrophysical model constraints. *Geophysics* 2020;**85**(2):R101–11.

57. Singh S, Tsvankin I, Naeini EZ. Full-waveform inversion with borehole constraints for elastic VTI media. *Geophysics* 2020;**85**(6):R553–63.

58. Li Y, Alkhalifah T, Zhang Z. High-resolution regularized elastic full waveform inversion assisted by deep learning 82nd EAGE Annual Conference & Exhibition volume 2020. European Association of Geoscientists & Engineers 2020:1–5.

59. AlRegib G, Deriche M, Long Z, Di H, Wang Z, Alau- dah Y et al. Subsurface structure analysis using computational interpretation and learning: a visual signal processing perspective. *IEEE Signal Process Mag* 2018;**35**(2):82–98.

60. Wu Y, Lin Y, Zhou Z, Bolton DC, Liu Ji, Johnson P. Deep-detect: a cascaded region-based densely connected network for seismic event detection. *IEEE Trans Geosci Remote Sens* 2018;**57**(1):62–75.

61. Zhu W, Beroza GC. PhaseNet: a deep-neural-network-based seismic arrival-time picking method. *Geophys J Int* 2019;**216**(1):261–73.

62. Ovcharenko O, Kazei V, Kalita M, Peter D, Alkhalifah T. Deep learning for low-frequency extrapolation from multioffset seismic data. *Geophysics* 2019;**84**(6):R989–1001.

63. Saad OM, Chen Y. Automatic waveform-based source-location imaging using deep learning extracted microseismic signals. *Geophysics* 2020;**85**(6):KS171–83.

64. Bergen KJ, Johnson PA, Maarten V, Beroza GC. Machine learning for data-driven discovery in solid Earth geoscience. *Science* 2019;**363**(6433).

65. Liu DC, Nocedal J. On the limited memory BFGS method for large scale optimization. *Math Program* 1989;**45**(1-3):503–28.

66. Wolfe P. Convergence conditions for ascent methods. *SIAM Rev* 1969;**11**(2):226–35.

67. Glorot X, Bordes A, Bengio Y. Deep sparse rectifier neural networksProceedings of the fourteenth international conference on artificial intelligence and statistics 2011:315–23.
68. Nair V, Hinton GE. Rectified linear units improve restricted boltzmann machines Proceedings of the 27th international conference on machine learning (ICML-10) 2010:807–14.
69. Srivastava N, Hinton G, Krizhevsky A, Sutskever I, Salakhutdinov R. Dropout: a simple way to prevent neural networks from overfitting. *J Mach Learn Res* 2014;**15**(1):1929–58.
70. Krizhevsky A, Sutskever I, Hinton GE. Imagenet classification with deep convolutional neural networks. *Advances in neural information processing systems* 2012:1097–105.
71. Naeini EZ, Exley R. Quantitative interpretation using facies-based seismic inversion. SEG Technical Program Expanded Abstracts 2016 Society of Exploration Geophysicists 2016:2906–10.
72. Duan Y, Sava P. Elastic wavefield tomography with physical model constraints. *Geophysics* 2016;**81**(6):R447–56.
73. Hansen PC, O'Leary DP. The use of the L-curve in the regularization of discrete ill-posed problems. *SIAM J Sci Comput* 1993;**14**(6):1487–503.
74. Versteeg R. The marmousi experience: velocity model determination on a synthetic complex data set. *The Leading Edge* 1994;**13**(9):927–36.
75. Kohavi R et al. A study of cross-validation and bootstrap for accuracy estimation and model selection. *Ijcai* 1995;**14**:1137–45.
76. Szydlik TJ, Way S, Smith P, Aamodt L, Friedrich C. 3d pp/ps prestack depth migration on the volve field 68th EAGE Conference and Exhibition incorporating SPE EUROPEC 2006 European Association of Geoscientists & Engineers 2006:2.
77. Guitton A, Alkhalifah T. A parameterization study for elastic vti full-waveform inversion of hydrophone components: synthetic and North Sea field data examples. *Geophysics* 2017;**82**(6):R299–308.
78. Berryman JG, Grechka VY, Berge PA. Analysis of Thomsen parameters for finely layered VTI media. *Geophys Prospect* 1999;**47**(6):959–78.
79. He W, Plessix R, Singh S. Modified boundary conditions for elastic inversion of active land seismic data in VTI media78th EAGE Conference and Exhibition 2016:2016.
80. Pica A, Diet JP, Tarantola A. Nonlinear inversion of seismic reflection data in a laterally invariant medium. *Geophysics* 1990;**55**(3):284–92.
81. Yoon K, Suh S, Cai J, Wang B. Improvements in time domain FWI and its applications SEG Technical Program Expanded Abstracts 2012. Society of Exploration Geophysicists 2012:1–5.
82. Oh Ju-W, Alkhalifah T. Optimal full-waveform inversion strategy for marine data in azimuthally rotated elastic orthorhombic media. *Geophysics* 2018;**83**(4):R307–20.
83. Alkhalifah T, Plessix Ře-É. A recipe for practical full-waveform inversion in anisotropic media: an analytical parameter resolution study. *Geophysics* 2014;**79**(3):R91–101.
84. Asnaashari A, Brossier R, Garambois St, Audebert F, Thore P, Virieux J. Regularized seismic full waveform inversion with prior model information. *Geophysics* 2013;**78**(2):R25–36.
85. Shen X, Jiang Li, Dellinger J, Brenders A, Kumar C, James M et al. High resolution full waveform inversion for structural imaging in explorationSEG Technical Program Expanded Abstracts 2018. Society of Exploration Geophysicists 2018:1098–102.
86. Abadi M′, Barham P, Chen J, Chen Z, Davis A, Dean J et al. Tensorflow: a system for large-scale machine learning. *OSDI* 2016;**16**:265–83.
87. Li W, Deffenbaugh M, Gillard DG, Chen G, Xu X. Method for estimating subsurface properties from geophysical survey data using physicsbased inversion, July 4 2017. US Patent 9,696,442.

CHAPTER 10

A holistic approach to computing first-arrival traveltimes using neural networks

Umair bin Waheed[a], Tariq Alkhalifah[b], Ehsan Haghighat[c], Chao Song[b]
[a]Department of Geosciences, King Fahd University of Petroleum and Minerals, Dhahran, Saudi Arabia
[b]Department of Physical Science and Engineering, King Abdullah University of Science and Technology, Thuwal, Saudi Arabia
[c]Department of Civil Engineering, Massachusetts Institute of Technology, MA, United States

Abstract

Since the original algorithm by John Vidale in 1988 to numerically solve the isotropic eikonal equation, widely used in seismic wave propagation studies, there has been tremendous progress on the topic addressing an array of challenges, including improvement of the solution accuracy, incorporation of surface topography, adding more accurate physics by accounting for anisotropy/attenuation in the medium, and speeding up computations using multiple CPUs and GPUs. Despite these advances, there is no mechanism in these algorithms to carry information gained by solving one problem to the next. Moreover, these approaches may breakdown for certain complex forms of the eikonal equation, requiring simplification of the equations to estimate approximate solutions. Therefore, we seek an alternate approach to address the challenge in a holistic manner, i.e., a method that not only makes it simpler to incorporate topography, allows accounting for any level of complexity in physics, benefiting from computational speedup due to the availability of multiple CPUs or GPUs, but also able to transfer knowledge gained from solving one problem to the next. We develop an algorithm based on the emerging paradigm of physics-informed neural network to solve various forms of the eikonal equation. We show how transfer learning and surrogate modeling can be used to speed up computations by utilizing information gained from prior solutions. We also propose a two-stage optimization scheme to expedite the training process in the presence of sharper heterogeneity in the velocity model and recommend using a locally adaptive activation function for faster convergence. Furthermore, we demonstrate how the proposed approach makes it simpler to incorporate additional physics and other features in contrast to conventional methods that took years and often decades to make these advances. Such an approach not only makes the implementation of eikonal solvers much simpler but also puts us on a much faster path to progress. The method paves the pathway to solving complex forms of the eikonal equation that have remained unsolved using conventional algorithms or solved using some approximation techniques at best; thereby, creating new possibilities for advancement in the field of numerical eikonal solvers.

Keywords

Anisotropy; Eikonal equation; Neural networks; Scientific machine learning; Traveltimes

Advances in Subsurface Data Analytics
DOI: https://doi.org/10.1016/B978-0-12-822295-9.00006-6

10.1 Introduction

The eikonal equation is a nonlinear partial differential equation (PDE) obtained from the first term of the Wentzel-Kramers-Brillouin expansion of the wave equation and represents a class of Hamilton-Jacobi equations[1]. It finds applications in multiple domains of science and engineering, including image processing[2], robotic path planning and navigation[3], computer graphics[4], and semi-conductor manufacturing[5]. In seismology, it used to compute first-arrival traveltimes, which are necessary for the success of a wide range of seismic processing and imaging tools including statics and moveout correction[6], traveltime tomography for initial velocity model building[7,8], microseismic source localization[9], and ray-based migration[10]. Ray tracing and finite-difference based solutions of the eikonal equation are the most popular approaches for computing traveltimes.

Ray tracing methods compute traveltimes along the characteristics of the eikonal equation by solving a system of ordinary differential equations[11]. The approach is generally efficient for a sparse source-receiver geometry, but the computational cost increases dramatically with the increase in the number of source-receiver pairs.

Moreover, for practical applications such as imaging and velocity model building, traveltime solutions need to be interpolated onto a regular grid. This requirement not only adds to the computational cost of the method but also poses a challenge, particularly in complex media where rays may diverge from one another, leading to large spatial gaps between rays, creating regions known as shadow zones[12]. Additionally, in strongly varying velocity models, multiple ray-paths may connect a source receiver pair, making it easy to miss the path with the minimum traveltime. Therefore, the numerical solution of the eikonal equation has been a topic of continued research interest over the years.

Vidale[13] led the development of numerical eikonal solvers by proposing an expanding box strategy to compute first-arrival traveltimes in heterogeneous media. Subsequently, the method was improved and extended to three dimensions[12], to incorporate anisotropy[14,15], and to high-order accurate solutions[16]. The instability of the expanding box method due to turning rays led to the development of the expanding wavefront scheme[17]. This was further improved to obtain maximum energy traveltimes[18], and to incorporate anisotropy in the model[19].

Another algorithm that became popular during the late 1990s was the fast marching method[20]. The popularity of the method was due to its accuracy, stability, and efficiency properties. The fast marching method saw great interest and development in the subsequent period. This included extension of the method to improve traveltime accuracy[21–23], incorporating anisotropy[24–26], parallelization for computational speedup using multiple CPUs[27], and even acceleration using GPUs[28].

Despite its success, the fast marching method was overtaken in popularity by the fast sweeping method[29] since the mid-2000s. This was mainly due to the flexibility and robustness of the fast sweeping method to various forms of the eikonal equation. Numerous advances to the fast sweeping method have since been proposed to improve the accuracy of the method[30,31], to incorporate anisotropy[32-35], to account for attenuation[36], to tackle surface topography[37], and parallelization for computational speedup[38,39].

Several other hybrid strategies have also been proposed to solve the eikonal equation. For a detailed review of these methods, we refer the interested reader to[40].

In light of these developments, it is beyond doubt that there has been tremendous progress since the original eikonal solver by Vidale[13]. This huge and growing body of literature, spanning over three decades, on the numerical solution of the eikonal equation, required significant research efforts to address an array of challenges, including improvement of the solution accuracy, incorporation of surface topography, adding more accurate physics by accounting for anisotropy/attenuation in the medium, and speeding up computations by using multiple CPUs and GPUs. Therefore, we seek an alternate approach that could address these challenges in a holistic manner – a method that makes it simpler to incorporate topography, allow accounting for more accurate physics, and benefit from computational speedup due to the availability of multiple CPUs or GPUs. Such an approach would not only make the implementation of eikonal solvers much simpler but also put us on a much faster path to progress in solving complex forms of the eikonal equation.

Furthermore, a major drawback of the conventional eikonal solvers is that there is no mechanism to utilize the information gained by solving one problem to the next. Therefore, the same amount of computational effort is needed even for a small perturbation in the source position and/or the velocity model. This can lead to a computational bottleneck, particularly for imaging/inversion applications that require repeated computations, often with thousands of source positions and multiple updated velocity models. Therefore, a method that could use information gained from one solution to the next to speed up computations can potentially remedy this situation. With these objectives in mind, we look into the machine learning literature for inspiration.

Having shown remarkable success across multiple research domains[41], machine learning has recently shown promise in tackling problems in scientific computing. The idea to use an artificial neural network for solving PDEs has been around since the 1990s[42]. However, due to recent advances in the theory of deep learning coupled with a massive increase in computational power and efficient graph-based implementation of new algorithms and automatic differentiation, we are witnessing a resurgence of interest in using neural networks to approximate solutions of PDEs.

Recently, Raissi et al.[43] developed a deep learning framework for the solution and discovery of PDEs. The so-called physics-informed neural network (PINN) leverages the capabilities of deep neural networks (DNNs) as universal function approximators. Contrary to the conventional deep learning approaches, PINNs restrict the space of admissible solutions by enforcing the validity of the underlying PDE governing the actual physics of the problem. This is achieved by using a simple feed-forward neural network leveraging automatic differentiation (AD) to compute the differential variables in the PDE. It is worth noting that PINNs do not require a labeled set to learn the mapping between inputs and outputs, rather learning is facilitated through the loss function formed by the underlying PDE. PINNs have already demonstrated success in solving forward and inverse problems in geophysics[44–48].

In this chapter, we present a neural network approach to solve various forms of the eikonal equation. We use the PINN framework, where the governing equation is incorporated into the loss function of the neural network. We also show how the proposed method addresses the highlighted challenges compared to conventional algorithms. Specifically, we show that by simply updating the loss function of the neural network, we can account for more accurate physics in the traveltime solution. Moreover, since the proposed method is mesh-free, we will observe that to incorporate topography, no special treatment is needed as opposed to conventional finite-difference methods. In addition, the use of computational graphs allows us to run the same piece of code on different platforms (CPUs, GPUs) and architectures (desktops, clusters) without worrying about the implementation details. Most importantly, the proposed method allows us to use information gained while solving for a particular source position and velocity model to speed up computations for perturbations in the velocity model and/or source position. We demonstrate this aspect through the use of machine learning techniques like transfer learning and surrogate modeling.

The rest of the chapter is organized as follows: We begin by presenting the theoretical foundations of the proposed method and discuss how it can be used to solve more complex forms of the eikonal equation. Next, we test the method on a diverse set of 2D and 3D benchmark synthetic models and compare its performance with the popular fast sweeping method. Finally, we conclude the chapter by discussing the strengths of the method and identifying future research opportunities.

10.2 Theory

In this section, we describe how neural networks can be used to compute traveltime solutions for eikonal equations corresponding to isotropic and anisotropic media. We do so by first introducing the different forms of the eikonal equation and the concept of factorization. Next, we outline the general mechanism of a feed-forward neural network followed by its capability as a function approximator. This is followed by a brief overview of the concept of automatic differentiation, which is used to compute the derivative of the networks' output with respect to the inputs. Finally, putting these concepts together, we will present the proposed algorithm for solving various forms of the eikonal equation.

10.2.1 Eikonal equations

The eikonal equation is a nonlinear first-order PDE that is, for an isotropic medium, given as:

$$\left(\frac{\partial T}{\partial x}\right)^2 + \left(\frac{\partial T}{\partial z}\right)^2 = \frac{1}{v(x,z)^2},$$ (10.1)

subject to a point-source boundary condition as:

$$T(x_s, z_s) = 0,$$ (10.2)

where $T(x, z)$ is the traveltime from the source point (x_s, z_s) to a point (x, z) in the computational domain, and $v(x, z)$ is the phase velocity of the isotropic medium. Since the curvature of the wavefront near the point-source is extremely large, previous studies[31,49] have shown that it is better to solve the factored eikonal equation instead of Eq. (10.1). The idea is to factor the unknown traveltime into two multiplicative factors, where one of the factors is specified analytically to capture the source-singularity such that the other factor is gently varying in the source neighborhood. Therefore, we factor $T(x, z)$ into two multiplicative functions:

$$T(x,z) = T_0(x,z) \cdot \tau(x,z),$$ (10.3)

where $T_0(x, z)$ is the known function and $\tau(x, z)$ is the unknown function. Plugging this into Eq. (10.1), we get the factored eikonal equation for an isotropic model as:

$$\left(T_0 \frac{\partial \tau}{\partial x} + \tau \frac{\partial T_0}{\partial x}\right)^2 + \left(T_0 \frac{\partial \tau}{\partial z} + \tau \frac{\partial T_0}{\partial z}\right)^2 = \frac{1}{v^2},$$ (10.4)

subject to the updated point-source condition:

$$\tau(x_s, z_s) = 1.$$ (10.5)

The known factor T_0 is the traveltime solution in a homogeneous isotropic model given as:

$$T_0(x,z) = \frac{\sqrt{(x-x_s)^2 + (z-z_s)^2}}{v_s},$$ (10.6)

where v_s is taken to be the velocity at the point-source location.

Eq. (10.4) is the factored eikonal equation for an isotropic medium; however, sedimentary rocks exhibit at least some degree of anisotropy due to a number of factors including thin layering and preferential alignment of grains cracks[50]. This results in the velocity being a function of the wave propagation direction, making the isotropic approximation of the Earth invalid. Therefore, traveltime computation algorithms must honor the anisotropic nature of the Earth for accurate subsurface imaging and other applications. Thus, we consider a realistic approximation of the subsurface anisotropy known as the tilted transverse isotropy (TTI) case. The factored eikonal equation for a TTI medium is considerably more complex than the isotropic case and is given, under the acoustic assumption, as[49]:

$$
(1+2\varepsilon)\left[\cos\theta\left(T_0\frac{\partial\tau}{\partial x}+\tau\frac{\partial T_0}{\partial x}\right)+\sin\theta\left(T_0\frac{\partial\tau}{\partial z}+\tau\frac{\partial T_0}{\partial z}\right)\right]^2
$$
$$
+\left[\cos\theta\left(T_0\frac{\partial\tau}{\partial z}+\tau\frac{\partial T_0}{\partial z}\right)-\sin\theta\left(T_0\frac{\partial\tau}{\partial x}+\tau\frac{\partial T_0}{\partial x}\right)\right]^2 \tag{10.7}
$$
$$
\times\left[1-\frac{2\eta v_t^2(1+2\varepsilon)}{1+2\eta}\left(\cos\theta\left(T_0\frac{\partial\tau}{\partial x}+\tau\frac{\partial T_0}{\partial x}\right)+\sin\theta\left(T_0\frac{\partial\tau}{\partial z}+\tau\frac{\partial T_0}{\partial z}\right)\right)^2\right]=\frac{1}{v_t^2},
$$

where $v_t(x, z)$ is the velocity along the symmetry axis, $\epsilon(x, z)$ and $\eta(x, z)$ are the anisotropy parameters, and $\theta(x, z)$ is the tilt angle that the symmetry axis makes with the vertical. The point-source condition is the same as the one given in Eq. (10.5). Again $\tau(x, z)$ is the unknown function we solve Eq. (10.7) for, whereas $T_0(x, z)$ is the known function which may be taken as the solution of a homogeneous, tilted elliptically isotropic medium, given as[32]:

$$
T_0(x, z)=\sqrt{\frac{b_s(x-x_s)^2+2c_s(x-x_s)(z-z_s)+a_s(z-z_s)^2}{a_s b_s-c_s^2}} \tag{10.8}
$$

with

$$
a_s=v_{ts}^2(1+2\varepsilon_s)\cos\theta_s^2+v_{ts}^2\sin\theta_s^2,
$$
$$
b_s=v_{ts}^2\cos\theta_s^2+v_{ts}^2(1+2\varepsilon_s)\sin\theta_s^2, \tag{10.9}
$$
$$
c_s=(v_{ts}^2-v_{ts}^2(1+2\varepsilon_s))\cos\theta_s\sin\theta_s.
$$

In the above expressions, v_{ts} and ϵ_s are the velocity along the symmetry axis and the anisotropy parameter, respectively, at the point-source location. Similarly, θ_s is the tilt angle taken at the source point.

It is worth highlighting that the isotropic and TTI cases represent mathematical approximations of the subsurface. An isotropic model considers the velocity to be invariant with respect to the direction of propagation, which is a crude representation of the Earth's crust. The simplest practical anisotropic symmetry system is axisymmetric anisotropy, commonly known as transverse isotropy (TI). A TI medium with a vertical axis of symmetry (VTI) is a good approximation for horizontally layering shale formation or thin-layering sediments. The factored eikonal equation for VTI media can be obtained by setting $\theta = 0$ in Eq. (10.7). For more complex geology, such as sediments near the flanks of salt domes and fold-and-thrust belts like the Canadian foothills, a TTI model represents the best approximation. Fig. 10.1 illustrates these approximations graphically.

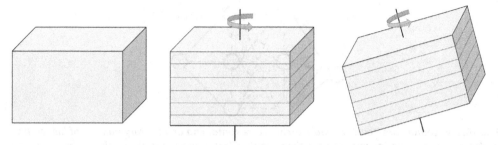

Fig. 10.1 Illustration of the different subsurface approximations: isotropic (*left*), vertically transversely isotropic (VTI) (*center*), and tilted transversely isotropic (TTI) (*right*). The solid lines indicate the direction of the symmetry axis for the VTI and TTI cases.

The reason for considering eikonal equations corresponding to different media is to highlight, in comparison with the conventional methods, how easy it is to adapt the proposed method to solve a relatively more complex eikonal equation (more on this in Section 10.2.4).

10.2.2 Approximation property of neural networks

A feed-forward neural network, also known as a multi-layer perceptron, is a set of neurons organized in layers in which evaluations are performed sequentially through the layers. It can be seen as a computational graph with an input layer, an output layer, and an arbitrary number of hidden layers. In a fully connected neural network, neurons in adjacent layers are connected with each other, but neurons within a single layer share no connections. It is called a feed-forward neural network because information flows from the input through each successive layer to the output. Moreover, there are no feedback or recursive loops in a feed-forward neural network.

Neural networks are well-known for their strong representational power. A neural network with n neurons in the input layer and m neurons in the output layer can be used to represent a function $u: \mathbb{R}^n \to \mathbb{R}^m$. In fact, it has been shown that a neural network with a finite number of neurons in the hidden layer can be used to represent any bounded continuous function to the desired accuracy. This is also known as the universal approximation theorem[51,52]. In addition, it was later shown that by using a deep network with multiple hidden layers and a nonlinear activation function, the total number of neurons needed to represent a given function could be significantly reduced[53]. Therefore, our goal here is to train a DNN that could represent the mapping between the spatial coordinates (x, z), as inputs to the network, and the unknown traveltime function $\tau(x, z)$ representing the output of the DNN. Fig. 10.2 illustrates this idea pictorially showing a neural network with input neurons for the spatial coordinates (x, z) that are passed through the hidden layers to the output layer for predicting the traveltime factor at the inputted spatial location.

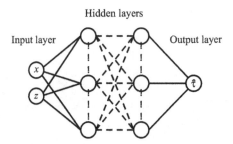

Fig. 10.2 *A feedforward neural network architecture containing an arbitrary number of hidden layers/neurons that is used to approximate the traveltime factor given spatial coordinates (x, z) for a 2D computational domain.*

We formulate here considering a 2D case for simplicity of illustration. In a 3D model, one would need a neural network with three input neurons, one for each spatial dimension. It must also be noted that while DNNs are, in theory, capable of representing very complex functions, finding the actual parameters (weights and biases) needed to solve a given PDE can be very challenging.

10.2.3 Automatic differentiation

Solving a PDE using neural networks requires a mechanism to accurately compute derivatives of the network's output(s) with respect to the input(s). There are multiple ways to compute derivatives including hand-coded analytical derivatives, symbolic differentiation, numerical approximation, and automatic differentiation (AD)[54]. While manually working out the derivatives is exact, it is often time consuming to code and it is error-prone. Symbolic differentiation, while also exact, may result in exponentially large expressions and, therefore, can be prohibitively slow and memory intensive. Numerical differentiation, on the other hand, is easy to implement but can be highly inaccurate due to round-off errors. Contrary to these approaches, AD uses exact expressions with floating-point values instead of symbolic strings and it involves no approximation errors. This results in an accurate evaluation of derivatives at machine precision. Therefore, we evaluate the partial derivatives of the unknown traveltime factor τ with respect to the inputs (x, z) using AD.

However, it must be noted that an efficient implementation of the AD algorithm can be non-trivial. Fortunately, many existing computational frameworks such as Tensorflow[55] and PyTorch[56] have made available efficiently implemented AD libraries.

10.2.4 Solving eikonal equations

We begin by considering how the different pieces of the puzzle outlined in the previous subsections can be combined to solve eikonal equations. First, we illustrate this using the factored isotropic eikonal Eq. (10.4) and then demonstrate how simple it is under the proposed framework to solve a more complex eikonal equation, such as the factored TTI eikonal Eq. (10.7).

To solve Eq. (10.4), we leverage the capabilities of neural networks as function approximators and define a loss function that minimizes the residual of the underlying PDE for a chosen set of training (collocation) points. This is achieved using the following components:

1. A DNN approximation of the unknown traveltime field variable $\tau(x, z)$,
2. A differentiation algorithm, that, AD in this case, to evaluate partial derivatives of $\tau(x, z)$ with respect to the spatial coordinates (x, z), and
3. A loss function incorporating the underlying eikonal equation, sampled on a collocation grid, and an optimizer to minimize the loss function by updating the neural network parameters.

To illustrate the idea, let us consider a two-dimensional domain $\Omega \in \mathbb{R}^2$. A point-source is located at coordinates (x_s, z_s), where $\tau(x_s, z_s) = 1$. The unknown traveltime factor $\tau(x, z)$ is approximated using a DNN, N_τ, such that:

$$\tau(x, z) \approx \hat{\tau}(x, z) = N_\tau(x, z; \theta),\tag{10.10}$$

where x, z are network inputs, $\hat{\tau}$ is the network output, and $\Omega \in \mathbb{R}^D$ represents the set of all trainable parameters of the network with D as the total number of parameters.

The loss function is given by the mean-squared error norm as

$$\Im = \frac{1}{N_I} \sum_{(x*, z*)\in I} \|\mathcal{L}\|^2 + \frac{1}{N_I} \sum_{(x*, z*)\in I} \left\|\mathcal{H}(-\hat{\tau})|\hat{\tau}|\right\|^2 + \left\|\hat{\tau}(x_s, z_s) - 1\right\|^2,\tag{10.11}$$

where $\mathcal{L}$ represents the residual of the factored isotropic eikonal Eq. (10.4), given by

$$\mathcal{L} = \left(T_0 \frac{\partial \hat{\tau}}{\partial x} + \hat{\tau} \frac{\partial T_0}{\partial x}\right)^2 + \left(T_0 \frac{\partial \hat{\tau}}{\partial z} + \hat{\tau} \frac{\partial T_0}{\partial z}\right)^2 - \frac{1}{v^2}.\tag{10.12}$$

The first term on the right side of Eq. (10.11) imposes validity of the factored eikonal Eq. (10.4) on a given set of training points $(x*, z*) \in I$, where N_I is the number of training samples. The second term forces the solution $\hat{\tau}$ to be positive by penalizing negative solutions using the Heaviside function $\mathcal{H}()$. The last term enforces the boundary condition by imposing the solution $\hat{\tau}$ to be unity at the source point (x_s, z_s). The set of network parameters θ^* that minimizes the loss function (11) on this set of training points, $(x*, z*) \in I$, is then identified by solving the optimization problem:

$$\theta^* = \arg\min_{\theta \in R^D} \Im(x^*, z^*; \theta).\tag{10.13}$$

Once the DNN is trained, we evaluate the network on a set of regular grid-points in the computational domain to obtain the unknown traveltime field. The final traveltime solution is obtained by multiplying it with the known traveltime part, i.e.,

$$\hat{T}(x, z) = T_0(x, z) \cdot \hat{\tau}(x, z)\tag{10.14}$$

This yields traveltimes corresponding to an isotropic approximation of the Earth. However, it is well-known that the subsurface is anisotropic in nature. Therefore, a significant amount of research effort has been spent over the years on extending numerical eikonal solvers to anisotropic media. The complication in numerically solving the eikonal equation arises due to anellipticity of the wavefront[57], resulting in high-order nonlinear terms in the eikonal equation. These high-order terms dramatically increase the complexity in solving the anisotropic eikonal equation and, therefore, have been a topic of immense research interest. On the contrary, the proposed neural network formulation allows solving for the anisotropic eikonal equation by simply replacing the residual in Eq. (10.11) with the one corresponding to the anisotropic eikonal equation. Therefore, to solve the factored TTI eikonal Eq. (10.7), we would, instead of Eq. (10.12), use the following:

$$
\begin{aligned}
\mathcal{L} = \left(1 + 2\varepsilon\right) & \left(\cos\theta \left(T_0 \frac{\partial \hat{\tau}}{\partial x} + \hat{\tau} \frac{\partial T_0}{\partial x} \right) + \sin\theta \left(T_0 \frac{\partial \hat{\tau}}{\partial z} + \hat{\tau} \frac{\partial T_0}{\partial z} \right) \right)^2 \\
& + v_t^2 \left(\cos\theta \left(T_0 \frac{\partial \hat{\tau}}{\partial z} + \hat{\tau} \frac{\partial T_0}{\partial z} \right) - \sin\theta \left(T_0 \frac{\partial \hat{\tau}}{\partial x} + \hat{\tau} \frac{\partial T_0}{\partial x} \right) \right)^2 \\
& \times \left(1 - \frac{2\eta v_t^2 (1 + 2\varepsilon)}{1 + 2\eta} \left(\cos\theta \left(T_0 \frac{\partial \hat{\tau}}{\partial x} + \hat{\tau} \frac{\partial T_0}{\partial x} \right) + \sin\theta \left(T_0 \frac{\partial \hat{\tau}}{\partial z} + \hat{\tau} \frac{\partial T_0}{\partial z} \right) \right)^2 \right) - \frac{1}{v_t^2}
\end{aligned}
\tag{10.15}
$$

This is a highly desirable feature of this approach because eikonal equations corresponding to models with even lower symmetry than TTI can be easily solved by simply using a different residual term. By contrast, conventional algorithms such as fast marching or fast sweeping methods would require significant effort to incorporate such changes, thereby resulting in much slower scientific progress.

A workflow summarizing the proposed solver for the factored TTI eikonal equation is shown in Fig. 10.3.

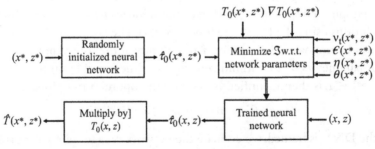

Fig. 10.3 A workflow of the proposed factored TTI eikonal solver in 2D: A randomly initialized neural network is trained on a set of randomly selected collocation points (x^*, z^*) in the model space with given model parameters $v_t(x^*, z^*)$, $\epsilon(x^*, z^*)$, $\eta(x^*, z^*)$, $\theta(x^*, z^*)$, the known traveltime function $T_0(x^*, z^*)$, and its spatial derivative $T_0(x^*, z^*)$ to minimize the loss function given in Eq. 10.11. Once the network is trained, it is evaluated on a regular grid of points (x, z) to yield an estimate of the traveltime field $\hat{\tau}$, which is then multiplied with the factored traveltime part T_0 to yield the estimated first-arrival traveltime solution $\hat{T}$.

10.3 Numerical tests

In this section, we test the neural network based eikonal solver and compare its performance with the first-order fast sweeping method, which is routinely used in geophysical (and other) applications for traveltime computations. We consider several isotropic and anisotropic 2D/3D models for these tests and also include a model with topography to demonstrate the flexibility of the proposed method.

For each of the examples presented below, we use a neural network having 20 hidden layers with 20 neurons in each layer and minimize the neural network's loss function using full-batch optimization. A locally adaptive inverse tangent activation function is used for all hidden layers except the final layer, which uses a linear activation function. Locally adaptive activation functions have been shown to achieve superior optimization performance and convergence speed over base methods. The introduction of a scalable parameter in the activation function for each neuron changes the slope of the activation function and, therefore, alters the loss landscape of the neural network for improved performance. For more information on locally adaptive activation functions, we refer the interested reader to[58].

We choose the afore-mentioned configuration of the neural network based on some initial tests and keep it fixed for the entire study to minimize the need for hyperparameter tuning for each new velocity model.

The following examples are prepared and trained using the neural network library, SciANN[59], a Keras/Tensorflow API that is designed and optimized for physics-informed deep learning. SciANN leverages the latest advancements of Tensorflow while keeping the interface close to the mathematical description of the problem.

Example 1: An isotropic model with constant vertically varying gradient

First, we consider a vertically varying 1×1 km isotropic model. The velocity at zero depth is 2 km/s and it increases linearly with a gradient of 1 s^{-1} to 3 km/s at a depth of 1 km. We compute traveltime solutions using the neural network and first-order fast sweeping method by considering a pointsource located at (0.5 km, 0.5 km). We compare their performance with a reference solution computed analytically[60]. The velocity model is shown in Fig. 10.4 and is discretized on a 101×101 grid with a 10 m grid interval along both axes.

For training the neural network, we begin with randomly initialized parameters and train on 50% of the total grid points selected randomly and use the Adam optimizer[61] with 10,000 epochs. Once the network is trained, we evaluate the trained network on the regularly sampled (101 101) grid to obtain the unknown traveltime field $\hat{\tau}$, which is then multiplied with the corresponding factored traveltime field T_0 to obtain the final traveltime solution. We compare the accuracy of the neural network solution and the

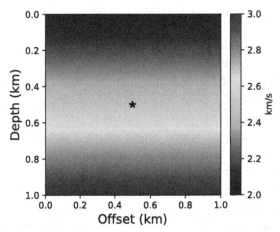

Fig. 10.4 *A vertically varying velocity model with a constant velocity gradient of 1 s⁻¹.* The velocity at zero depth is equal to 2 km/s and it increases linearly to 3 km/s at a depth of 1 km. The black star indicates the point-source location used for the test.

first-order fast sweeping solution, computed on the same regular grid in Fig. 10.5. We observe significantly better accuracy for the neural network based solution despite using only half of the total grid points for training. In Fig. 10.6, we confirm this observation by plotting the corresponding traveltime contours.

Example 2: Smoothly varying TTI model.

Next, we train the neural network to solve the TTI eikonal equation. Compared to the fast sweeping method that requires significant modifications to the isotropic eikonal solver, the neural network based approach requires only an update to the loss function by incorporating the appropriate residual based on the TTI eikonal equation. For the velocity parameter v_p, we consider a linear velocity model with a vertical gradient of

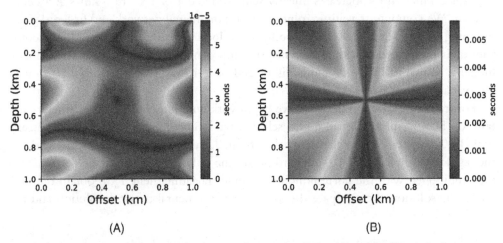

(A)   (B)

Fig. 10.5 Absolute traveltime errors for the neural network solution (A) and the first-order fast sweeping solution (B) for the isotropic model and the source location shown in Fig. 10.4.

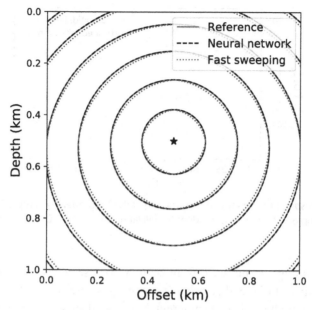

Fig. 10.6 Traveltime contours for the reference solution (*solid red*) computed analytically, neural network solution (dashed black), and the first-order fast sweeping solution (*dotted blue*) for the vertically varying isotropic model. The black star shows the location of the point source.

1.5 s^{-1} and a horizontal gradient of 0.5 s^{-1} as shown in Fig. 10.7. We use homogeneous models for the anisotropy parameters with $\epsilon = 0.2$ and $\eta = 0.083$. We also consider a homogeneous tilt angle of $\theta = 30°$. These models are also discretized on a 101 × 101 grid with 10 m grid interval along both axes.

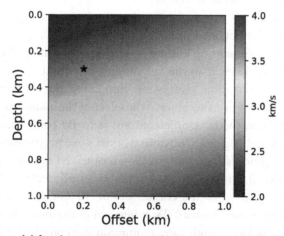

Fig. 10.7 *A velocity model for the parameter vt with a constant vertical gradient of 1.5s^{-1} and a horizontal velocity gradient of 0.5 s^{-1}.* A homogeneous model is used for the anisotropy parameters ($\epsilon = 0.2, \eta = 0.083$) and for the tilt angle ($\theta = 30°$). The black star indicates the point-source location used for the test.

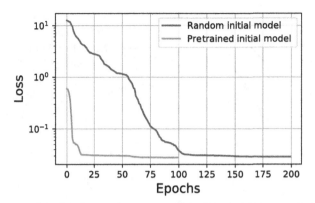

Fig. 10.8 A comparison of the loss history for training of the TTI model in example 2 using pre-trained weights from example 1 (*orange*) and random initialization (*blue*). An L-BFGS-B optimizer is used for both cases.

Instead of training the network from scratch, we use transfer learning, which is a machine learning technique that relies on storing knowledge gained while solving one problem and applying it to a different but related problem. Starting with a pre-trained network from example 1, we fine-tune the neural network parameters for the TTI model using 50% of the total grid points, selected randomly, using the L-BFGS-B solver[62] for 100 epochs. Starting with a pre-trained network allows us to use the L-BFGS-B method for faster convergence as opposed to starting with the Adam optimizer and then switching to L-BFGS-B as suggested by previous studies[43]. For comparison, we also train a neural network from scratch and the convergence history, shown in Fig. 10.8, confirms that the solution converges much faster when using transfer learning.

Fig. 10.9 compares absolute traveltime errors computed using the neural network and the first-order fast sweeping method using the iterative solver of Waheed et al.[33]. The reference solution is obtained using a high-order fast sweeping method on a finer grid. We observe that, despite using transfer learning, the accuracy of the neural network solution is considerably better than the fast sweeping method.

We confirm this observation visually by comparing the corresponding traveltime contours in Fig. 10.10. One can also observe the effect of the additional anisotropy parameters and the tilt angle on the traveltime contours here. By comparing the shapes of the contours in Figs. 10.6 and 10.10, it is obvious that the wave propagation speed varies with the direction of propagation. A faster propagation is observed orthogonal to the symmetry direction, given by the tilt angle, compared to the propagation along the symmetry axis.

It is worth noting that while the complexity of the fast sweeping solvers and their computational cost increases dramatically when switching from an isotropic to a TTI model, for the neural network both cases require similar complexity and computational cost. Therefore, the proposed method is particularly suited to model complex physics involving media with anisotropy, attenuation, etc.

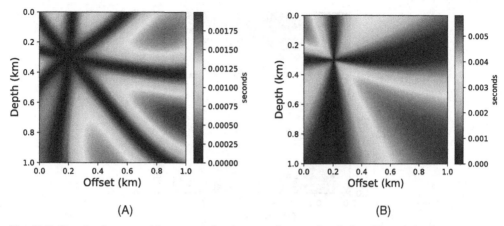

Fig. 10.9 The absolute traveltime errors for the neural network solution (A) and the fast sweeping solution (B) for the TTI model considered in example 2.

One of the main challenges in seismic imaging and inversion is the need for repeated traveltime computations for thousands of source locations and multiple updated velocity models. Unfortunately, conventional techniques do not allow the transfer of information from one solution to the next and, therefore, the same amount

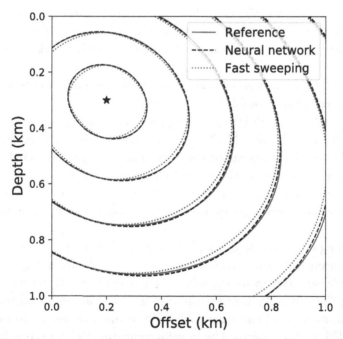

Fig. 10.10 The traveltime contours for the reference solution (*solid red*), neural network solution (dashed black), and the first-order fast sweeping solution (*dotted blue*) for the smoothly varying TTI model. The black star shows the location of the point source.

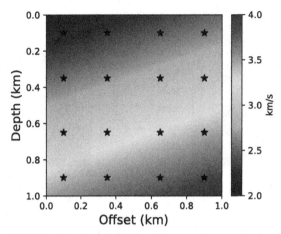

Fig. 10.11 *A velocity model for the parameter vt with a constant vertical gradient of 1.5 s⁻¹ and a horizontal velocity gradient of 0.5 s⁻¹.* A homogeneous model is used for the anisotropy parameters ($\epsilon = 0.2$, $\eta = 0.083$) and for the tilt angle ($\theta = 30°$). Black stars indicate locations of sources used to train the network as a surrogate model.

of computational effort is needed for even small perturbations in the source location and/or the velocity model. We noted above how transfer learning can be used to speed up convergence for a new velocity model and source position. This could be further extended by adding source location (x_s, z_s) as input to the network and training a surrogate model.

To do so, we train a neural network on solutions computed for 16 sources located at regular intervals in the considered TTI model as shown in Fig. 10.11. Through the training process, the network learns the mapping between a given source location and the corresponding traveltime field. Once the surrogate model is trained, traveltime fields for additional source locations can be computed instantly by using a single evaluation of the trained network. This is similar to obtaining an analytic solver as no further training is needed for computing traveltimes corresponding to new source locations. This feature is particularly advantageous for large 3D models that need thousands of such computations.

After training the surrogate model, we test its performance by computing the traveltime field corresponding to a randomly chosen source location. Fig. 10.12 compares the absolute traveltime errors for the solution predicted by the surrogate model and the fast sweeping TTI solver. We observe that even without any additional training for this new source position, we obtain remarkably high accuracy compared to the fast sweeping method. This is also confirmed by visually comparing the corresponding traveltime contours in Fig. 10.13.

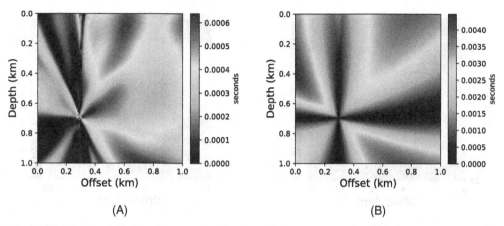

(A)   (B)

Fig. 10.12 The absolute traveltime errors for the solution computed using the surrogate model (a) and the fast sweeping solution (b) for the TTI model considered in example 2 for a randomly chosen source location.

Example 3: VTI SEAM model.

Next, we test the performance of the proposed method on a portion of the VTI SEAM model, shown in Fig. 10.14. The model parameters are extracted from the 3D SEG advanced modeling (SEAM) phase I subsalt earth model[63]. This is a particularly

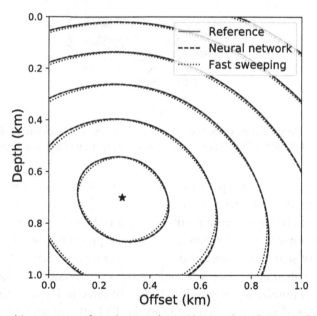

Fig. 10.13 The traveltime contours for solutions obtained using the reference solution (*solid red*), the neural network surrogate model (*dashed black*), and the first-order fast sweeping solver (*dotted blue*) for a randomly chosen source point in a smoothly varying TTI model.

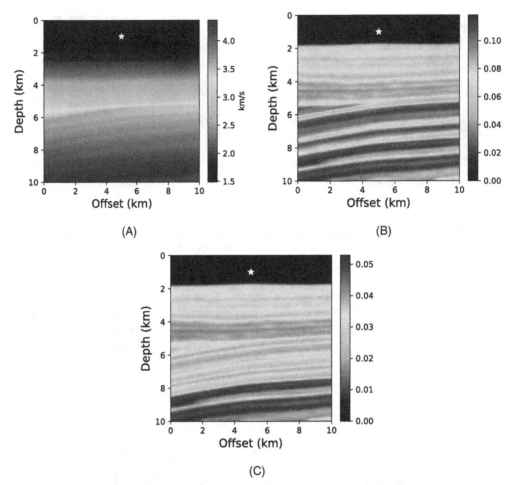

Fig. 10.14 (A) The vertical velocity $v_{t'}$ (B) the ϵ parameter, and (C) the η parameter for the considered portion of the VTI SEAM model. The white star indicates the position of the point-source.

interesting example due to sharper variations in the velocity model and the anisotropy parameters. The model is discretized on a 101×101 grid with a grid spacing of 100 m along both axes. Based on our recent efforts in using the neural network solver[44,47], we observe that the convergence of the neural network approach slows down considerably in the presence of sharp variations in the velocity model. We have already seen above that using a pre-trained neural network yields faster convergence by allowing the use of the second-order optimization method (L–BFGS–B) directly. Therefore, in this example, we use the pretrained network obtained from the TTI eikonal solver in example 2.

For further speedup, we propose a two-stage training scheme to obtain an accurate traveltime solution. In the first stage, when the neural network is learning a smooth representation of the underlying function, we use only a small percentage of grid points for training. In this case, we use only 1% of the total grid points chosen randomly for the first 200 epochs and then switch to using 50% of the total grid points in stage 2 for another 1000 epochs to update the learned function in better approximating sharp features in the resulting traveltime field. Again, since we start training with a pre-trained network, we use the L–BFGS–B optimizer for faster convergence.

Fig. 10.15 shows traveltime contours for a point-source located at (5 km, 1 km) for the reference solution, the neural network solution, and the firstorder fast sweeping solution. In Fig. 10.15A, we show the neural network solution at the end of stage 1. It can be seen that the solution is quite smooth and misses sharp features visible in the reference solution. In Fig. 10.15B, we observe that the neural network solution captures these features as additional training points are added in the second stage of training. Therefore, using a small number of training points in stage 1 reduces the training cost without compromising on solution accuracy. By comparing the absolute traveltime errors in Fig. 10.16, we observe that the neural network solution after stage 2 is considerably more accurate than the first-order fast sweeping method, even for a realistic VTI model.

Example 4: BP TTI model.

We have already seen how the neural network based approach is flexible in incorporating complex physics compared to conventional techniques. In this example, we will see how incorporating irregular topography is straight-forward using the proposed

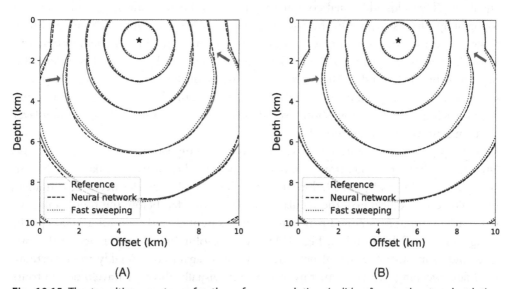

(A)   (B)

Fig. 10.15 The traveltime contours for the reference solution (*solid red*), neural network solution (*dashed black*), and the first-order fast sweeping solution (*dotted blue*) for the VTI SEAM model. Red arrows indicate the improvement of accuracy due to the additional training points used in the neural network solution in going from stage 1 (A) to stage 2 (B) of the proposed training process. The black star shows the location of the point-source.

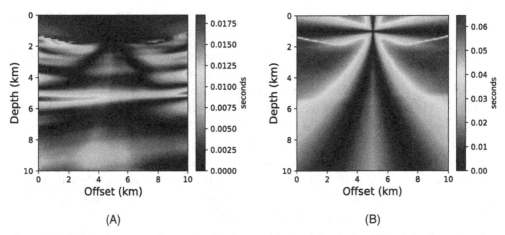

Fig. 10.16 The absolute traveltime errors for the neural network solution (A) and the fast sweeping solution (B) for the SEAM VTI model.

approach. The free-surface encountered in land seismic surveys is often non-flat and requires taking this into account for accurate traveltime computation. One approach to tackle this is to transform from Cartesian to curvilinear coordinate system to mathematically flatten the free-surface and solve the resulting topography-dependent eikonal equation[37]. This adds additional computational cost and may result in instabilities when topography varies sharply. On the contrary, the neural network approach outlined here is mesh-free and doesn't require any modification to the algorithm. We demonstrate this through a test on a portion of the BP TTI model, shown in Fig. 10.17. The model was developed by Shah[64] and publicly provided courtesy of the BP Exploration Operating Company. The considered portion of the model is discretized on a 161 × 161 grid using a grid spacing of 6.25 m along both axes. Points above the considered topography layer are then removed from the model.

For a point-source located at (5 km, 0.5 km), we compare the neural network and fast sweeping traveltime solutions. For training the neural network, we take into account about 50% of the grid points below the topography and once the network is trained, we evaluate the solution on the regular grid points that fall below the topography. We start training using pre-trained parameters from example 2 and train for 10,000 epochs using an L–BFGS-B optimizer. Fig. 10.18 shows absolute traveltime errors for the two cases, indicating that the neural network solution is again considerably more accurate than fast sweeping. We confirm this observation visually through traveltime contours plotted in Fig. 10.19.

Example 5: 3D TTI model.

Finally, we show an example of extending the proposed method to a 3D TTI model. The model for the velocity parameter v_t is shown in Fig. 10.20. A homogeneous model

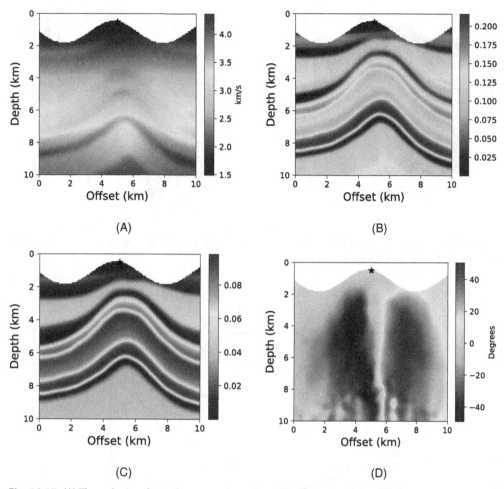

Fig. 10.17 (A) The velocity along the symmetry axis v_t, (B) the ϵ parameter, (C) the η parameter, and (D) the tilt angle θ for the considered portion of the BP TTI model with a layer of non-flat topography.

is used for the anisotropy parameters with $\epsilon = 0.2$ and $\eta = 0.1$. We also consider a homogeneous tilt angle of $30°$. The model is discretized on a $101 \times 101 \times 51$ grid with a grid spacing of 50 m along each axis. The workflow for obtaining the neural network solution is essentially the same. In this case, the neural network takes three input parameters corresponding to the spatial axes (x, y, z) and outputs the corresponding unknown traveltime field $\hat{\tau}(x, y, z)$. Similar to before, the neural network output is multiplied by the known traveltime factor $T_0(x, y, z)$ to obtain the final traveltime solution.

Starting with a pretrained neural network on a smoothly varying TTI model, we train the network using 50% of the total grid points chosen randomly. We use 20,000 L-BFGS-B epochs during the training process. For a point-source located at

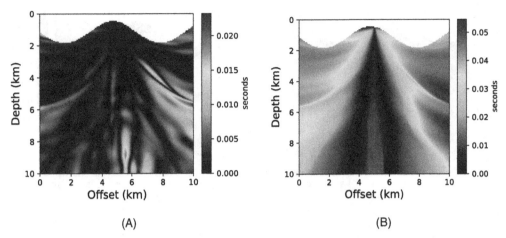

Fig. 10.18 The absolute traveltime errors for the neural network solution (A) and the fast sweeping solution (B) for the BP TTI model.

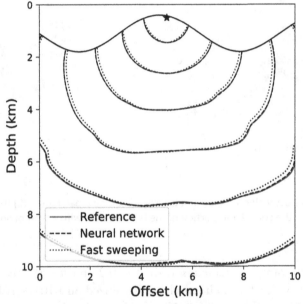

Fig. 10.19 The traveltime contours for the reference solution (*solid red*), neural network solution (*dashed black*), and the first-order fast sweeping solution (*dotted blue*) for the BP TTI model. The black star shows the location of the point source and the solid black curve indicates the topography layer.

$(x, y, z) = (2\text{ km}, 2\text{ km}, 1\text{ km})$, we compare the accuracy of the neural network solution and the first-order fast sweeping solution in Fig. 10.21. We observe that the proposed method is capable of computing accurate traveltimes for 3D TTI models as well without requiring any major alteration to the underlying algorithm.

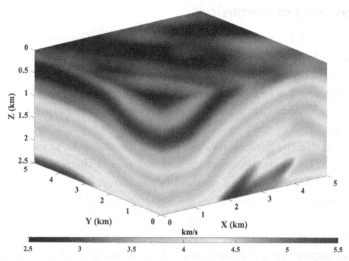

Fig. 10.20 *A velocity model for the parameter v$_t$ for the 3D TTI model test.* A homogeneous model is used for the anisotropy parameters ($\varepsilon = 0.2$, $\eta = 0.1$) and for the tilt angle ($\theta = 30°$).

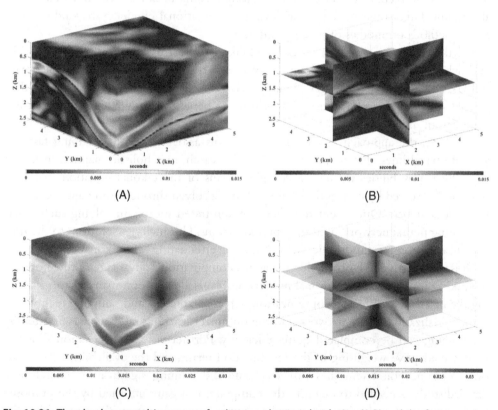

Fig. 10.21 The absolute traveltime errors for the neural network solution (A,B) and the fast sweeping solution (C,D) for the 3D TTI model.

10.4 Discussion and conclusions

We proposed a neural network approach to computing first-arrival traveltimes based on the framework of physics-informed neural networks. By leveraging the capabilities of neural networks as universal function approximators, we define a loss function to minimize the residual of the governing eikonal equation at a chosen set of training points. This is achieved with a simple feed-forward neural network using automatic differentiation.

We demonstrated the flexibility of the proposed framework in incorporating anisotropy in the model simply by updating the loss function of the neural network according to the underlying PDE. Since the method is mesh-free, we also saw how easy it is to incorporate non-flat topography into the solver compared to conventional methods. Another attractive feature, due to this mesh-free nature of the algorithm, is that sources and receivers do not have to lie on a regular grid as required by conventional finite-difference techniques. We also demonstrated that by using machine learning techniques like transfer learning and surrogate modeling, we could transfer information gained by solving one problem to the next – a key feature missing in our conventional numerical algorithms. This would be key in achieving computational efficiency beyond conventional methods on models of practical interest.

Furthermore, the neural network based eikonal solver uses Tensorflow at the back-end, which allows for easy deployment of computations across a variety of platforms (CPUs, GPUs) and architectures (desktops, clusters). On the contrary, significant effort is needed in adapting conventional algorithms to benefit from different computational platforms or architectures.

In short, the approach tackles many problems associated with obtaining fast and accurate traveltimes, that required decades of research, in a holistic manner. In fact, it opens new possibilities in solving complex forms of the eikonal equation that have remained unsolved by conventional algorithms or solved through some approximation techniques at best. Our recent tests have demonstrated success in solving such equations using neural networks without approximations. This includes solutions to the qSV eikonal equation for anisotropic media[65] and the attenuating VTI eikonal equation[66].

It is also worth emphasizing that the actual computational advantage of the proposed method, compared to conventional numerical solvers, depends on many factors including the network architecture, optimization hyper-parameters, and sampling techniques. If the initialization of the network and the optimizer learning rate are chosen carefully, the training can be completed quite efficiently. Furthermore, the activation function used, the adaptive weighting of the loss function terms, and the availability of second-order optimization techniques can accelerate the training significantly. Therefore, a detailed study is needed to quantify the computational gains afforded by the proposed neural network-based solver compared to conventional algorithms by considering the aforementioned factors. A rudimentary analysis performed in[44] indicates that once the surrogate model is obtained, the PINN eikonal solver is computationally faster by more than an order of magnitude than the fast sweeping method. Since the computational complexity of solving the anisotropic eikonal equation using conventional methods increases dramatically compared to the isotropic case[33], the proposed framework is computationally more attractive for traveltime modeling in anisotropic media.

Nevertheless, there are a few challenges associated with the method that requires further research. Chief among them is the slow convergence of the solution in the presence of sharp heterogeneity in the velocity and/or anisotropy models. We proposed a two-stage optimization process in this chapter that alleviates part of the problem by using only a small fraction of the training points during the initial training phase. Since at this stage, the network is learning a smooth representation of the underlying function, we could save some computational cost by using a small number of training points initially. We also used a locally adaptive activation function that has been shown to achieve faster convergence. Other possible solutions may include an adaptive sampling of the velocity model by using denser sampling of training points around parts of the model with large velocity gradients. Another challenge concerns the optimal choice of the neural networks' hyper-parameters. In this study, we alleviated the problem by choosing them through some quick initial tests and keeping them fixed for all the test examples. Recent advances in the field of metalearning may, potentially, enable automated selection of these parameters in the future.

References

1. Crandall MG, Lions P-L. Viscosity solutions of Hamilton-Jacobi equations. *Trans Am Math Soc* 1983;**277**:1–42.
2. Alvino C, Unal G, Slabaugh G, Peny B, Fang T. Efficient segmentation based on eikonal and diffusion equations. *Int J Comput Math* 2007;**84**:1309–24.
3. Garrido S, Álvarez D, Moreno L. Path planning for mars rovers using the fast marching method Robot 2015: Second Iberian Robotics Conference 2016 Springer, pp.93–105.
4. Raviv D, Bronstein AM, Bronstein MM, Kimmel R, Sochen N. Affine-invariant geodesic geometry of deformable 3D shapes. *Comput Graph* 2011;**35**:692–7.
5. Helmsen JJ, Puckett EG, Colella P, Dorr M. Two new methods for simulating photolithography development in 3D Optical Microlithography IX 2726 International Society for Optics and Photonics 1996:253–61.
6. Lawton DC. Computation of refraction static corrections using first-break traveltime differences. *Geophysics* 1989;**54**:1289–96.
7. Hole J, Zelt B. 3-D finite-difference reflection traveltimes. *Geophys J Int* 1995;**121**:427–34.
8. Taillandier C, Noble M, Chauris H, Calandra H. First-arrival traveltime tomography based on the adjoint-state method. *Geophysics* 2009;**74**:WCB1–10.
9. Grechka V, De La Pena A, Schisselé-Rebel E, Auger E, Roux P-F. Relative location of microseismicity. *Geophysics* 2015;**80**:WC1–9.
10. Lambare G, Operto S, Podvin P, Thierry P. 3D ray+ born migration/inversion—part 1: theory. *Geophysics* 2003;**68**:1348–56.
11. Cerveny V. Seismic Ray Theory 2001 Cambridge University Press.
12. Vidale JE. Finite-difference calculation of traveltimes in three dimensions. *Geophysics* 1990;**55**:521–6.
13. Vidale J. Finite-difference calculation of travel times. *Bull Seismol Soc Am* 1988;**78**:2062–76.

14. Dellinger J. Anisotropic finite-difference traveltimes, in: SEG Technical Program Expanded Abstracts 1991. *Soc Explor Geophys* 1991:1530–3.
15. Dellinger J, Symes W. Anisotropic finite-difference traveltimes using a Hamilton-Jacobi solver SEG Technical Program Expanded Abstracts 1997 1997 Society of Exploration Geophysicists: 1786–9.
16. Kim S, Cook R. 3-D traveltime computation using second-order ENO scheme. *Geophysics* 1999;**64**:1867–76.
17. Podvin P, Lecomte I. Finite difference computation of traveltimes in very contrasted velocity models: a massively parallel approach and its associated tools. *Geophys J Int* 1991;**105**:271–84.
18. Nichols DE. Maximum energy traveltimes calculated in the seismic frequency band. *Geophysics* 1996;**61**:253–63.
19. Wang Y, Nemeth T, Langan RT. An expanding-wavefront method for solving the eikonal equations in general anisotropic media. *Geophysics* 2006;**71**:T129–35.
20. Sethian JA, Popovici AM. 3-D traveltime computation using the fast marching method. *Geophysics* 1999;**64**:516–23.
21. Rickett J, Fomel S. A second-order fast marching eikonal solver. *Stanford Exploration Project Report* 1999;**100**:287–93.
22. Alkhalifah T, Fomel S. Implementing the fast marching eikonal solver: spherical versus cartesian coordinates. *Geophys Prospect* 2001;**49**:165–78.
23. Popovici AM, Sethian JA. 3-D imaging using higher order fast marching traveltimes. *Geophysics* 2002;**67**:604–9.
24. Sethian JA, Vladimirsky A. Ordered upwind methods for static Hamilton–Jacobi equations: theory and algorithms. *SIAM J Numer Anal* 2003;**41**:325–63.
25. Cristiani E. A fast marching method for Hamilton-Jacobi equations modeling monotone front propagations. *J Sci Comput* 2009;**39**:189–205.
26. bin Waheed U, Alkhalifah T, Wang H. Efficient traveltime solutions of the acoustic TI eikonal equation. *J Comput Phys* 2015;**282**:62–76.
27. Breuß M, Cristiani E, Gwosdek P, Vogel O. An adaptive domaindecomposition technique for parallelization of the fast marching method. *Appl Math Comput* 2011;**218**:32–44.
28. Monsegny J, Monsalve J, Léon K, Duarte M, Becerra S, Agudelo W, Arguello H. Fast marching method in seismic ray tracing on parallel GPU devices Latin American High Performance Computing Conference Springer 2018: 101–11.
29. Zhao H. A fast sweeping method for eikonal equations. *Math Comput* 2005;**74**:603–27.
30. Zhang Y-T, Zhao H-K, Qian J. High order fast sweeping methods for static hamilton–jacobi equations. *J Sci Comput* 2006;**29**:25–56.
31. Fomel S, Luo S, Zhao H. Fast sweeping method for the factored eikonal equation. *J Comput Phys* 2009;**228**:6440–55.
32. Luo S, Qian J. Fast sweeping methods for factored anisotropic eikonal equations: multiplicative and additive factors. *J Sci Comput* 2012;**52**:360–82.
33. Waheed UB, Yarman CE, Flagg G. An iterative, fast-sweeping- based eikonal solver for 3D tilted anisotropic media. *Geophysics* 2015;**80**:C49–58.
34. Han S, Zhang W, Zhang J. Calculating qP-wave traveltimes in 2D TTI media by high-order fast sweeping methods with a numerical quartic equation solver. *Geophys J Int* 2017;**210**:1560–9.
35. Bouteiller PLe, Benjemaa M, Métivier L, Virieux J. A discontinuous galerkin fast-sweeping eikonal solver for fast and accurate traveltime computation in 3D tilted anisotropic media. *Geophysics* 2019;**84**:C107–18.
36. Hao Q, Waheed U, Alkhalifah T. A fast sweeping scheme for P-wave traveltimes in attenuating VTI media 80th EAGE Conference and Exhibition 2018 Volume 2018 European Association of Geoscientists & Engineers 2018:1–5.
37. Lan H, Zhang Z. A high-order fast-sweeping scheme for calculating first-arrival travel times with an irregular surface. *Bull Seismol Soc Am* 2013;**103**:2070–82.
38. Zhao H. Parallel implementations of the fast sweeping method. *J Comput Math* 2007:421–9.
39. Detrixhe M, Gibou F, Min C. A parallel fast sweeping method for the eikonal equation. *J Comput Phys* 2013;**237**:46–55.
40. Gómez JV, Álvarez D, Garrido S, Moreno L. Fast methods for eikonal equations: an experimental survey. *IEEE Access* 2019;**7**:39005–29.

41. Jordan MI, Mitchell TM. Machine learning: trends, perspectives, and prospects. *Science* 2015;**349**:255–60.
42. Lagaris IE, Likas A, Fotiadis DI. Artificial neural networks for solving ordinary and partial differential equations. *IEEE Trans Neural Networks* 1998;**9**:987–1000.
43. Raissi M, Perdikaris P, Karniadakis GE. Physics-informed neural networks: a deep learning framework for solving forward and inverse problems involving nonlinear partial differential equations. *J Comput Phys* 2019;**378**:686–707.
44. Waheed U, Haghighat E, Alkhalifah T, Song C, Hao Q. PINNeik: Eikonal solution using physics-informed neural networks. *Comput. Geosci.* 2020;**155**:104833.
45. Smith JD, Azizzadenesheli K, Ross ZE. Eikonet: solving the eikonal equation with deep neural networks. *IEEE Trans Geosci Remote Sens* 2020;**59**(12):10685–10696.
46. Moseley B, Markham A, Nissen-Meyer T. Solving the wave equation with physics-informed deep learning. *arXiv preprint arXiv* **2020**;2006:11894.
47. Song C, Alkhalifah T, Waheed UB. Solving the frequency-domain acoustic vti wave equation using physics-informed neural networks. *Geophys J Int* 2021;225(2):846–859.
48. Waheed UB, Alkhalifah T, Haghighat E, Song C, Virieux J. PINNtomo: Seismic tomography using physics-informed neural networks. *arXiv preprint arXiv* 2021;2104:01588.
49. Waheed UB, Alkhalifah T. A fast sweeping algorithm for accurate solution of the tilted transversely isotropic eikonal equation using factorization. *Geophysics* 2017;**82**(6):WB1–8.
50. Thomsen L. Weak elastic anisotropy. *Geophysics* 1986;**51**:1954–66.
51. Cybenko G. Approximation by superpositions of a sigmoidal function. *Math Control Signals Systems* 1989;**2**:303–14.
52. Hornik K, Stinchcombe M, White H. Multilayer feedforward networks are universal approximators. *Neural Netw* 1989;**2**:359–66.
53. Lu Z, Pu H, Wang F, Hu Z, Wang L. The expressive power of neural networks: a view from the widthAdvances in Neural Information Processing Systems 2017:6231–9.
54. Baydin AG, Pearlmutter BA, Radul AA, Siskind JM. Automatic differentiation in machine learning: a survey. *J Mach Learn Res* 2018;**18**:1–43.
55. Abadi M, Agarwal A, Barham P, Brevdo E, Chen Z, Citro C, et al., Tensor-flow: large-scale machine learning on heterogeneous systems, 2015. Avaiable from: https://www.tensorflow.org/, software available from tensorflow.org.
56. Paszke A, Gross S, Chintala S, Chanan G, Yang E, DeVito Z et al. Automatic differentiation in Py-Torch Proceedings of Neural Information Processing Systems 2017.
57. Alkhalifah T. An acoustic wave equation for anisotropic media. *Geo-physics* 2000;**65**:1239–50.
58. Jagtap AD, Kawaguchi K, Karniadakis GEm. Locally adaptive activation functions with slope recovery for deep and physics-informed neural networks. *Proc Royal Soc A* 2020;**476**:20200334.
59. Haghighat E, Juanes R. Sciann: a keras/tensorflow wrapper for scientific computations and physics-informed deep learning using artificial neural networks. *Comput Meth Appl Mech Eng* 2021;**373**:113552.
60. Slotnick M. Lessons in seismic computing. *Soc. Expl. Geophys* 1959;**268**.
61. Kingma DP, Ba J, Adam: a method for stochastic optimization, (2014).
62. Zhu C, Byrd RH, Lu P, Nocedal J. Algorithm 778: L-BFGS-B: Fortran subroutines for large-scale bound-constrained optimization. *ACM Trans Math Softw* 1997;**23**:550–60.
63. Fehler M, Keliher PJ. SEAM phase 1: challenges of subsalt imaging in tertiary basins, with emphasis on deepwater Gulf of Mexico. *Soc Explor Geophys* 2011.
64. Shah H. The 2007 BP anisotropic velocity-analysis benchmark 70th Anuual EAGE meeting, workshop 2007.
65. Waheed U, Alkhalifah T, Li B, Haghighat E, Stovas A, Virieux J. Traveltime computation for qSV waves in TI media using physicsinformed neural networks In: 82nd EAGE Conference and Exhibition, Submitted 2021.
66. Taufik MH, Waheed U, Hao Q, Alkhalifah T. The eikonal solution for attenuatingVTI media using physics-informed neural networks In: 82nd EAGE Conference and Exhibition, Submitted 2021.

PART 4

New directions

CHAPTER 11

Application of artificial intelligence to computational fluid dynamics

Shahab D. Mohaghegh[a], Ayodeji Aboaba[a], Yvon Martinez[a], Mehrdad Shahnam[b], Chris Guenther[b], Yong Liu[b,c]
[a]West Virginia University, United States
[b]Department of Energy, National Energy Technology Laboratory, United States
[c]Leidos Research Support Team, United States

Abstract

Smart proxy technology leverages the art and science of artificial intelligence and machine learning in order to build accurate and very fast proxy models for highly complex numerical simulation models. The main characteristics of the smart proxy modeling are (a) physics of the numerical simulation is not reduced or modified, (b) resolution of the numerical simulation is not reduced or modified, (c) detail cell results of the numerical simulation is replicated with high accuracy, and (d) deployment of the smart proxy to provide numerical simulation results for every cell will take few minutes using a laptop or desktop workstation. In this project, smart proxy technology is used to simulate the combustion of natural gas under various conditions such as varying natural gas composition and flow rate, inlet air flow rate and temperature, and outlet pressure in a high-pressure combustion facility (B6 combustor) with more than four million simulation cells.

Keywords

Artificial intelligence; CFD; Machine learning; Smart proxy

11.1 Introduction

Artificial intelligence and machine learning represent a highly advanced incorporation of science and technology with capabilities that continue to surprise the scientific community with its accomplishments[1]. The idea behind this technology was initiated soon after the World War II, involving Alan Turing (1950), Warren McCulloch, and Walter Pitts (1943) but did not take its realistic first step of using data instead of Aristotelian logic rules until 1957 and 1960 (Frank Rosenblatt in 1957 and Widrow and Hoff in 1960). It took a considerable amount of time before this technology took some important steps (Hopfield in 1982, Rumelhart in 1986) towards what today is known as artificial intelligence and machine learning.

Artificial intelligence: the definition of artificial intelligence is "imitating life to solve complex and dynamic problems." Based on this definition, "artificial intelligence" is an umbrella term that covers a series of new and innovative scientific methods that mimic human brain when performing analysis, modeling, and decision-making. As one

Advances in Subsurface Data Analytics
DOI: https://doi.org/10.1016/B978-0-12-822295-9.00001-7

becomes more interested in this technology, developing serious understanding of topics such as philosophy, neuroscience, and biology, as well as evolution and physics become more and more important.

"Artificial intelligence" is currently impacting the petroleum industry and not too far from now will be seriously influencing everything in our lives, including our health, age, jobs, economy, politics, environment, etc. Not taking "artificial intelligence" seriously, can tell much about a person, especially a scientist.

Machine learning: machine learning includes series of tools, techniques, and algorithms that make "artificial intelligence" a possibility. The definition of "machine learning" is using open computer algorithms to learn from experience (in form of data) rather than detail and explicit programming for the computer to perform certain tasks. In the context human species, our DNA is strictly coded in detail, but our mind and how our brain works, is not.

Smart proxy modeling is one of the major contributions of artificial intelligence and machine learning to numerical simulation. Historically, one of the major issues associated with numerical simulation has been extensive computational footprint. The amount of time that is required to deploy complex and high-fidelity numerical simulations are very high. For example, the single run of the numerical simulation model (computational fluid dynamics–CFD) that is the subject of this chapter take about 24 hours on the DOE's HPC. Traditionally, reduced order models (ROM) and response surface methods (RSM) were developed to minimize the computational footprint of the numerical simulation models.

These traditional technologies either simplify the physics of the numerical simulation models (ROM) or simplify the overall characteristics of "how" and "why" the numerical simulation model (RSM) is used in order to achieve their objectives. While replicating the results of the numerical simulation model with high accuracy (as it is covered in this chapter) without modification of its physics or simplification of its approach and space, time resolution, a single run of the smart proxy modeling a laptop only takes a few minutes. May be not too far from today, deployment of smart proxy models can be done on a tablet or a smart phone.

Smart proxy modeling is a technology that was first developed and generated in the oil and gas industry. Smart proxy modeling was developed for numerical reservoir simulations and was successfully applied in multiple cases[2-5]. This AI and machine learning related technology was then applied to computational fluid dynamics (CFD).

DOE-NETL is supporting projects to develop technologies that will improve the efficiency, cost, and environmental performance of complex power generation systems such as gas turbines and coal-fired power plants. In-situ monitoring of combustion phenomena is a critical need for optimal operation and control of such systems. CFD is an important tool currently being used to investigate and understand the dynamics of the combustion process in these systems. Gas turbine combustion is a complex process, and it can be a challenge to achieve accurate and reliable CFD simulation results at a reasonable computational cost.

The challenge in CFD simulation of complex reaction flows is to adequately resolve the structures that exist at different spatial and temporal scales in an inherently transient flow. Additionally, in reacting gas-solid flow simulations, small time steps are needed in order to not only resolve the temporal scales of the flow, but also ensure numerical stability of the solution. A rule of thumb for adequate spatial resolution is for the grid spacing to be about 10 times the particle diameter[6]. The grid requirement for maintaining such a ratio of grid size to particle diameter for smaller size particles makes such simulations computationally costly and impractical[7]. Recent work at NETL[7] has shown the number of simulations, which is required for non-intrusive uncertainty quantification, can easily exceed many tens of simulations. The spatial and temporal resolution requirements for multiphase flows make CFD simulations computationally expensive and potentially beyond the reach of many design analysts[8].

The goal of this research work is to develop a data driven predictive model capable of replicating the thermal-flow pattern and species distribution results of CFD simulation of natural gas combustion in a high-pressure combustion facility (B6 combustor). Achieving this goal will significantly reduce the typical long time-to-solution characteristics of CFD simulations while preserving traditional CFD solver accuracy for the CFD simulation model under study. The developed smart proxy will contribute greatly to the development of technologies that improve the efficiency, cost, and environmental performance of complex power generation systems.

11.1.1 Structure of the work

The study details the research work performed in building a data driven predictive model that replicates transport variables in the CFD simulation of the combustion of natural gas in a high-pressure combustion facility (B6 combustor). The data driven modeling framework presented for the B6 Combustor model in this study will ultimately be applied to a more complex system (the tristate coal-fired boiler) in a separate project.

In section one (this chapter), the problem was defined, and the final objective of the research was articulated. In section two, a brief background information is provided on key elements of the research work. A brief description of the reaction flow and governing equations used in the numerical CFD simulation software is provided in order to lay the groundwork for understanding the engineering and scientific details associated with the CFD model being studied. A summary of the different machine learning and computing techniques used on the project is included in section two to provide a background to the solution methodology utilized. Also, a brief literature review on the use of AI and machine learning relating to fluid dynamics is provided.

In section three, a general overview of the end-to-end workflow of the smart proxy development process is introduced with a description of the design and implementation of the framework used for the development of the smart proxy model. This section provides a description of B6 combustor CFD simulation model, a detailed description of the simulation data received including the boundary conditions, a brief description of steps taken to develop the smart proxy model and a detailed description of the different machine learning algorithms used in the development process.

Section four provides a lot more detail on every step taken to meet the objective of the research. Detail information is provided regarding the input training dataset and the neural network setup at each step of the development process. This section includes results and discussions for each step taken towards building the smart proxy model.

11.2 Background

This section provides some basic but necessary background on some key components of this research work.

11.2.1 NETL's high-pressure combustor facility (B6 combustor)

Unconventional gas supplies, like shale gas, are expected to grow which will make US natural gas composition more variable and the composition of fuel sources may vary significantly from existing domestic natural gas supplies. The effect of gas composition on combustion behavior is of interest to allow end-use equipment to accommodate the widest possible gas composition. The B6 combustor is a high-pressure facility used at NETL to study both research and commercial gas turbine fuel injectors. B6 combustor experiments have been conducted at NETL to investigate the effect of varying fuel composition on combustion dynamics.

The CFD simulation model was first validated with available experimental data which had been collected from the pressurized single injector combustion test rig. The tests were conducted at 7.5 atm with a 589K preheated air. A propane blending facility was used to vary the site natural gas composition. The CFD simulation model predicted results were within the experimental error bar. After the CFD simulation model was validated with experimental results, a much wider range of gas composition was simulated to investigate the effect of gas composition on combustion. Another purpose of the high-pressure combustion facility is to develop a combustion control and diagnostic sensor (CCADS) to in-situ monitor the combustion phenomena which is based on the mechanisms for ion formation and electrical properties of a flame.

11.2.2 Ansys fluent

The CFD model is based on the mass, momentum and energy balance equations with some other constitutive equations such as the equation of state to calculate the gas phase density.

$$\rho_g = \frac{PM_{wg}}{RT_g} \tag{11.1}$$

Continuity equation:

$$\frac{\partial}{\partial t}\left(\rho_g\right) + \nabla \cdot \left(\rho_g \overrightarrow{u_g}\right) = \sum_{i=1}^{N} R_{gi} \tag{11.2}$$

Momentum equation:

$$\frac{\partial}{\partial t}\left(\rho_g \vec{u}_g\right) + \nabla \cdot \left(\rho_g \vec{u}_g \vec{u}_g\right) = -\nabla P_g + \nabla \cdot \bar{\bar{\tau}}_g + \rho_g \vec{g} \tag{11.3}$$

$$\bar{\bar{\tau}}_g = \mu_e\left[\left(\nabla \vec{u}_g + \nabla \vec{u}_g^{-T}\right) - \frac{2}{3}\nabla \cdot \vec{u}_g I\right] \tag{11.4}$$

As the gas phase is composed of several components such as the O_2, N_2, CO_2 etc., the species transport equation

$$\frac{\partial}{\partial t}\left(\rho_g Y_i\right) + \nabla \cdot \left(\rho_g \vec{u}_g Y_i\right) = -\nabla \cdot \vec{J}_i + R_{gi} \tag{11.5}$$

$$\vec{J}_i = -\left(\rho_g D_{im} + \frac{\mu_t}{Sc_t}\right)\nabla Y_i \tag{11.6}$$

Where:

- ρ_g = gas density (kg/m³).
- P = operating gas pressure (outlet pressure).
- M_w = average molecular weight of gas.
- R = universal gas constant (8.314 J/mol/K)
- T = gas phase temperature.
- $\vec{u}_g$ = gas phase velocity in x, y and z direction respectively.
- $\bar{\bar{\tau}}_g$ = stress tensor.
- Y_i = fraction of species i in the gas phase.
- R_{gi} = net rate of production of species i by chemical reaction.
- $\vec{g}$ = gravity.
- $\vec{J}_i$ = diffusion flux of species due to the gradients of concentration.
- D_{im} = mass diffusion coefficient for species j in the mixture.
- μ_e = effective viscosity ($\mu_e = \mu_t + \mu$).
- μ_t = turbulent viscosity $\left(\mu_t = \rho C_\mu \dfrac{k^2}{\varepsilon}\right)$.

**Subscript "g" means the gas phase, subscript "i" means the species i.

Incompressible ideal gas law is used to calculate the gas density as the temperature changes a lot but the pressure changes little. The operating pressure is the pressure at the coal boiler outlet. For turbulent flows, the molecular viscosity is much smaller than the turbulent viscosity.

11.2.2.1 Turbulence model

Realizable k-ε with standard wall functions as realizable k-ε model is more suitable for flow with swirling[9–15]. The realizable k-ε model differs from the standard k-ε model in two important ways: the realizable model contains an alternative formulation for the turbulent viscosity. A modified transport equation for the dissipation rate, has been derived from an exact equation for the transport of the mean-square vorticity fluctuation. The term "realizable" means that the model satisfies certain mathematical constraints on the Reynolds stresses, consistent with the physics of turbulent flows. Neither the standard k-ε model nor the RNG k-ε model is realizable.

The difference between the realizable k-ε model and the standard and RNG k-ε models is that C_μ is no longer constant but a function of the mean strain and rotation rates, the angular velocity of the system rotation, and the turbulence fields.

k is the turbulence kinetic energy

$$\frac{\partial}{\partial t}\left(\rho_g k\right)+\nabla\cdot\left(\rho_g k \overrightarrow{u_g}\right)=\nabla\cdot\left[\left(\mu+\frac{\mu_t}{\sigma_k}\right)\nabla k\right]+G_k+G_b-\rho\varepsilon-Y_m+S_k \quad (11.7)$$

ε is the dissipation rate of turbulence kinetic energy

$$\frac{\partial}{\partial t}\left(\rho_g\varepsilon\right)+\nabla\cdot\left(\rho_g\varepsilon\overrightarrow{u_g}\right)=\nabla\cdot\left[\left(\mu+\frac{\mu_t}{\sigma_\varepsilon}\right)\nabla\varepsilon\right]+\rho C_1 S\varepsilon-\rho C_2\frac{\varepsilon^2}{k+\sqrt{\upsilon\varepsilon}}+C_{1\varepsilon}\frac{\varepsilon}{k}C_{3\varepsilon}G_b+S_\varepsilon \quad (11.8)$$

Where:
- G_k is generation of turbulence kinetic energy due to the mean velocity gradients.
- G_b is the generation of turbulence kinetic energy due to buoyancy.
- Y_m is the contribution of the fluctuating dilatation in compressible turbulence to the overall dissipation rate.
- C_2 (1.9) and $C_{1\varepsilon}$ (1.44) are constants. σ_k (1.0) and σ_ε (1.2) are the turbulent Prandtl numbers for k and ε respectively.
- S_k and S_ε are user-defined source terms.

11.2.2.2 CFD reaction Eddy-dissipation model

Most fuels are fast burning, and the overall rate of reaction is controlled by turbulent mixing. The net rate of production of species due to reaction R_i, is given by the smaller (that is, limiting value) of the two expressions below:

$$R_{i,r}=\upsilon'_{v,r}M_{w,i}4.0\rho\frac{\varepsilon}{k}min_R\left(\frac{Y_R}{\upsilon'_{R,r}M_{w,R}}\right) \quad (11.9)$$

$$R_{i,r}=\upsilon'_{i,r}M_{w,i}2.0\rho\frac{\varepsilon}{k}\frac{\sum_P Y_P}{\sum_j^N \upsilon''_{j,r}M_{w,j}} \quad (11.10)$$

Where:
- Y_P = mass fraction of any product species, P
- Y_R = mass fraction of a reactant, R
- $\upsilon'_{i,r}$ − stoichiometric coefficient for reactant i in reaction r
- $\upsilon''_{i,r}$ = stoichiometric coefficient for product i in reaction r

11.2.2.3 CFD heat transfer model

$$\frac{\partial}{\partial t}\left(\rho_g H\right) + \nabla \cdot \left(\rho_g H \overrightarrow{u_g}\right) = \nabla \cdot \left(\frac{k_t}{C_p}\nabla H\right) + S_h \tag{11.11}$$

$$H_j = \int_{T_{ref}}^{T} C_{p,j}dT + H_j^0\left(T_{ref,j}\right) \tag{11.12}$$

H is the total enthalpy defined as $H = \sum_j Y_j H_j$ where Y_j is the mass fraction of species j and H_j is the enthalpy of species j. The heat capacity $C_{p,j}$ is defined as a function of temperature for each species. When the radiation model is being used, the source term S_h includes radiation source terms.

Both conduction and convection require matter to transfer heat. Radiation is a method of heat transfer that does not rely upon any contact between the heat source and the heated object. Thermal radiation (often called infrared radiation) is a type of electromagnetic radiation (or light). Radiation is a form of energy transport consisting of electromagnetic waves traveling at the speed of light. No mass is exchanged, and no medium is required for radiation.

11.2.2.3.1 Radiation model
Discrete ordinates (DO) model is used as DO model needs more computational resource than other radiation model but the DO model is more complete[9–12,14–18].

DO is recommended by fluent. Emissivity of gas can be calculated from weighted-sum-of-grey-gases model (WSGGM), which has been widely used in computational fluid dynamics and reached good balance between calculating efficiency and accuracy[19,20]. WSGGM assumed that the emissivity of flue gas was decided by local temperature and partial pressure of gas species.

11.2.3 Machine learning
Artificial intelligence and machine learning are widely known technologies that aim to teach machines to learn from input data. Machine learning algorithms can be classified mainly into supervised and unsupervised learning algorithms. Supervised learning algorithms learn a function that, given a sample of data and desired outputs, best approximates the relationship between input features and output (also known as ground truth) observable in the data. Unsupervised learning algorithms, on the other hand, do not have labeled outputs; so, the goal is to infer the natural structure or underlying pattern present within a set of data points.

11.2.3.1 Fuzzy clustering
Clustering is a form of unsupervised learning technique which involves assigning data points (or objects) to clusters (groups) such that points in the same cluster are as similar

as possible. The simplest form of cluster analysis is the hard clustering in which a data point exclusively belongs to a single cluster. Fuzzy clustering is useful in avoiding the arbitrariness of assigning an object or data point to only one cluster when it may be close to several. In fuzzy clustering (also called soft clustering), every object or data point belongs to every cluster with a membership weight that is between 0 (absolutely does not belong) and 1 (absolutely belongs)[21]. Cluster membership weights for any data point must sum up to 1.

In this project, the skfuzzy package from a popular open-source machine learning library called Scikit-learn is used in performing fuzzy clustering tasks.

11.2.3.2 Artificial neural networks

One of the most common supervised learning algorithms is the artificial neural network (ANN). An ANN is a simple mathematical computational algorithm that is capable of learning from input data (machine learning) as well as discovering patterns (pattern recognition)[22].

ANN is biologically inspired by the interconnections that take place between neurons in a human brain. Neurons carry and pass information from one neuron to another via synapse. The architecture of artificial neural networks consists of an input layer, one or more hidden layers, and an output layer. The input layer contains the information provided to the neural network in the form of attributes. The hidden layer is responsible for translating the information from the input layer to the output layer by a system of weighted connections and non-linear activation functions[22]. Fig. 11.1 shows a typical ANN with four input attributes, three neurons in the hidden layer and a single neuron in the output layer. The strength of information passed from one artificial neuron to another is assigned by its "weight." Optimization of these weights is crucial in the development of a well-trained neural network.

In this study, a machine learning library in Python called Keras is used in modeling artificial neural networks[23]. Keras is an open source high-level neural networks API

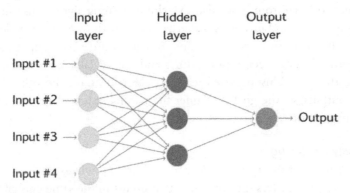

Fig. 11.1 *Artificial neural network architecture.*

written in Python and capable of running on top of TensorFlow, CNTK, or Theano. In this project, the TensorFlow backend is used.

11.2.3.3 Artificial neural network performance evaluation metrics

During the ANN training process, the objective function (also called loss function) is used in updating the weights of the neurons of the neural network. The performance of a neural network is often evaluated in the context of minimizing the total error between the neural network predicted values and the actual output values (ground truth). The most commonly used objective function (especially for regression tasks) is the mean square error (MSE) which is the sum of squared differences between the predicted values and the actual output values, as shown by Eq. 11.13.

$$MSE = \frac{1}{N}\sum_{i=1}^{N}\left(y_{actual} - y_{predicted}\right)^2 \tag{11.13}$$

In practice, while a portion of the entire dataset (called test or blind dataset) is set aside prior to training, to test the performance of the neural network on unseen data, a portion of the remaining data is usually set aside for calibration of the neural network as it is being trained. An optimization of the loss function on the calibration set helps to ensure that the neural network is not over-trained (overfitting) or under-trained. In this project, we have taken a more unique approach to data partitioning for the development of neural networks.

There are various ways to evaluate the performance of a neural network model. In machine learning, a very common metric is the percentage error which expresses the difference between the predicted value and actual value as a percentage of the actual value as shown in Eq. 11.14.

$$\% \ Error = \frac{CFD_{value} - Smart\ Proxy_{value}}{CFD_{value}} \ x \ 100 \tag{11.14}$$

11.2.3.4 Data batching

Artificial neural networks are trained using the gradient descent optimization algorithm in which the difference between the ANN predicted values and the actual values are used in estimating the error gradient. The error gradient is then used to update the model weights and the process is repeated. The estimate of the loss gradient is usually calculated based on all or a subset of the training dataset. The number of training samples used in estimating the error gradient is called batch size and it is an important hyper parameter that influences how the neural network learns. We refer to the concept of controlling the batch size as data batching.

The error gradient is a statistical estimate. The more training examples used in the estimate, the more accurate this estimate will be and the more likely that the weights of the network will be adjusted in a way that will improve the performance of the ANN model in fewer number of iterations[24]. However, computing the gradient over a very large number of examples could be very computationally expensive especially when dealing with a large amount of data. However, using too few examples from the training data could result in less accurate and noisy estimates of the gradient. Nevertheless, these noisy updates can result in faster learning and sometimes a more robust model[24]. Data batching can be used to manage the tradeoff between computational cost and neural network performance.

To manage the large amount of dataset involved during the training of neural networks, we have used a combination of computing techniques in python programming called data generators and memory mapping. Memory-mapping is a python computing technique used for accessing small segments of large files on disk, without reading the entire file into memory[25]. When it is not practical to load entire training dataset into the machine learning library due to memory limitation, data generators could be used to generate data in batches and continuously feed the data to the machine learning algorithm. In this project, we have memory-mapped very large files containing training datasets and used custom built-in functions called data generators to feed the data in batches to the artificial neural networks.

11.2.4 Previous work

The idea of using artificial intelligence in solving petroleum engineering problems was first introduced by Mohaghegh and Ameri[26]. They showed that an ANN can be used to automate the task of determining formation permeability based on geological well logs, thereby eliminating the need to perform the task repeatedly by log analysts. Mohaghegh and Ameri[26] also stated that neural network can handle far more complex tasks. Mohaghegh et al.[27] used ANN for predicting gas storage well performance after hydraulic fracture in their later investigations. This technology has been used to implement a unique approach to petroleum reservoir modeling by constructing top-down reservoir models ("TDM") that use field measurements (i.e., production history, hydraulic fracturing, well logs, etc.) to predict oil and gas production from shale reservoirs[28].

Data-driven smart proxy models have been used to take advantage of the "big data" solutions (machine learning and pattern recognition) to develop highly accurate replicas of numerical models with very fast response time[29]. Data-driven smart proxies implement machine learning and pattern recognition techniques, using generated numerical simulation data with efforts to significantly reduce the computational footprint and the time spent to conduct large complex numerical simulation runs[30].

Boosari[31] in a study developed a smart proxy model to predict the unsteady state behavior of fluid flow resulting from wall collapse in a two-dimensonal rectangular water tank, using dataset generated from OpenFOAM CFD simulations. The results from the study showed that a smart proxy model can predict the CFD simulation results with less than 10% error, within a significantly reduced amount of time compared to the large computational footprint of the CFD simulations. Ansari et al.[8] used AI and machine learning to construct a smart proxy model to replicate the flow and particle behavior for a gas-solid multiphase flow in a non-reacting rectangular fluidized bed. The smart proxy was developed using cell-level data generated from CFD simulation runs using MFiX CFD simulation software. The work performed by Ansari et al.[8] showed that this technology can reproduce CFD simulation results with less than 10% error within just a few seconds.

11.3 B6 combustor model

The objective of this section of the chapter is to provide a general overview of the end-to-end workflow involved in developing the smart proxy model for the B6 combustor simulation models. This section provides a detailed description of the simulation data received including the boundary conditions, a brief description of descriptive analytical techniques used to develop the smart proxy model and a detailed description of the different machine learning algorithms used in the development process. A more detailed description of specific steps taken to develop the smart proxy model is provided in the next section, together with the results of the smart proxy model at each step of the development process.

11.3.1 B6 combustor problem definition

It is known that relatively small changes in turbine engine ambient conditions and fuel composition can affect the combustion dynamics of operating engines. Combustion dynamics are a form of oscillating combustion that produces pressure oscillations at hundreds of cycles per second. If left uncontrolled, these oscillations can be very damaging, cracking metal combustion liners, triggering flashback, and producing thermal failure from enhanced heat transfer. A combination of numerical models and experimental testing have been used at NETL to investigate the effects of changing ambient conditions and fuel interchangeability (the volume fraction of propane addition to natural gas) on combustion dynamics (instabilities)[32].

The B6 combustor has been used to study the effect of fuel composition on premix turbine combustion as well as to develop a combustion control and diagnostic sensor (CCADS) to in-situ monitor the combustion phenomena. The experimental test rig has a propane blending facility which is used to vary the natural gas composition. The resultant pollutant and dynamic response from lean-premixed gas turbine systems relies heavily on adequate mixing of the fuel and air prior to reaching the reacting zones within the combustor[31]. The objective is to develop a smart proxy model capable of replicating the B6 combustor CFD simulation results of pressure, temperature, nitrogen, carbon-di-oxide, and oxygen distribution in the system, for varying composition and flow rates of air and fuel. Fig. 11.2 shows a picture of the combustor rig[32].

Fig. 11.2 *NETL's B6 combustion rig.*

11.3.2 B6 combustor CFD simulation model

NETL's high-pressure combustion facility test rig is roughly 17.8 cm (7.0 in) diameter combustion chamber and the length of combustion zone is 0.91 m (36 in). It has been used to investigate the effect of fuel interchangeability on combustor performance (the volume fraction of propane addition to the natural gas) by testing different fuel-air equivalence ratios (0.42 ~ 0.48) on combustion performances.

The B6 CFD simulation model is 17.8 cm in diameter, and 0.91 m in length, with a total of 9,291,712 grid cells. Fig. 11.3 shows the schematic of the simulation model. The B6 combustor is a steady state single phase (gas only) combustion model. Preheated air and premixed fuel (main and pilot fuel at room temperature) is injected through the inlet. A set of nozzles at the inlet provide swirling effect for turbulent mixing of the gas as it enters the combustion chamber. A total of eight CFD simulation runs were initially generated with which the smart proxy model was developed. An additional two simulation runs were provided as a blind test following the development of the smart proxy model. CFD simulation runs were generated using the Ansys FLUENT software. Tables 11.1 and 11.2 show the percent fuel composition, and other boundary conditions respectively for the initial 8 CFD simulation runs. Four of the eight simulation runs have methane composition ranging from 88.96% to 89.13%, these were tagged as the "base" composition simulation cases. The other four cases were blended with more propane and have methane composition ranging from

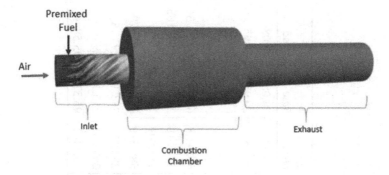

Fig. 11.3 *Schematic of the B6 combustion simulation model.*

Table 11.1 Boundary conditions for development cases-inlet fuel composition.

Case Number	1	2	3	4	5	6	7	8
	Base vol.%	Blend vol.%	Base vol.%	Base vol.%	Blend vol.%	Blend vol.%	Base vol.%	Blend vol.%
CH_4	89.025	84.425	89.113	89.130	84.283	84.153	88.961	84.235
C_2H_6	7.727	7.488	7.636	7.622	7.450	7.510	7.728	7.500
C_3H_8	1.223	6.415	1.197	1.182	6.273	6.329	1.213	6.264
C_4H_{10}	0.468	0.455	0.471	0.472	0.457	0.462	0.479	0.463
C_5H_{12}	0.130	0.125	0.132	0.132	0.125	0.127	0.135	0.128
N_2	1.061	1.006	1.067	1.075	1.047	1.060	1.106	1.051
CO_2	0.366	0.356	0.384	0.387	0.364	0.358	0.378	0.359

84.15% to 84.425%, and these were tagged as the "Blend" cases. The gas composition of air at inlet and the fuel temperature at inlet remained the same across all cases.

Tables 11.3 and 11.4 show the percent fuel composition and other boundary conditions respectively for the additional two CFD simulation runs. One of the simulation runs is a "base" case and the other a "blind" case. These two simulation runs are sometimes referred to as "extra base" and "extra blind" respectively in this study. The gas composition of air at inlet and the fuel temperature at inlet used are the same as was used in the development simulation case runs.

The simulation results (solution data) generated from fluent for each CFD run includes cell-level distribution of five attributes of interest: pressure, temperature, nitrogen, carbon dioxide, and oxygen.

11.3.3 B6 smart proxy development overview

Development of the smart proxy model for the simplified B6 combustor model was carried out in three stages with multiple tasks performed in each stage. The general

Table 11.2 Other boundary conditions for development cases.

Boundary condition	Description	Unit	1 Base	2 Blend	4 Base	4 Base	5 Blend	6 Blend	7 Base	8 Blend
N2_Air_Inlet		%	78.08	78.08	78.08	78.08	78.08	78.08	78.08	78.08
O2_Air_Inlet	Inlet Air Gas	%	20.95	20.95	20.95	20.95	20.95	20.95	20.95	20.95
Ar_Air_Inlet	Composition	%	0.93	0.93	0.93	0.93	0.93	0.93	0.93	0.93
CO2_Air_Inlet		%	0.04	0.04	0.04	0.04	0.04	0.04	0.04	0.04
Air_Flow_Rate	Air Flow Rate	scf/hr	66269.58	66222.6	66272.97	66315.72	66241.71	66241.83	66244.22	66253.56
Main_Fuel_Flow	Fuel Flow Rate	scf/hr	2592.92	2783.134	2948.073	2592.51	2783.444	2448.18	2947.512	2448.165
Pilot_Fuel_Flow			137.099	146.068	155.573	136.815	146.725	129.074	155.804	128.781
Air_Inlet_Temp	Air Temp. at Inlet	deg. K	589.046	588.824	588.615	588.543	588.666	588.579	588.579	588.579
Main_Fuel_Inlet_Temp	Fuel Temp. at Inlet		293.15	293.15	293.15	293.15	293.15	293.15	293.15	293.15
Pilot_Fuel_Inlet_Temp		deg.k	293.15	293.15	293.15	293.15	293.15	293.15	293.15	293.15
Pressure_Outlet	Pressure at Outlet	kPa	95.4662	95.4687	95.5355	95.4484	95.5262	95.4461	95.5731	95.5846

Table 11.3 Boundary conditions for blind validation cases-inlet fuel composition.

Case number	9 Blind validation run Base vol.%	10 Blind validation run Blend vol.%
CH_4	89.122	84.225
C_2H_6	7.600	7.600
C_3H_8	1.200	6.200
C_4H_{10}	0.471	0.468
C_5H_{12}	0.132	0.130
N_2	1.102	1.009
CO_2	0.373	0.368

Table 11.4 Other boundary conditions for blind validation cases.

Boundary condition	Description	Unit	9 Blind base	10 Blind blend
N2_Air_Inlet		%	78.08	78.08
O2_Air_Inlet	Inlet air gas	%	20.95	20.95
Ar_Air_Inlet	Composition	%	0.93	0.93
CO2_Air_Inlet		%	0.04	0.04
Air_Flow_Rate	Air flow rate	scf/hr	66280	66240
Main_Fuel_Flow		2750	2599	
Pilot_Fuel_Flow	Fuel flow rate	scf/hr	142	135
Air_Inlet_Temp	Air temp. at inlet	deg. K	588.772	588.66
Main_Fuel_Inlet_Temp		293.15	293.15	
Pilot_Fuel_Inlet_Temp	Fuel temp. at inlet	deg.K	293.15	293.15
Pressure_Outlet	Pressure at outlet	kPa	95.5	95.5

overview of these stages is shown in Fig. 11.4 below. The first stage involves performing data quality check and building a data visualization tool for the project. In the second stage, a detailed exploratory and descriptive analysis of the data was performed, and the predictive model developed in the last stage.

It is important to note that the descriptive and predictive modeling tasks of the framework were carried out in multiple steps, such that the modeling approach was continuously refined based on the resulting performance of the smart proxy model at each step. At each step, new features were generated and added to the database. A summary of the features that were added at each modeling step is provided in the descriptive analytics section of this section (3.3.3). A more detail description of these features with the corresponding smart proxy model results at each step is provided in the next section.

11.3.3.1 Data received from CFD simulation runs

In addition to the B6 model boundary condition information described in Section 11.3.2, other data received from NETL include the solution data and model geometry

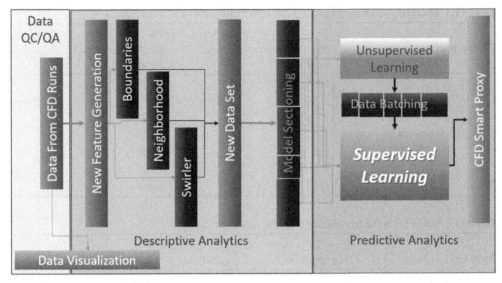

Fig. 11.4 *Smart proxy development framework for the B6 combustion model.*

data. The solution data contains cell level information on the distribution of each of the five transport variables of interest. The model geometry data contained the cell id, cell center coordinates (x, y, z), adjacent cell ids (i.e. ID of neighboring cells), and node ids of each cell in the simulation model. Data received for each simulation run contained a total of 21 files in *.txt* file format with a total hard disk storage size of about 1.7 GB per simulation run. The boundary condition data was received in a single *.csv* file.

11.3.3.2 Data visualization tool
In order to check the quality of data received and for the purpose of validating the results of the B6 smart proxy model to be developed, a python script was developed to transform the data received (originally stored in *.txt* file format) into a file format in which the model results can be visualized in three dimensions. As illustrated in Fig. 11.5 below, the data was processed into *.vtk* (visualization tool kit) format files which were then imported into ParaView for visualization. ParaView is an open-source data analysis and visualization application which allows for 3D interactive data exploration.

The B6 combustor model simulation mesh contains a total of 9,291,712 elements (i.e., cells) with about eight million tetrahedral cells and approximately 950,000 wedge cells. Fig. 11.6 provides a brief description of these cell types in terms of the number of faces and points.

Based on the entire simulation dataset collected from Fluent and the model boundary conditions described in Section 11.3.2, a structured database was generated for each CFD simulation case (for a total of eight development cases). The following provides a description of the data received for each case run as shown in Fig. 11.5.

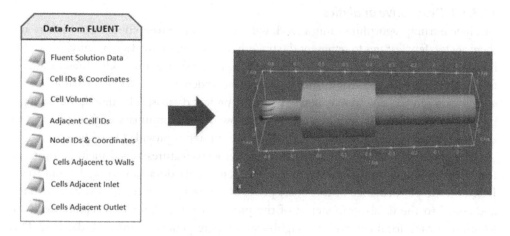

Fig. 11.5 *ParaView image generated from *.vtk files built from CFD simulation data.*

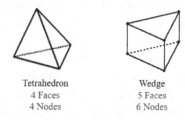

Fig. 11.6 *B6 combustor simulation mesh cell types.*

1. Fluent solution data: These set of files contain the output of CFD simulation calculations performed by Fluent for the five different transport variables of interest, at the cell level: pressure (P), temperature (T), oxygen concentration (O_2), nitrogen concentration (N_2), and carbon dioxide concentration (CO_2).
2. Cell IDs and coordinates: These files contain the ID and cell center coordinates (x, y, z) for each element (i.e., focal cell) in the mesh.
3. Cell volume: File contains the volume of each focal cell in the mesh.
4. Adjacent cell IDs: These files contain the ID of face-bounding (adjacent by face) cells for every focal cell in the mesh.
5. Node IDs and coordinates: These files contain the ID and coordinates of points (nodes) defining each focal cell in the mesh.
6. Cells adjacent to walls: These files contain the ID of cells that are adjacent to the external walls of the model.
7. Cells adjacent to inlet: These files contain the ID of cells that are adjacent to the inlet wall.
8. Cells adjacent to outlet: These files contain ID of cells that are adjacent to the outlet wall.

11.3.3.3 Descriptive analytics

Machine learning algorithms cannot work without data. Little to nothing can be achieved if there are too few features to represent the underlying pattern in the data to a machine learning algorithm. Comprehensive descriptive analytics of the simulation model dataset was performed and more features that further represents the underlying physics and dynamics of the combustion phenomena were generated into the project database. The descriptive analytics process was completed in four iterative steps. Below is a brief summary of features added at each step. More detail information and results at each step is provided in section four.

In the first step of the descriptive analytics process, features added include information that represent the geometry of the cells and their distances to wall boundaries. Based on the performance of the smart proxy at this step, more features were generated and added to the database in step 2 of the process. In step 2, features representing the location of each focal cell and its neighborhood were generated into the database. This included information on the volumes of neighboring (face-bounding) cells, and information indicating if the focal cell was bounding a wall or not. In a realistic combustor, the overall rate of reaction is controlled by turbulent mixing. Mixing does not only occur as a result of a jet crossflow of the preheated air and fuel at the inlet but typically through some additional mechanism such as swirling flow which increases the turbulence intensity[32]. Features generated in step 3 of the process included a representation of this swirling effect on the combustion process. These include distances of each focal cell in the system to each of the swirler nozzles (see detail in section four). In step 4, an unsupervised machine learning technique called fuzzy clustering was applied to the entire system in order to group the over 9 million cells into classes based on the distribution of each transport variable of interest across all development simulation cases. This technique helps to provide more information to the training algorithm regarding the behavior of the B6 combustor simulation models. More detail regarding fuzzy clustering is provided in the predictive analytics section (Section 11.3.3.4) of this chapter.

11.3.3.3.1 B6 combustor model sectioning

The computer machine available to the research team at the time of the project had 500GB hard disk storage, and 24GB of memory. Generating dataset required to train the smart proxy from the database required a significant amount of memory larger than the computer memory that was available to the research team. In fact, in steps 1 and 2 of the descriptive analytics process, the smart proxy model was developed for a small section of the B6 Combustor model at each step. This made it possible to complete the training of the model in a shorter length of time and quickly assess the contribution of newly added features at each step of the process. Due to this limitation in compute resource (memory and speed), the B6 Combustor simulation model was divided into seven sections in step 3, based on a detailed analysis of the distribution of the transport variables across all simulation runs. Fig. 11.7 shows the model sectioning and naming convention for each section. Table 11.5 shows the number of cells by section for a single

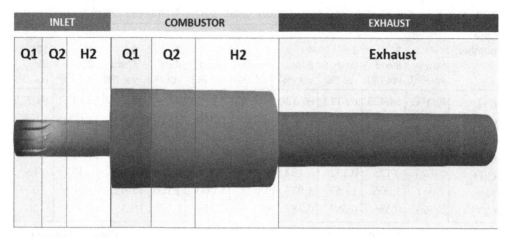

Fig. 11.7 *B6 combustor model sectioning.*

Table 11.5 Total number of cells by model section.

Total no. of cells by model section

Section name	Convention name	No. cells	Z-axis BGN	Z-axis END
1st quarter inlet	Q1_Inlet	1,623,745	-0.83058	-0.78868
2nd quarter inlet	Q2_Inlet	1,662,119	-0.78868	-0.74678
2nd half inlet	H2_Inlet	344,757	-0.74678	-0.66298
1st quarter combustor	Q1_Combustor	962.001	-0.66298	-0.58981
2nd quarter combustor	Q2_Combustor	971,256	-0.58981	-0.51664
2nd half combustor	H2_Combustor	1,898,924	-0.516642	-0.37030
Exhaust	Exhaust	1,903,951	-0.37030	0.0762
Entire mopdel	CFD	9,291,712	-0.83058	0.0762

CFD simulation run. For each transport variable of interest, an artificial neural network was trained for each section of the model and these networks were later combined to form the smart proxy model for the B6 combustor. Tables 11.6 and 11.7 show boundary conditions for B6 combustor model. More information is provided in Section 11.3.3.4 regarding the setup of neural networks that were developed.

11.3.3.4 Predictive analytics

This section presents the details of the machine learning algorithms used in learning the underlying pattern in the dataset, and the approach taken to validate the performance of the developed smart proxy model.

Both supervised and unsupervised learning methods were used to meet the objective of the project. The unsupervised learning technique involves a cluster analysis of the dataset based on the cell-level distribution of each CFD simulation output of interest

Table 11.6 Boundary conditions for B6 combustor simulation model-fuel composition.

Case number	1 Base vol. (%)	2 Blend vol. (%)	3 Base vol. (%)	4 Base vol. (%)	5 Blend vol. (%)	6 Blend vol. (%)	7 Base vol. (%)	8 Blend vol. (%)	9 Base vol. (%)	10 Blend vol. (%)
CH_4	89.025	84.425	89.113	89.130	84.283	84.153	88.961	84.235	89.122	84.225
C_2H_6	7.727	7.488	7.636	7.622	7.450	7.510	7.728	7.500	7.600	7.6
C_3H_8	1.223	6.415	1.197	1.182	6.273	6.329	1.213	6.264	1.200	6.2
C_4H_{10}	0.468	0.455	0.471	0.472	0.457	0.462	0.479	0.463	0.471	0.468
C_5H_{12}	0.130	0.125	0.132	0.132	0.126	0.127	0.135	0.128	0.132	0.130
N_2	1.061	1.006	1.067	1.075	1.047	1.060	1.106	1.051	1.102	1.009
CO_2	0.366	0.356	0.384	0.387	0.364	0.358	0.378	0.359	0.373	0.368

▀▀▀ Validation case   ▀▀▀ Blind validation case

across multiple simulation runs, while the supervised learning technique involves the use of artificial neural networks to learn the pattern in the dataset.

11.3.3.4.1 Data partitioning

Before any learning (that is model training) was performed (whether supervised or unsupervised), data partitioning approach was decided and carefully implemented to avoid data leakage and ensure true confidence in the performance of the smart proxy to be developed. The importance of data partitioning has been highlighted in Section 11.2.3.1 in the background section of this study.

Fig. 11.8 illustrates the data partitioning approach used in the development of the smart proxy. We have taken a "double blind" approach. To successfully validate the performance of the proposed smart proxy, the two blind validation runs (referenced as blind validation base–case 9 and blind validation blend – case 10 in Section 11.3.2) were completely kept out for double blind validation of the neural networks to be developed. During the development phase (i.e., neural network training), two (base case 1 and blend case 5) out of the eight development simulation cases were used as validation to assess the performance of the neural network before adjudging it as being completely developed. So only a total of six CFD simulation runs was used in training the final smart proxy model. The smart proxy was deployed on the blind validation cases 9 and 10 only after completing all of the four development steps identified in the Descriptive Analytics section. The detail smart proxy results for the blind validation cases (9 and 10) is provided in section four, as well as the results of performance on the validation cases (1 and 5) at every step of the development process.

Table 11.7 Additional boundary conditions for B6 combustor simulation model.

Case numer	1	2	3	4	5	6	7	8	9	10
	Base	Blend	Base	Base	Blend	Blend	Base	Blend	Blind base	Blind blend
N2 air inlet (%)	78.08	78.08	78.08	78.08	78.08	78.08	78.08	78.08	78.08	78.08
O2 air inlet (%)	20.95	20.95	20.95	20.95	20.95	20.95	20.95	20.95	20.95	20.95
Argon air inlet (%)	0.93	0.93	0.93	0.93	0.93	0.93	0.93	0.93	0.93	0.93
CO2 air inlet (%)	0.04	0.04	0.04	0.04	0.04	0.04	0.04	0.04	0.04	0.04
Air flow rate (scf/h)	66269.58	66222.59	66272.97	66315.71	66241.711	66241.8	66244.22	66252.56	66280.00	66240.00
Main fuel flow rate (scf/h)	2592.92	2783.134	2948.073	2592.51	2783.444	2448.18	2947.512	2448.165	2750	2599
Pilot fuel flow rate (scf/h)	137.099	146.068	155.573	136.815	146.725	129.074	155.804	128.781	142	135
Air inlet temperature (°K)	589.046	588.824	588.615	588.543	588.666	588.579	588.579	588.579	588.772	588.66
Main fuel inlet temperature (°K)	293.15	293.15	293.15	293.15	293.15	293.15	293.15	293.15	293.15	293.15
Pilot fuel inlet temperature (°K)	293.15	293.15	293.15	293.15	293.15	293.15	293.15	293.15	293.15	293.15
Pressure outlet (psi)	95.4662	95.4687	95.5355	95.4484	95.5262	95.4461	95.5731	95.5846	95.5	95.5

Validation case

Blind validation case

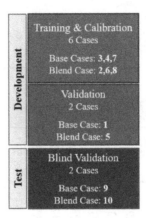

Fig. 11.8 *Data partitioning.*

11.3.3.4.2 Fuzzy clustering

A background to fuzzy clustering is already provided in Section 11.2.3.1 (in the background section of this study). Fuzzy clustering helps to discover distribution of patterns in datasets. When meaningful patterns are identified, it can be very valuable input for a neural network to learn. The objective of the fuzzy clustering analysis performed in this project is to find the distribution pattern of each transport variable (i.e., CFD simulation output) of interest (for example, Pressure) across the simulation case runs at the cell level in the system. Clustering involves assigning the cells to clusters (groups) such that cells in the same cluster are as similar as possible, in terms of the distribution of each transport variable of interest. Table 11.8 shows an example of the inputs and output information used in the analysis for n number of cells, where n is the total number of cells in the specific model section being developed or trained. The input to the clustering algorithm is the number of clusters to group the cells into, and cell-level values of the attribute of interest across the training and calibration cases. For example, $P_{3,2}$ refers to the pressure

Table 11.8 Input and output data to the fuzzy clustering algorithm.

| Cell ID | Cell pressure values by CFD case | | | | | | Cluster memberships & ID | | | |
	2	3	4	6	7	8	CM1	CM2	CM3	Cluster ID
1	$P_{1,2}$	$P_{1,3}$	$P_{1,4}$	$P_{1,6}$	$P_{1,7}$	$P_{1,8}$	$CM_{1,1}$	$CM_{1,2}$	$CM_{1,3}$	$f(Max(CM_{1,1},CM_{1,2},CM_{1,3})$
2	$P_{2,2}$	$P_{2,3}$	$P_{2,4}$	$P_{2,6}$	$P_{2,7}$	$P_{2,8}$	$CM_{2,1}$	$CM_{2,2}$	$CM_{2,3}$	$f(Max(CM_{2,1},CM_{2,2},CM_{2,3})$
3	$P_{3,2}$	$P_{3,3}$	$P_{3,4}$	$P_{3,6}$	$P_{3,7}$	$P_{3,8}$	$CM_{3,1}$	$CM_{3,2}$	$CM_{3,3}$	$f(Max(CM_{3,1},CM_{3,2},CM_{3,3})$
•	•	•	•	•	•	•	•	•	•	•
•	•	•	•	•	•	•	•	•	•	•
•	•	•	•	•	•	•	•	•	•	•
•	•	•	•	•	•	•	•	•	•	•
n	$P_{n,2}$	$P_{n,3}$	$P_{n,4}$	$P_{n,6}$	$P_{n,7}$	$P_{n,8}$	$CM_{n,1}$	$CM_{n,2}$	$CM_{n,3}$	$f(Max(CM_{n,1},CM_{n,2},CM_{n,3})$

value in cell 3, simulation case 2. For this problem we have used three clusters and only the six training and calibration cases identified in Data Partitioning section above as input. The output is the degree of membership (membership weight) of each cell in each of the clusters (shown in cluster membership columns CM1, CM2, CM3 in Table 11.7). In practice, fuzzy clustering is often converted to an exclusive clustering by assigning each data point to the cluster in which its membership weight is highest. For this problem, we have added the exclusive clustering information (cluster ID column in Table 11.8) to the fuzzy clustering information (CM1, CM2, and CM3 columns) by assigning each cell to the cluster in which it has the highest membership. As mentioned in the Descriptive Analytics section (Section 11.3.3.3), clustering analysis was applied in the last step of the development process and so, these two pieces of information (total four attributes) were included as input to every neural network model that was developed for the final B6 smart proxy model.

11.3.3.4.3 Artificial neural network setup

A neural network was trained, calibrated and validated for each model section and transport variable (a total of 35 neural networks). These neural networks were then coupled together to form the B6 smart proxy model. 80% of the samples (randomly selected) from the six training and calibration cases (base cases 3,4,7 and blend cases 2,6,8) identified in the data partitioning section was used to train the neural networks while the remaining 20% was used to continuously calibrate the neural network and check when to stop training in order to avoid overfitting. Neural network hyper-parameters were continuously tuned for better performance on validation cases 1 and 5.

All neural networks trained were single hidden layer, one output networks. The rectilinear activation function was used in all neural network models. The number of hidden neurons used varied from 15 to 2500 and the total number of training epochs completed for any attribute and model section varied from 15 to 100. The smart proxy model was developed for pressure, temperature, carbon dioxide, oxygen, and nitrogen.

11.3.3.4.4 Data batching

Though dividing the model into seven smaller sections helped in managing the compute resource limitation by preparing the training dataset section by section, the total number of training samples by section was still too large to fit into the computer memory to train a neural network. This problem was addressed by using a combination of computing techniques called memory-mapping and data generators. The basis for these techniques has been described in Section 11.2.3.4 in the background section of this study. For each section of the model, file containing the training dataset was memory-mapped to the virtual memory of the machine and the data fed in batches to the neural network for training. Memory-mapping is a python computing technique used for accessing small segments of large files on disk, without reading the entire file into memory[25]. When it is not practical to load entire training dataset into the machine learning library due to memory limitation, data generators could be used to generate data in batches and continuously feed the data to the machine learning algorithm. Different training data batch sizes (ranging from 5,000 to about 200,000 samples) were tested in other to optimize the performance of the neural networks.

11.4 Smart proxy development steps

In the previous section, we presented a general overview of the tasks performed at different stages in the project – Data quality check and visualization, Descriptive analytics, and Predictive analytics. Data received directly from the CFD simulation runs does not have enough information for the neural network to train as-is in the predictive analytics stage. Multiple steps were taken to generate features that represent the flow, reaction, and heat transfer phenomenon occurring in the combustion chamber, so as to assist the neural network to learn the underlying pattern in the distribution of the transport and species variables (pressure, temperature, nitrogen, oxygen, and carbon dioxide) throughout the combustor system.

The descriptive and predictive modeling stages of the process were carried out in multiple steps, such that the modeling approach was continuously refined based on the resulting performance of the smart proxy model following the inclusion of newly generated features at each step. These steps are summarized into four main steps listed below and are labelled based on the description of what features were generated into the database and included in the neural network training at each step.

- Step 1 – cell geometry and distances to wall boundaries
- Step 2 – cell neighborhood, location and euclidian distances to wall boundaries
- Step 3 – swirler distances
- Step 4 – fuzzy clustering

Detail description of work performed and smart proxy results for each step are presented in this section.

11.4.1 Model development step 1: Cell geometry and distances to wall boundaries

In gas combustion chambers such as the B6 combustor, a large proportion of the total heat flux to the walls of the combustor is by radiation from the flame[33]. Radiation exchange between surfaces in addition to their radiative properties and temperatures strongly depends on the surface's geometries, orientations, and separations distance[34]. The closer a cell is to the source of heat in the combustor, the greater the intensity of radiation energy received.

In step 1 of the development process, features representing the geometry and location of each cell were generated into the database. The cell geometry attribute is represented by the number of nodes bounding each focal cell in the model, which specifies if a cell is a tetrahedron or a wedge. The cell location attribute specifies in which section of the system each focal cell is located (whether in inlet, combustor, or exhaust) and additionally specifies how close to the wall the focal cell is. To generate these features, four boundaries were identified as inlet, outlet, inlet-to-combustor, combustor-to-exhaust. The inlet and outlet boundaries refer to the walls bounding the inlet and outlet of the system respectively. The inlet-to-combustor refers to the boundary where the system geometry transitions from the inlet into the combustion chamber while the combustor-to-exhaust refers to the boundary where the system geometry transitions from the combustion chamber into the exhaust section. Since the three main sections of the system have different diameters, the radial distance of each focal cell to the radial boundary must be accounted for. The distances of each focal cell to the nearest radial boundary and farthest radial boundary was calculated and added to the database. A total of seven new features were added to the database in step 1.

Figs. 11.9-11.11 illustrate how the distances of cells to the boundaries, were calculated for arbitrary cells located at the inlet, combustion chamber, and exhaust, respectively.

A 500 GB storage, 4 cores, and 24 GB RAM desktop computer was used in preparing the training dataset and training the neural networks. In order to address the limitations in compute memory and speed considering the large amount of data being processed, only the inlet section of the system (as shown in Fig. 11.12) was used in model development in step 1. The total number of cells at the inlet section is approximately

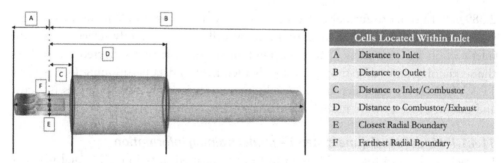

Cells Located Within Inlet	
A	Distance to Inlet
B	Distance to Outlet
C	Distance to Inlet/Combustor
D	Distance to Combustor/Exhaust
E	Closest Radial Boundary
F	Farthest Radial Boundary

Fig. 11.9 *Calculated distances of cells located at the inlet.*

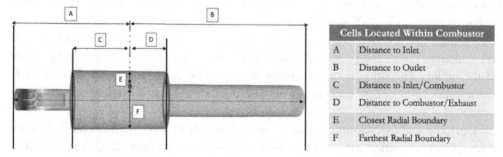

Cells Located Within Combustor	
A	Distance to Inlet
B	Distance to Outlet
C	Distance to Inlet/Combustor
D	Distance to Combustor/Exhaust
E	Closest Radial Boundary
F	Farthest Radial Boundary

Fig. 11.10 *Calculated distances of cells located at the combustion chamber.*

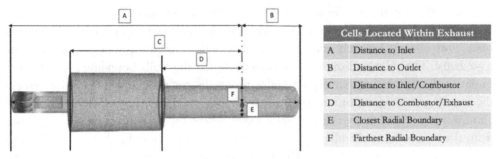

Cells Located Within Exhaust	
A	Distance to Inlet
B	Distance to Outlet
C	Distance to Inlet/Combustor
D	Distance to Combustor/Exhaust
E	Closest Radial Boundary
F	Farthest Radial Boundary

Fig. 11.11 *Calculated distances of cells located at the exhaust.*

Fig. 11.12 *Step 1 – target section for training and development of the neural network.*

2,389,668. In order to further address the memory limitation problem and increase the processing and development time, two million cells were randomly selected from the inlet section for processing. The same random set of cells were selected for each of the six training and calibration cases already identified in data partitioning (See Section 11.3.3.4). As a first trial step in the development process, only two attributes (pressure and carbon dioxide) were modeled and analyzed.

11.4.1.1 Model development step 1 – model training information

For each attribute, a total of 23 input features were used in training a neural network for the randomly selected cells at the inlet section of the system. Table 11.9 shows the list of input attributes. 16 out of the 23 features were originally provided as part of data exported from FLUENT software. These includes focal cell location and geometry information (X, Y, and Z coordinates and cell volume), and simulation model boundary conditions (fuel composition, air and fuel flow rates and temperature, and pressure at the outlet). The other seven attributes were generated in step 1, and these include the number of nodes on each focal cell, and distances to the boundaries described in Section 11.4.1.

Only the eight development cases (cases 1 through 8) were used in step 1. The six training and calibration cases (base cases 3, 4, 7 and blended cases 2, 6, 8) identified in Section 11.3.3.4 were used in training the neural network. A total of 12 million samples

Table 11.9 Training input attributes in development step 1.

Attributes provided from FLUENT	Attributes generated in step 1
X_Coord	Node_Count
Y_Coord	Dist_To_Inlet
Z_Coord	Dist_To_Outlet
volume	Dist_To_Inlet_Combustor
CH4_Inlet	Dist_To_Combustor_Exhaust
C2H6_Inlet	Closest_Radial_Boundary
C3H8_Inlet	Farthest_Radial_Boundary
C4H10_Inlet	
C5H12_Inlet	
N2_Inlet	
CO2_Inlet	
Air_Flow_Rate	
Main_Fuel_Flow	
Pilot_Fuel_Flow	
Air_Inlet_Temp	
Pressure_Outlet	
Total = 16	Total = 7

(2 million cells from inlet section across all six cases) were used in training and calibration. Specifically, 80% of the samples (selected at random) were used in training while the remaining 20% were used as calibration samples. The base case 1 and blind case 5 were used for validating each neural network after each training attempt. The neural network hyper-parameters were continuously tuned for better performance on the validation cases 1 and 5.

For each attribute, a single hidden-layer neural network with one output was built. The size of the training dataset (approximately 6.39 GB) could not fit in memory of the desktop machine being used to train the neural networks and therefore, computing techniques mentioned in Section 11.2.3.4 (data batching) in the background section of this study, were used to address this challenge. File containing the training dataset (for an attribute) was memory-mapped to the virtual memory of the machine and the data fed in batches to the neural network for training. Different training data batch sizes were tested in other to optimize the performance of the neural networks. Batch sizes tested range from about 10,000 samples/batch to 100,000 samples/batch). The rectilinear activation function was used in the hidden and output layers. Different number of hidden neurons was also tested to optimize neural network performance. Number hidden neurons tested ranged from about 50–200 neurons.

11.4.1.2 Model development step 1 – presentation of results

In this section, the smart proxy results for the distribution of carbon-dioxide in the inlet section for the two validation cases (base case 1 and blend case 5) are presented.

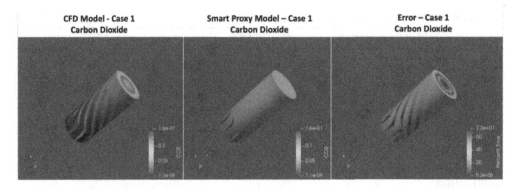

Inlet Exterior

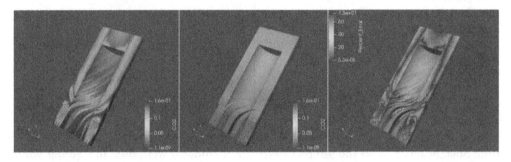

Inlet Interior

Fig. 11.13 *Step 1 smart proxy results for validation case 1 – carbon dioxide.*

Two figures (Figs. 11.13 and 11.14) are displayed in which the first figure shows the distribution of carbon-dioxide in the exterior of the inlet section while the second figure shows a cross-section of the inlet when cut in half. The second figure shows the distribution of carbon dioxide in the interior parts of the inlet section. In both figures, the first image (on left) shows the result from the CFD simulation model as obtained from FLUENT, followed by the results of the smart proxy model (middle) while the last image (on right) is the percent error plot comparing the CFD simulation result with the smart proxy result. For the carbon-dioxide and other gas species, percent error is calculated by taking the absolute difference between the CFD simulation result and the smart proxy result.

11.4.2 Model development step 2: Cell neighborhood, location, and Euclidian distances to wall boundaries

Results obtained in step 1 showed that there was still not enough information for the neural network to learn the underlying pattern in the system. In FLUENT software, as with many other numerical simulation applications, the value of a dynamic variable or

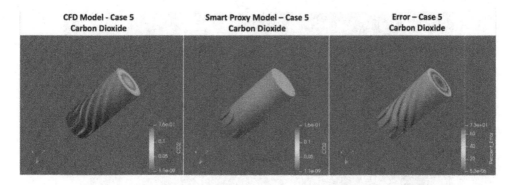

Inlet Exterior

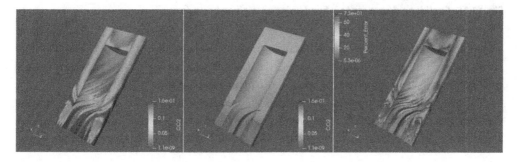

Inlet Interior

Fig. 11.14 *Step 1 smart proxy results for validation case 5 – carbon dioxide.*

properties in any given grid or cell is mostly impacted by the value of the variable in neighboring cells. In addition to already existing features, more features further representing the location of each cell and its neighborhood were therefore generated into the database.

In step 1, the calculated distances to the inlet, inlet-combustor, combustor-exhaust and outlet boundaries were straight horizontal distances. In order to further communicate focal cell locations to the neural network, the Euclidean distances to a fixed point (center) on these boundaries were calculated and added to the database. Fig. 11.15 is an illustration of how the Euclidean distance was calculated for an arbitrary cell at position x, y, z (represented by the yellow circle) to the center (represented by the red circle) of the inlet-combustor boundary. The closest and farthest radial distances were calculated in step 1, the ratio of these distances was calculated and added to the database in step 2.

In order to represent focal cell neighborhood information, the volume of up to any four face surrounding cells was included for each focal cell. In addition to the volume of surrounding cell attributes, the total number of adjacent (face-bounding) cells was

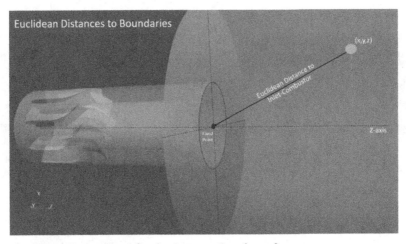

Fig. 11.15 *Euclidean distances to a fixed point on a given boundary.*

added and features representing whether a focal cell is bounding the inlet, outlet or any surrounding wall were also added to the database. A total of 13 new features (as shown in Table 11.10) were generated and added to the database in step 2.

Recall that random sampling of cells in the inlet section was necessary to address the compute resource limitation challenge in order to train the neural networks in step 1. In order to avoid having to train with a sampled dataset, it was decided to perform the

Table 11.10 Training input attributes in development step 2.

Attributes provided from FLUENT	Attributes generated in step 1	Attributes generated in step 2
X_Coord	Node_Count	Euclidean_Dist_Inlet
Y_Coord	Dist_To_Inlet	Euclidean_Dist_Inlet_Comb
Z_Coord	Dist_To_Outlet	Euclidean_Dist_Comb_Exhaust
volume	Dist_To_Inlet_Combustor	Euclidean_Dist_Outlet
CH4_Inlet	Dist_To_Combustor_Exhaust	Radial_Boundary Ratio
C2H6_Inlet	Closest_Radial_Boundary	Adjacent_Cell_Volume_1
C3H8_Inlet	Farthest_Radial_Boundary	Adjacent_Cell_Volume_2
C4H10_Inlet		Adjacent_Cell_Volume_3
C5H12_Inlet		Adjacent_Cell_Volume_4
N2_Inlet		Total_Adjacent_Cells
CO2_Inlet		Cell_Adjacent_Inlet
Air_Flow_Rate		Cell_Adjacent_Outlet
Main_Fuel_Flow		Cell_Adjacent_Wall
Pilot_Fuel_Flow		
Air_Inlet_Temp		
Pressure_Outlet		
Total = 16	Total = 7	Total = 13

Half Combustor
Used in training

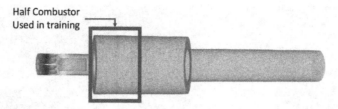

Fig. 11.16 *Step 2 – target section for training and development of the neural network.*

next development analysis with a section of the model where a sizeable portion of the system can be analyzed without exceeding the memory capacity of the machine being used. The first half combustion chamber portion of the system has approximately two million cells while it has about the same length along the Z-axis as the inlet section. As shown in Fig. 11.16, the first half of the combustion chamber was therefore selected for analysis in step 2 of the development process.

11.4.2.1 Model development step 2 – model training information
In step 2 of the development process, only the pressure distribution was modeled. The data partitioning approach is as described in the model development step 1. Six cases (2, 3, 4, 6, 7, 8) were used for training and calibration of the neural network while two cases (base case 1 and blend case 5) were used as validation after each training attempt.

The total number of training and calibration samples was 12,034,230 (6 cases x 2,005,705) while 4,011,410 (2 cases x 2,005,705) samples was used to validate the neural network. As mentioned in Step 1, the data batching technique was used at this step to feed data in batches to the training algorithm, and a single hidden–layer neural network with one output was built. A total of 36 input attributes were used in training the neural network as shown in Table 11.10.

11.4.2.2 Model development step 2 – presentation of results
The following figures (Figs. 11.17 and 11.18) show the results of the smart proxy model developed in step 2. The results presented include the pressure distribution for the validation cases (1 and 5) and two of the training cases (2 and 6).

Each figure contains a total of 12 images: the first six images show an exterior view of the model for two cases while the following six images show an interior view of the model for the same two cases. Moreover, the left-hand side images represent the actual CFD model simulation generated from Ansys Fluent, the middle images represent the smart proxy model, and the right-hand side images show the error difference between the actual CFD model and the smart proxy model.

Results for the remaining attributes of interest were not generated as the objective was to evaluate the performance of the developed neural network.

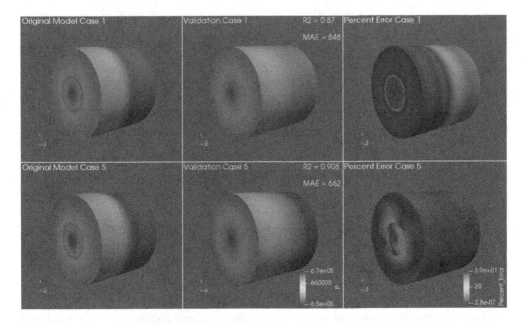

Half Combustor Exterior

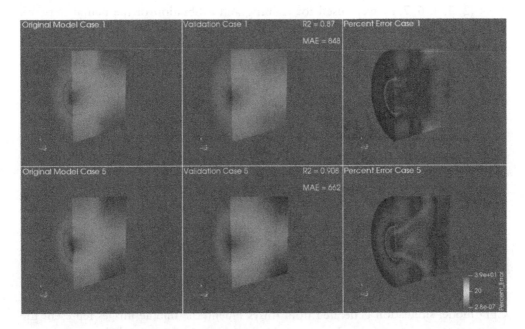

Half Combustor Exterior

Fig. 11.17 *Step 2 smart proxy results for training cases 1 and 5 – pressure.*

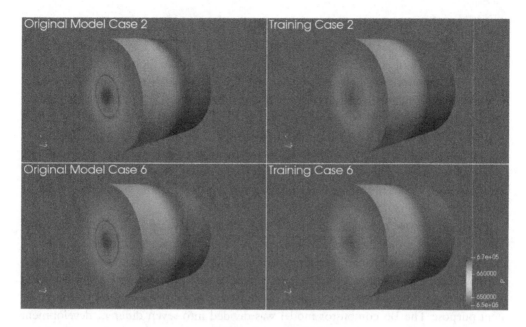

Half Combustor Exterior

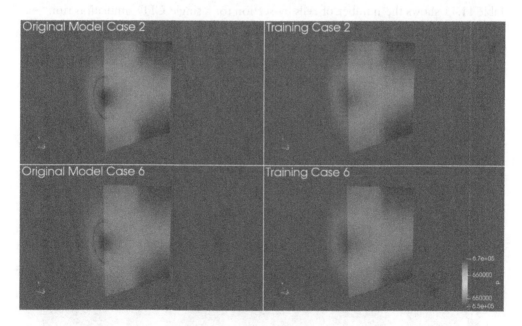

Half Combustor Interior

Fig. 11.18 *Step 2 smart proxy results for training cases 2 and 6 – pressure.*

11.4.3 Model development step 3: Swirler distances

In a realistic combustor, the overall rate of reaction is controlled by turbulent mixing. Mixing does not only occur as a result of a jet crossflow of the preheated air and fuel at the inlet but typically through some additional mechanism such as swirling flow which increases the turbulence intensity[32]. The results obtained in Step 2 showed that the neural network was missing the reaction flow pattern (especially at the inlet–combustor boundary) created by the air-fuel mixing effect of the swirlers at the inlet. The newly generated features in Step 3 included a representation of this swirling effect on the combustion process. These includes calculating the distances of each focal cell in the system to each of the thirteen swirler nozzles. Fig. 11.19 illustrates how these distances are calculated for an arbitrary cell in the system.

Based on what was learned in steps 1 and 2 of the development process, the total number of samples that the available desktop machine could handle in development was already known. This information, coupled with a detail analysis of the distribution of the transport variables of interest (pressure, temperature, oxygen, nitrogen and carbon-dioxide) across all development cases was used in dividing the system into sections for development purpose. The B6 combustor model was divided into seven different development sections. Fig. 11.20 shows the model sectioning and naming convention for each section. Table 11.11 shows the number of cells by section for a single CFD simulation run.

Once the model was properly sectioned, the section with the smallest number of cells, the first Quarter Combustor (Q1-Combustor) as shown in Fig. 11.21 was selected as the section to be modeled in the third step of the development process. The number

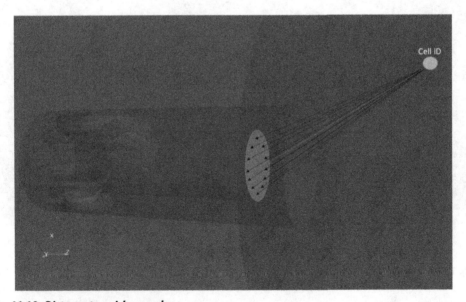

Fig. 11.19 *Distance to swirler nozzles.*

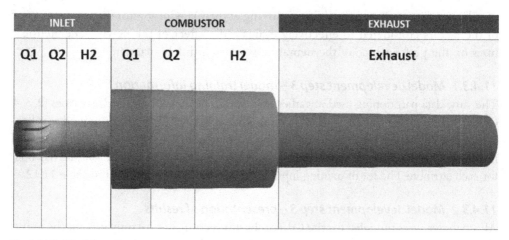

Fig. 11.20 *Model sectioning.*

Table 11.11 Total number of cells by model section.

Total no. of cells by model section

Section name	Convention name	No. of cells	Z-axis BGN	Z-axis END
1st quarter inlet	Q1_Inlet	1,623,745	−0.83058	−0.78868
2nd quarter inlet	Q2_Inlet	1,662,119	−0.78868	−0.74678
2nd half inlet	H2_Inlet	344,757	−0.74678	−0.66298
1st quarter combustor	Q1_Combustor	962,001	−0.66298	−0.58981
2nd quarter combustor	Q2_Combustor	971,256	−0.58981	−0.51664
2nd half combustor	H2_Combustor	1,898,924	−0.516642	−0.37030
Exhaust	Exhaust	1,903,951	−0.37030	0.0762
Entire mode	CFD	9,291,712	−0.83058	0.0762

First Quarter Combustor
Used in training

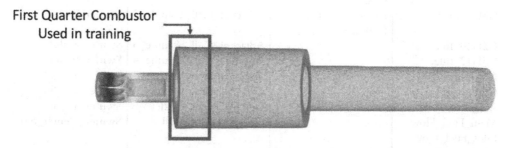

Fig. 11.21 *Step 2 – target section for training and development of the neural network.*

of cells and the combustion activity occurring at the Q1-Combustor section were some of the factors considered to effectively evaluate the effect of the newly generated features on the performance of the neural network in a timely manner.

11.4.3.1 Model development step 3 – model training information

The same data partitioning used in earlier steps was implemented in step 3. Six cases (2, 3, 4, 6, 7, 8) were used for training and calibration while two cases (1 and 5) were used as validation after each training attempt. All five attributes of interest (P, T, N_2, O_2, CO_2) were modeled in this step. Furthermore, a single hidden-layer neural network with one output was built for each attribute. The list of training input attributes in step 3 is provided in Table 11.12.

11.4.3.2 Model development step 3 – presentation of results

All attributes were modeled for the Q1 combustor section, but we only present the results for pressure, temperature, and carbon dioxide distributions for the two validation cases base case 1 and blend case 5 (Figs. 11.22–11.33). Similar quality of results was obtained for nitrogen and oxygen and these are presented in Appendix 11.7 (Figs. 11.34–11.63).

Table 11.12 Training input attributes in development step 3.

Attributes provided from Fluent	Attributes generated in step 1	Attributes generated in step 2	Attributes generated in step 3
X_Coord	Node_Count	Exclidean_Dist_Inlet	Swirler_1_dist
Y_Coord	Dist_To_Inlet	Euclidean_Dist_Inlet_Comb	Swirler_2_dist
Z_Coord	Dist_To_Outlet	Euclidean_Dist_Comb_Exhaust	Swirler_3_dist
volume	Dist_To_Inlet_Combustor	Euclidean_Dist_Outlet	Swirler_4_dist
CH4_Inlet	Dist_To_Combustor_Exhaust	Radial_Boundary_Ratio	Swirler_5_dist
C2H6_Inlet	Closest_Radial_Coundary	Adjacent_Cell_Volume_1	Swirler_6_dist
C3H8_Inlet	Farthest_Radial_Boundary	Adjacent_Cell_Volume_2	Swirler_7_dist
C2H10_Inlet		Adjacent_Cell_Volume_3	Swirler_8_dist
C5H12_Inlet		Adjacent_Cell_Volume_4	Swirler_9_dist
N2_Inlet		Total_Adjacent_Cells	Swirler_10_dist
CO2_Inlet		Cell_Adjacent_Inlet	Swirler_11_dist
Air_Flow_Rate		Cell_Adjacent_Outlet	Swirler_12_dist
Main_Fuel_Flow		Cell_Adjacent_Wall	Swirler_Center_dist
Pilot_Fuel_Flow			
Air_Inlet_Temp			
Pressure_Outlet			
Total = 16	Total = 7	Total = 13	Total = 13

Pressure - Case 1

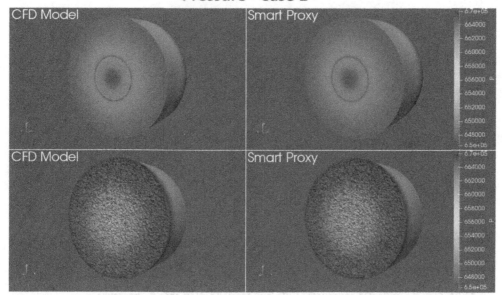

Fig. 11.22 *Step 3 results (Q1 combustor full view) for validation case 1 – pressure.*

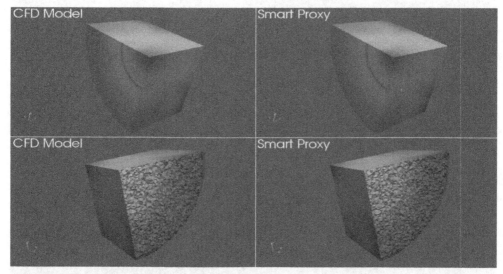

Fig. 11.23 *Step 3 results (Q1 combustor quarter view) for validation case 1 – pressure.*

Pressure - Case 5

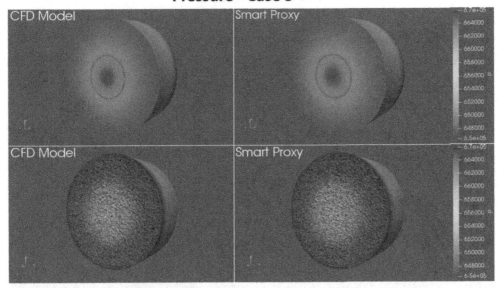

Fig. 11.24 *Step 3 results (Q1 combustor full view) for validation case 5 – pressure.*

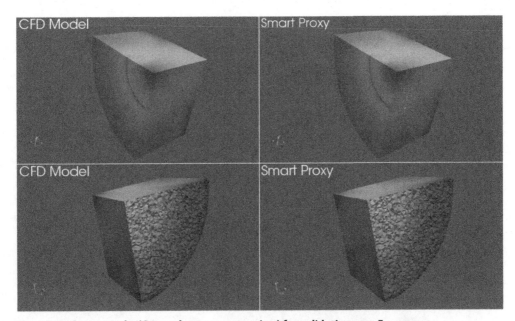

Fig. 11.25 *Step 3 results (Q1 combustor quarter view) for validation case 5 – pressure.*

Temperature - Case 1

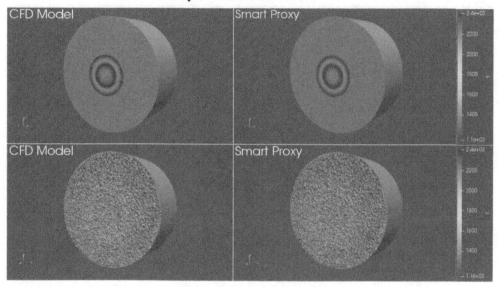

Fig. 11.26 *Step 3 results (Q1 combustor full view) for validation case 1 – temperature.*

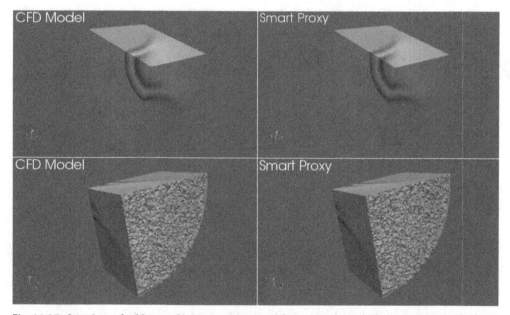

Fig. 11.27 *Step 3 results (Q1 combustor quarter view) for validation case 1 – temperature.*

Temperature - Case 5

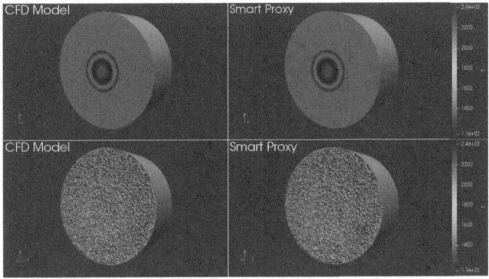

Fig. 11.28 *Step 3 results (Q1 combustor full view) for validation case 5 – temperature.*

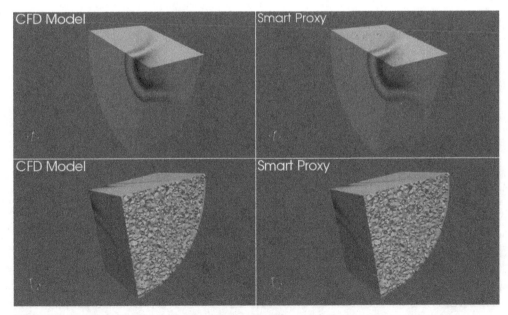

Fig. 11.29 *Step 3 results (Q1 combustor quarter view) for validation case 5 – temperature.*

Carbon Dioxide - Case 1

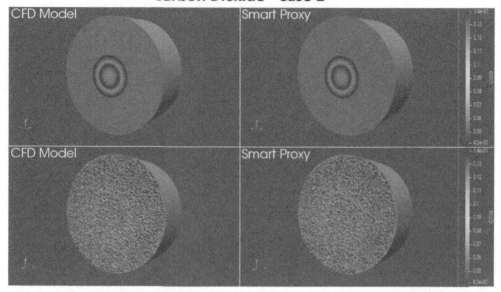

Fig. 11.30 *Step 3 results (Q1 combustor full view) for validation case 1 – carbon dioxide.*

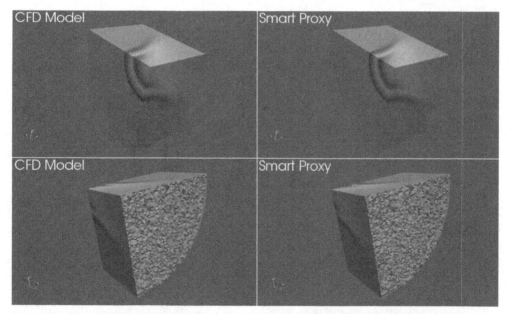

Fig. 11.31 *Step 3 results (Q1 combustor quarter view) for validation case 1 – carbon dioxide.*

Carbon Dioxide - Case 5

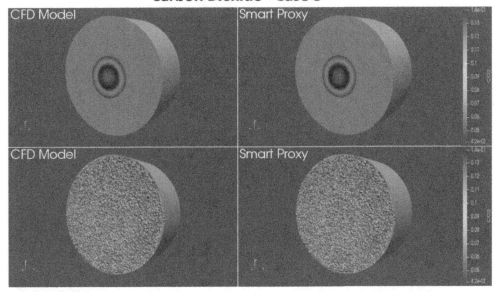

Fig. 11.32 *Step 3 results (Q1 combustor full view) for validation case 5 – carbon dioxide.*

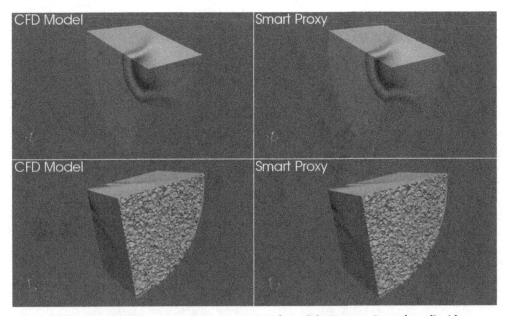

Fig. 11.33 *Step 3 results (Q1 combustor quarter view) for validation case 5 – carbon dioxide.*

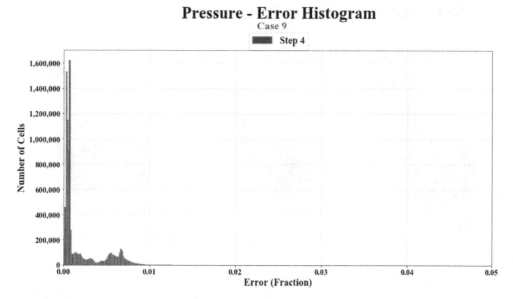

Fig. 11.34 *Step 4 error histogram for blind validation case 9 – pressure.*

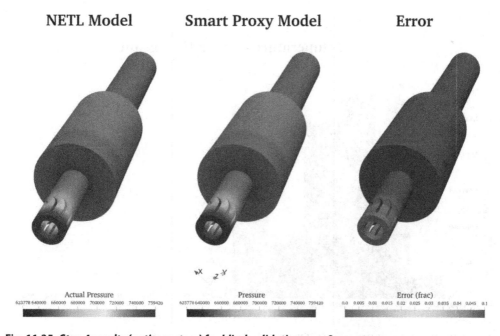

Fig. 11.35 *Step 4 results (entire system) for blind validation case 9 – pressure.*

NETL Model   **Smart Proxy Model**   **Error**

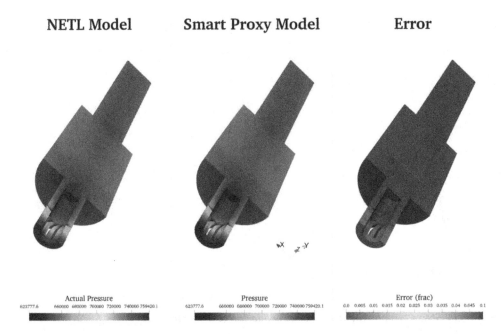

Fig. 11.36 *Step 4 results (half view) for blind validation case 9 – pressure.*

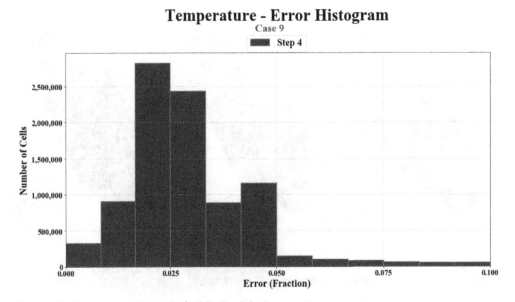

Fig. 11.37 *Step 4 error histogram for blind validation case 9 – temperature.*

NETL Model   **Smart Proxy Model**   **Error**

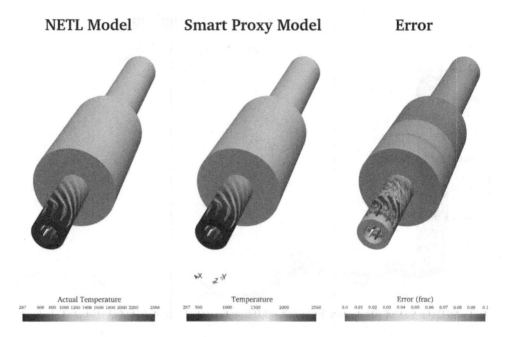

Actual Temperature
287   600 800 1000 1200 1400 1600 1800 2000 2200   2568

Temperature
287 500   1000   1500   2000   2568

Error (frac)
0.0   0.01 0.02 0.03 0.04 0.05 0.06 0.07 0.08 0.09   0.1

Fig. 11.38 *Step 4 results (entire system) for blind validation case 9 – temperature.*

NETL Model   **Smart Proxy Model**   **Error**

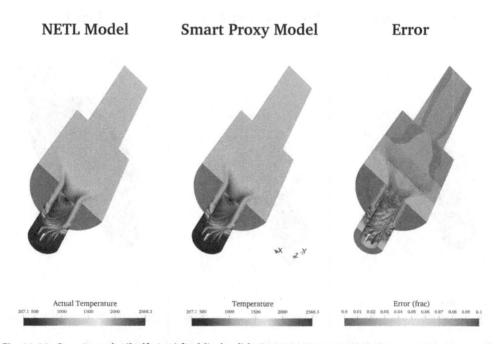

Actual Temperature
287.1 500   1000   1500   2000   2568.3

Temperature
287.1 500   1000   1500   2000   2568.3

Error (frac)
0.0   0.01 0.02 0.03 0.04 0.05 0.06 0.07 0.08 0.09   0.1

Fig. 11.39 *Step 4 results (half view) for blind validation case 9 – temperature.*

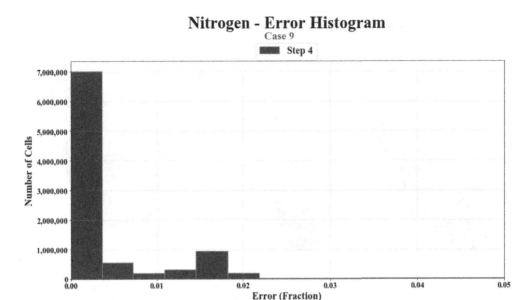

Fig. 11.40 *Step 4 error histogram for blind validation case 9 – carbon dioxide.*

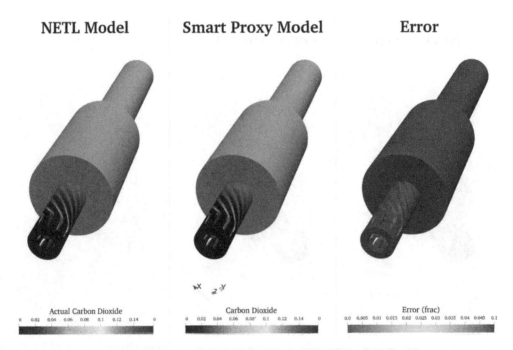

Fig. 11.41 *Step 4 results (entire system) for blind validation case 9 – carbon dioxide.*

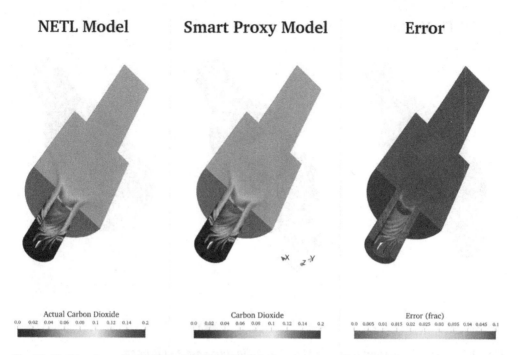

Fig. 11.42 *Step 4 results (half view) for blind validation case 9 – carbon dioxide.*

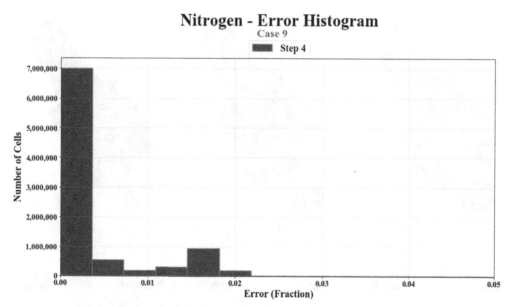

Fig. 11.43 *Step 4 error histogram for blind validation case 9 – nitrogen.*

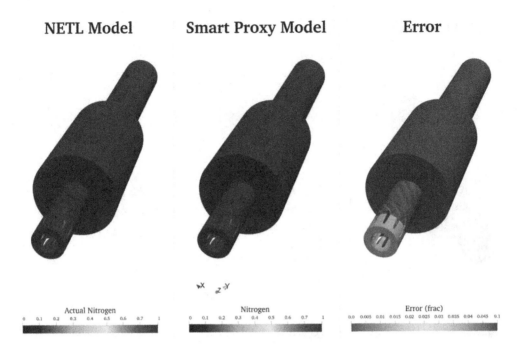

Fig. 11.44 *Step 4 results (entire system) for blind validation case 9 – nitrogen.*

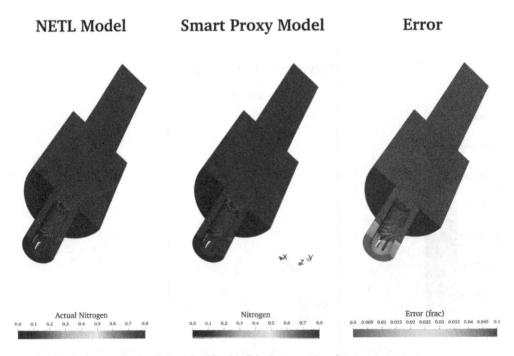

Fig. 11.45 *Step 4 results (half view) for blind validation case 9 – nitrogen.*

Oxygen - Error Histogram
Case 9

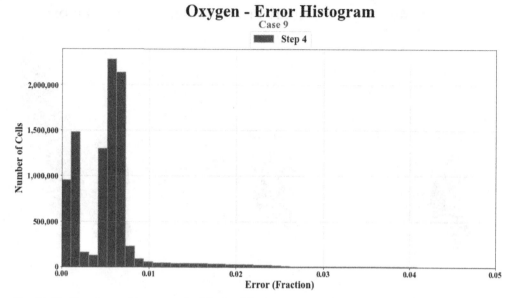

Fig. 11.46 *Step 4 error histogram for blind validation case 9 – oxygen.*

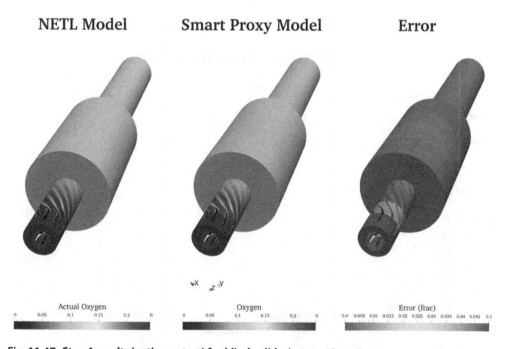

Fig. 11.47 *Step 4 results (entire system) for blind validation case 9 – oxygen.*

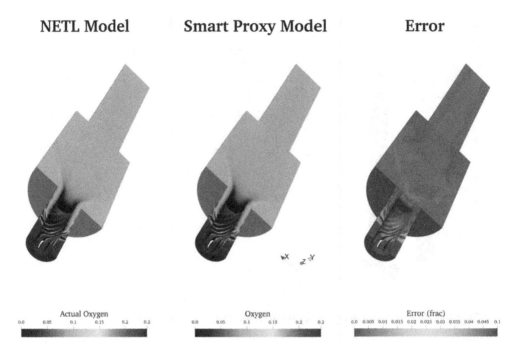

Fig. 11.48 *Step 4 results (half view) for blind validation case 9 – oxygen.*

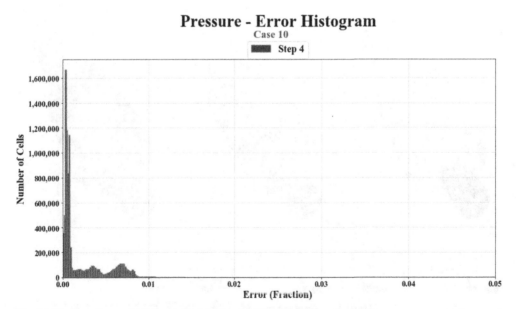

Fig. 11.49 *Step 4 error histogram for blind validation case 10 – pressure.*

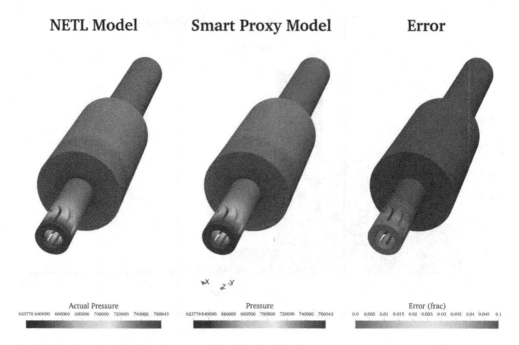

Fig. 11.50 *Step 4 results (entire system) for blind validation case 10 – pressure.*

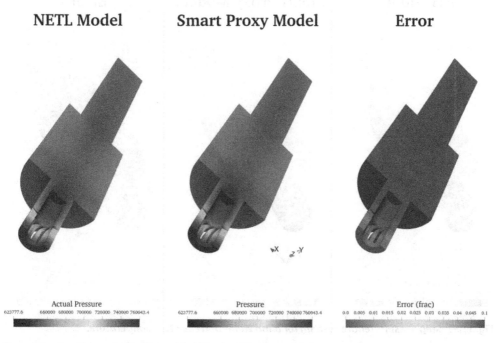

Fig. 11.51 *Step 4 results (half view) for blind validation case 10 – pressure.*

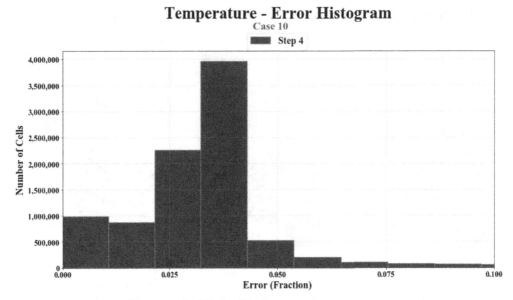

Fig. 11.52 *Step 4 error histogram for blind validation case 10 – temperature.*

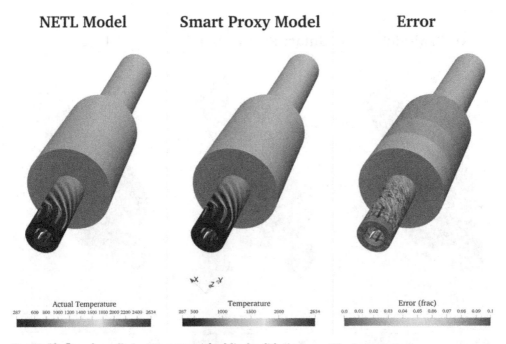

Fig. 11.53 *Step 4 results (entire system) for blind validation case 10 – temperature.*

* Application of artificial intelligence to computational fluid dynamics   333

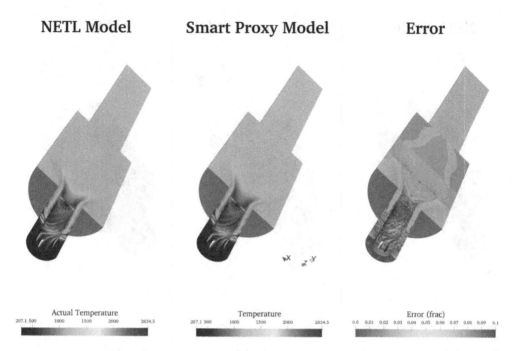

Fig. 11.54 *Step 4 results (half view) for blind validation case 10 – temperature.*

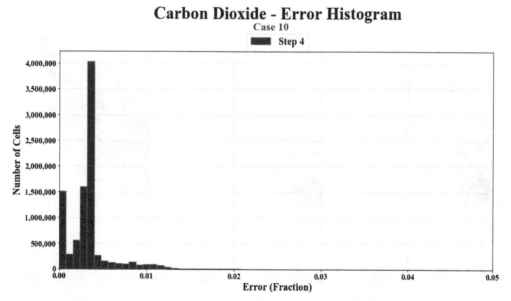

Fig. 11.55 *Step 4 error histogram for blind validation case 10 – carbon dioxide.*

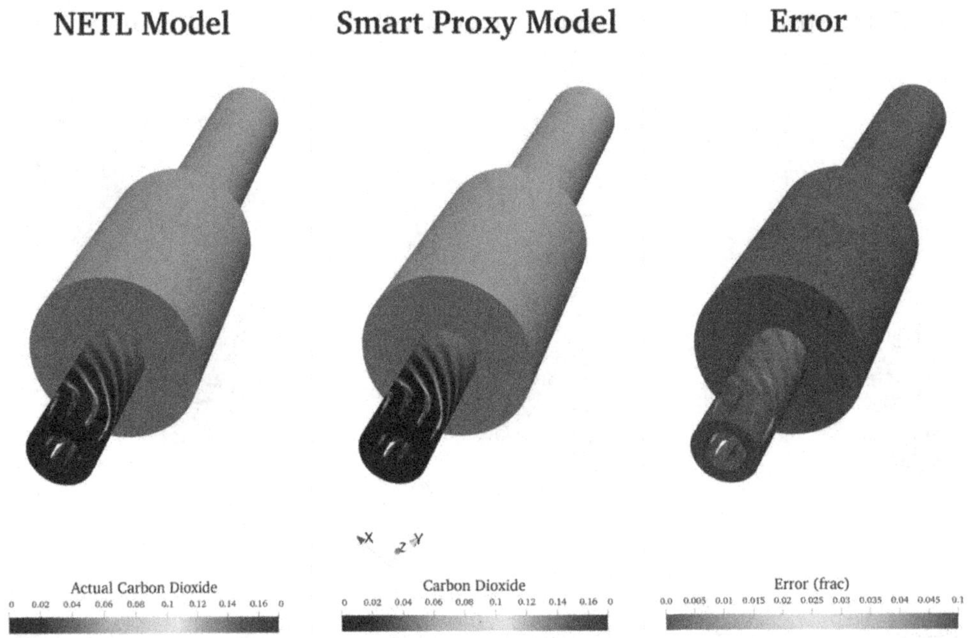

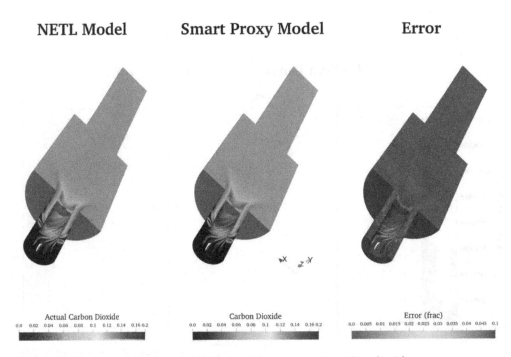

Fig. 11.56 *Step 4 results (entire system) for blind validation case 10 – carbon dioxide.*

Fig. 11.57 *Step 4 results (half view) for blind validation case 10 – carbon dioxide.*

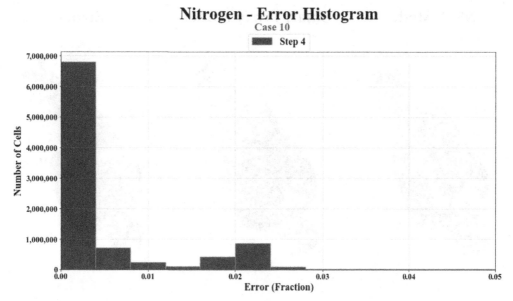

Fig. 11.58 *Step 4 error histogram for blind validation case 10 – nitrogen.*

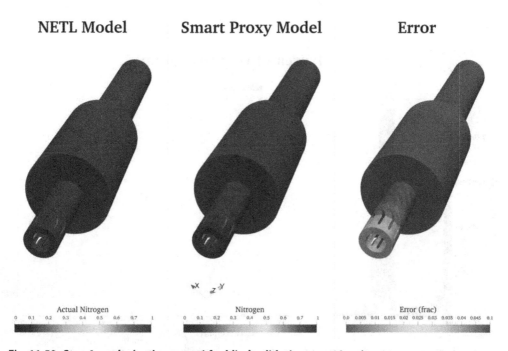

Fig. 11.59 *Step 4 results (entire system) for blind validation case 10 – nitrogen.*

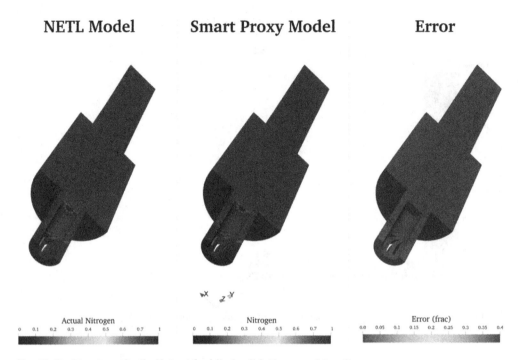

Fig. 11.60 *Step 4 results (half view) for blind validation case 10 – nitrogen.*

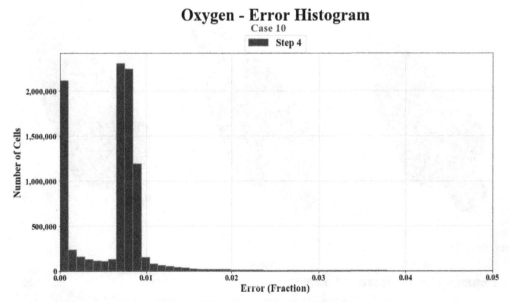

Fig. 11.61 *Step 4 error histogram for blind validation case 10 – oxygen.*

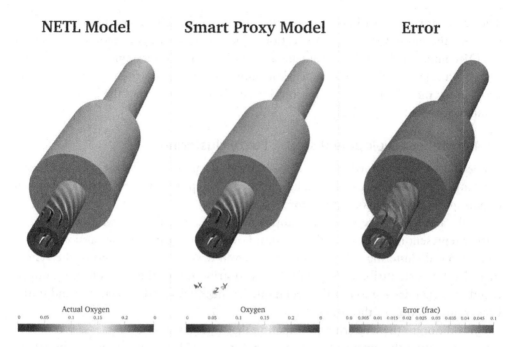

Fig. 11.62 *Step 4 results (entire system) for blind validation case 10 – oxygen.*

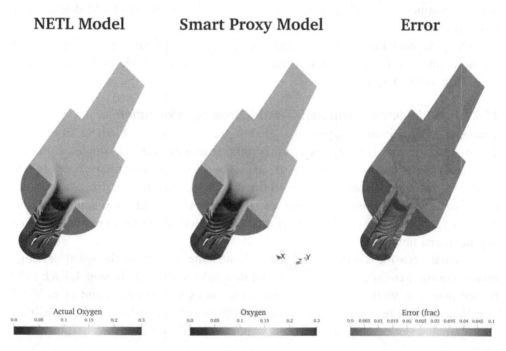

Fig. 11.63 *Step 4 results (half view) for blind validation case 10 – oxygen.*

The results for each attribute are presented in two figures. The first figure shows the result for the entire Q1 combustor section while the second figure shows a quarter of the Q1 combustor. Each figure contains a total of 4 images for a single case. The images on the left represent the CFD model simulation. The images on the right represent the smart proxy model. The top and bottom images show the front and the back of the Q1 combustor section, respectively.

11.4.4 Model development step 4: Fuzzy clustering

While the results obtained in step 3 showed significant improvement in smart proxy performance compared to the results obtained in step 1 and step 2, there were still opportunities to be explored for improvement.

In the last step of the model development process, new features were generated to further represent the physics of the reaction flow occurring in the combustion process. This involved clustering cells in each of the seven development sections of the system such that for each attribute, cells with similar distribution of the attribute are grouped together. Detail description of this technique has been provided in section 3 and in the background section of this study.

In addition to the clustering information that was added to the database, a new feature representing the extent of propane blend was added. This was specified by taking the ratio of methane flow rate to propane flow rate in the simulation model. A total of five new features were added in step 4. Table 11.13 shows the list of added features in this step.

In Step 4a, neural networks were trained, calibrated, and validated for all sections of the B6 combustor. Details of each section are summarized in Table 11.11 and graphically illustrated in Fig. 11.20.

11.4.4.1 Model development step 4 – model training information

The same data partitioning approach as used in previous steps was used. Six cases (cases 2, 3, 4, 6, 7, and 8) were used for training (80% randomly selected samples in the six cases) and calibration (remaining 20% of samples in the six cases) of the neural network. Base case 1 and blend case 5 were used for validation of each neural network at the end of each training attempt. The size of each development dataset is summarized in Table 11.11. As shown in Table 11.13, a total of 54 input attributes were used in training the neural networks.

A neural network was developed for each attribute in each of the seven development sections; therefore, a total of 35 neural networks was trained in step 4. Each fully trained neural network was finally deployed on the extra base case 9 and extra blend case 10, as a blind test of the performance of the neural networks. For each attribute, results for all seven development sections were combined for better visualization of the entire system.

Table 11.13 Training input attributes in development step 4.

Attributes provided from FLUENT	Attributes generated in step 1	Attributes generated in step 2	Attributes generated in step 3	Attributes generated in step 4
X_Coord	Node_Count	Euclidean_Dist_Inlet	Swirler_1_dist	CM_1
Y_Coord	Dist_To_Inlet	Euclidean_Dist_Inlet_Comb	Swirler_2_dist	CM_2
Z_Coord	Dist_To_Outlet	Euclidean_Dist_Comb_Exhaust	Swirler_3_dist	CM_3
volume	Dist_To_Inlet_Combustor	Euclidean_Dist_Outlet	Swirler_4_dist	Cluster_Id
CH4_Inlet	Dist_To_Combustor_Exhaust	Radial_Boundary_Ratio	Swirler_5_dist	Blend Ratio
C2H6_Inlet	Closest_Radial_Boundary	Adjacent_Cell_Volume_1	Swirler_6_dist	
C3H8_Inlet	Farthest_Radial_Boundary	Adjacent_Cell_Volume_2	Swirler_7_dist	
C4H10_Inlet		Adjacent_Cell_Volume_3	Swirler_8_dist	
C5H12_Inlet		Adjacent_Cell_Volume_4	Swirler_9_dist	
N2_Inlet		Total_Adjacent_Cells	Swirler_10_dist	
CO2_Inlet		Cell_Adjacent_Inlet	Swirler_11_dist	
Air_Flow_Rate		Cell_Adjacent_Outlet	Swirler_12_dist	
Main_Fuel_Flow		Cell_Adjacent_Wall	Swirler_Center_dist	
Pilot_Fuel_Flow				
Air_Inlet_Temp				
Pressure_Outlet				
Total = 16	Total = 7	Total = 13	Total = 13	Total = 5

11.4.4.2 Model development step 4 – presentation of results

This section presents the smart proxy results of the distribution of all attributes in the entire B6 combustor system (Figs. 11.64–11.71). Each attribute result is graphically and numerically presented by (a) a histogram describing the percent error distribution (all attributes shown under a 5% error with the exception of Temperature shown in a 10% error scale), (b) two tables; one table suggesting the number of cells within a 100% error scale and the second table providing a more detailed insight of the error distribution below 10% (Tables 11.14–11.33), and (c) two three-dimensional images showing distribution of the attribute in the entire system and a half cross-section view of the system.

Table 11.14 No of cells under 100% error for blind validation case 9 – pressure.

Percent ranges	Number of cells	Pere. cells
<10%	9,291,730	100.0%
10%–20%	0	0.0%
20%–30%	0	0.0%
30%–40%	0	0.0%
40%–50%	0	0.0%
50%–60%	0	0.0%
60%–70%	0	0.0%
70%–80%	0	0.0%
80%–90%	0	0.0%
>90%	0	0.0%

Table 11.15 No of cell under 10% error for blind validation case 9 – pressure.

Percent ranges	Number of cells	Pere. cells
<2%	9,291,730	100.0%
2%–4%	0	0.0%
4%–6%	0	0.0%
6%–8%	0	0.0%
8%–10%	0	0.0%
>10%	0	0.0%

Table 11.16 No of cells under 100% error for blind validation case 9 – temperature.

Percent ranges	Number of cells	Pere. cells
<10%	9,021,129	97.088%
10%–20%	222,291	2.392%
20%–30%	38,699	0.416%
30%–40%	8,919	0.096%
40%–50%	625	0.007%
50%–60%	51	0.001%
60%–70%	13	0.0%
70%–80%	2	0.0%
80%–90%	1	0.0%
>90%	0	0.0%

Table 4.17 No of cells under 10% error for blind validation cases 9 – temperature.

Percent ranges	Number of cells	Pere. cells
<2%	2,579,897	27.766%
2%–4%	4,465,385	48.058%
4%–6%	1,647,540	17,731%
6%–8%	199,023	2.142%
8%–10%	129,284	1.391%
>10%	270,601	2.912%

Table 11.18 No of cells under 100% error for blind validation case 9 – carbon dioxide.

Percent ranges	Number of cells	Pere. cells
<10%	9,291,730	100.0%
10%–20%	0	0.0%
20%–30%	0	0.0%
30%–40%	0	0.0%
40%–50%	0	0.0%
50%–60%	0	0.0%
60%–70%	0	0.0%
70%–80%	0	0.0%
80%–90%	0	0.0%
>90%	0	0.0%

Table 11.19 No of cells under 10% error for blind validation case 9 – carbon dioxide.

Percent ranges	Number of cells	Pere. cells
<2%	9,222,507	99.255%
2%–4%	53,275	0.573%
4%–6%	13,526	0.146%
6%–8%	2,422	0.026%
8%–10%	0	0.0%
>10%	0	0.0%

Table 11.20 No of cells under 100% error for blind validation case 9 – nitrogen.

Percent ranges	Number of cells	Pere. cells
<10%	9,252,113	99.574%
10%–20%	35,122	0.378%
20%–30%	4,429	0.048%
30%–40%	66	0.001%
40%– 50%	0	0.0%
50%–60%	0	0.0%
60%–70%	0	0.0%
70%–80%	0	0.0%
80%–90%	0	0.0%
>90%	0	0.0%

Table 11.21 No of cells under 10% error for blind validation case 9 – nitrogen.

Percent ranges	Number of cells	Pere. cells
<2%	9,130,895	98.269%
2%–4%	67,842	0.73%
4%–6%	28,232	0.304%
6%–8%	15,280	0.164%
8%–10%	9,864	0.106%
>10%	39,617	0.426%

Table 11.22 No of cells under 100% error for Blnd validation case 9 – oxygen.

Percent ranges	Number of cells	Pere. Cells
<10%	9,291,728	100.0%
10%–20%	2	0.0%
20%–30%	0	0.0%
30%–40%	0	0.0%
40%–50%	0	0.0%
50%–60%	0	0.0%
60%–70%	0	0.0%
70%–80%	0	0.0%
80%–90%	0	0.0%
>90%	0	0.0%

Table 11.23 No of cells under 10% error for blind validation case 9 – oxygen.

Percent ranges	Number of cells	Pere. cells
<2%	9,113,681	98.084%
2%–4%	144,830	1.559%
4%–6%	25,853	0.278%
6%–8%	7,246	0.078%
8%–10%	118	0.001%
>10%	2	0.0%

Table 11.24 No of cells under 100% error for blind validation case 10 – pressure.

Percent ranges	Number of cells	Pere. cells
<10%	9,291,730	100.0%
10%–20%	0	0.0%
20%–30%	0	0.0%
30%–40%	0	0.0%
40%–50%	0	0.0%
50%–60%	0	0.0%
60%–70%	0	0.0%
70%–80%	0	0.0%
80%–90%	0	0.0%
>90%	0	0.0%

Table 11.25 No of cells under 10% error for blind validation case 10 – pressure.

Percent ranges	Number of cells	Pere. cells
<2%	9,291,730	100.0%
2%–4%	0	0.0%
4%–6%	0	0.0%
6%–8%	0	0.0%
8%–10%	0	0.0%
>10%	0	0.0%

Table 11.26 No of cells under 100% error for blind validation case 10 – temperature.

Percent ranges	Number of cells	Pere. cells
<10%	9,053,161	97.432%
10%–20%	187,203	2.015%
20%–30%	40,293	0.434%
30%–40%	9,600	0.103%
40%–50%	1,052	0.011%
50%–60%	308	0.003%
60%–70%	78	0.001%
70%–80%	20	0.0%
80%–90%	9	0.0%
>90%	6	0.0%

Table 11.27 No of cells under 10% error for blind validation case 10 – temperature.

Percent ranges	Number of cells	Pere. cells
<2%	1,748,053	18.813%
2%–4%	4,973,509	53.526%
4%–6%	2,002,154	21.548%
6%–8%	207,036	2.228%
8%–10%	122,409	1.317%
>10%	238,569	2.568%

Table 11.28 No of cells under 100% error for blind validation case 10 – carbon dioxide.

Percent ranges	Number of cells	Pere. cells
<10%	9,291,730	100.0%
10%–20%	0	0.0%
20%–30%	0	0.0%
30%–40%	0	0.0%
40%–50%	0	0.0%
50%–60%	0	0.0%
60%–70%	0	0.0%
70%–80%	0	0.0%
80%–90%	0	0.0%
>90%	0	0.0%

Table 11.29 No of cells under 10% error for blind validation case 10 – carbon dioxide.

Percent ranges	Number of cells	Pere. cells
<2%	9,252,938	99.583%
2%–4%	33,072	0.356%
4%–6%	5,699	0.061%
6%–8%	20	0.0%
8%–10%	1	0.0%
>10%	0	0.0%

Table 11.30 No of cells under 100% error for blind validation case 10 – nitrogen.

Percent ranges	Number of cells	Pere. cells
<10%	9,256,463	99.62%
10%–20%	27,707	0.298%
20%–30%	7,209	0.078%
30%–40%	350	0.004%
40%–50%	1	0.0%
50%–60%	0	0.0%
60%–70%	0	0.0%
70%–80%	0	0.0%
80%–90%	0	0.0%
>90%	0	0.0%

Table 11.31 No of cells under 10% error for blind validation case 10 – nitrogen.

Percent ranges	Number of cells	Pere. cells
<2%	8,222,525	88.493%
2%–4%	982,487	10.574%
4%–6%	25,745	0.277%
6%–8%	15,571	0.168%
8%–10%	10,135	0.109%
>10%	35,267	0.38%

Table 11.32 No of cells under 100% error for blind validation case 10 – oxygen.

Percent ranges	Number of cells	Pere. cells
<10	9,292,730	100.0%
10%–20%	0	0.0%
20%–30%	0	0.0%
30%–40%	0	0.0%
40%–50%	0	0.0%
50%–60%	0	0.0%
60%–70%	0	0.0%
70%–80%	0	0.0%
80%–90%	0	0.0%
>90%	0	0.0%

Table 11.33 No of cells under 10% error for blind validation case 10 – oxygen.

Percent ranges	Number of cells	Pere. cells
<2%	9,178,770	98.784%
2%–4%	86,990	0.936%
4%–6%	23,215	0.25%
6%–8%	2,726	0.029%
8%–10%	29	0.0%
>10%	0	0.0%

11.5 Conclusions

It has been successfully demonstrated that a data driven predictive model can reproduce the results of CFD simulation of natural gas combustion in a high-pressure combustor system. The developed smart proxy replicates the thermal-flow patterns of pressure, temperature, and species concentrations (nitrogen, oxygen and carbon-dioxide) with a percent error of not more than 10%, and a faster execution time compared to the numerical CFD simulation approach. A single CFD simulation run of the B6 Combustor model takes about 24 hours or more to complete on the NETL HPC with single node 196GB RAM, 40 cores while the smart proxy generates results in about 5–6 minutes when executed on the same HPC configuration. This time can be further reduced if the compute resource limitations (computer memory and speed) are properly addressed and the system is modeled without having to divide into smaller development sections.

More importantly, the smart proxy model achieved this level of accuracy using a very minimal amount of data; only six CFD simulation runs of the B6 combustor model was used in developing the B6 smart proxy model. While CFD simulations require extensive compute resources, the developed smart proxy can be deployed on commodity computers (inexpensive laptop or desktop machines). This proves that this technology can contribute significantly to research studies that are targeted at determining optimal design and operating conditions that would maximize the efficiency of complex power generation systems.

11.5.1 Recommendations

The final modeling results shown in Section 11.4.4.2 of this study shows that the smart proxy development framework applied can replicate the results of CFD simulation of a single-phase reaction flow in a high-pressure combustor to a reasonable degree of accuracy. The results however could be improved, especially for a thermal flow field variable such as temperature. Apart from the simulation model boundary conditions, no reaction data from FLUENT software was used in developing the smart proxy model. It might be worth identifying potential contribution of the other successfully modeled transport and specie variables (pressure, oxygen, nitrogen and carbon-dioxide) to the modeling of the thermal flow field. Perhaps, the neural networks may further learn the reaction flow characteristics if transport data such as the turbulent kinetic energy and dissipation rate are also modeled.

Training a single neural network on the entire system for each attribute will increase the amount of information provided to any single neural network in the smart proxy model and improve performance as opposed to a total of seven networks currently built for each attribute. In order to efficiently fulfill the objective of the next phase of the project, which is a more complex multi-phase reaction flow in a coal-fired boiler, an access to a high-performance-computing (HPC) facility is highly recommended.

11.6 Appendix

11.6.1 Model development step 3: Q1 combustor results – nitrogen

Nitrogen – Case 1

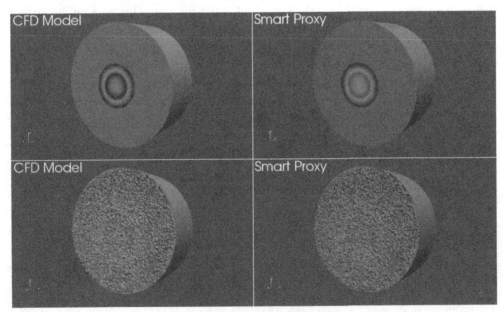

Fig. 11.64 *Step 3 results (Q1 combustor full view) for validation case 1 – nitrogen.*

Fig. 11.65 *Step 3 results (Q1 combustor full view) for validation case 1 – nitrogen.*

Nitrogen – Case 5

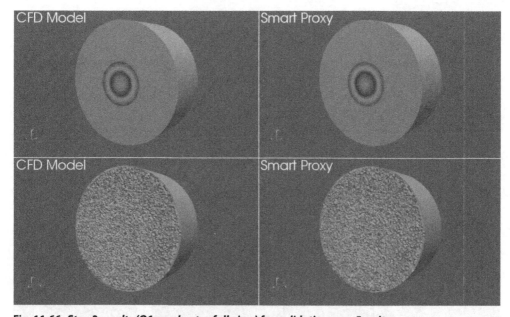

Fig. 11.66 *Step 3 results (Q1 combustor full view) for validation case 5 – nitrogen.*

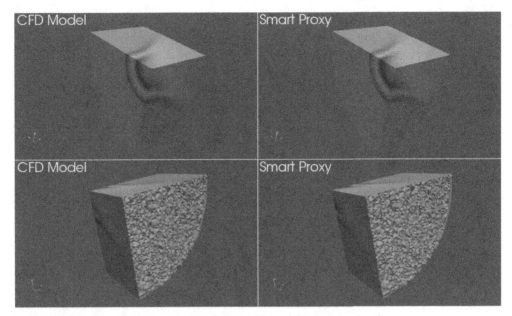

Fig. 11.67 *Step 3 results (Q1 combustor full view) for validation case 5 – nitrogen*

11.6.2 Model development step 3: Q1 combustor results – oxygen

Oxygen – Case 1

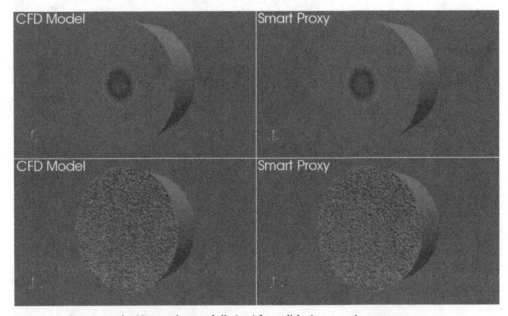

Fig. 11.68 *Step 3 results (Q1 combustor full view) for validation case 1 – oxygen.*

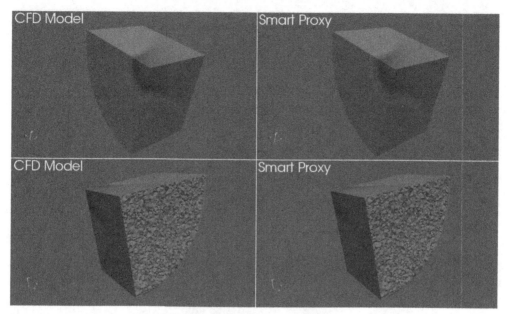

Fig. 11.69 *Step 3 results (Q1 combustor full view) for validation case 1 – oxygen.*

Oxygen – Case 5

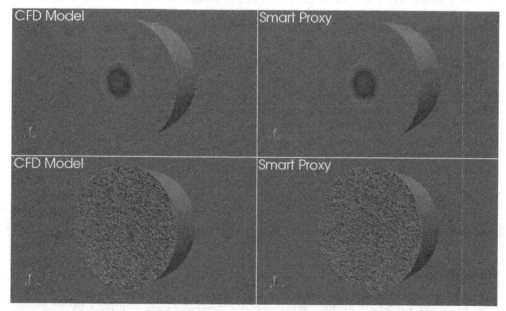

Fig. 11.70 *Step 3 results (Q1 combustor full view) for validation case 5 – oxygen.*

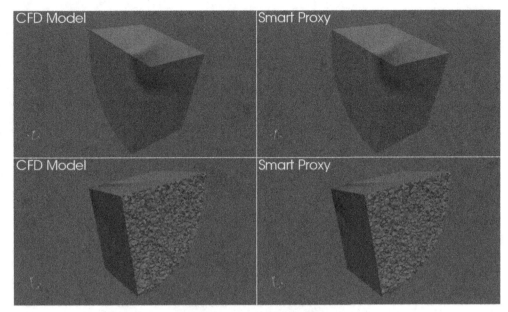

Fig. 11.71 *Step 3 results (Q1 combustor full view) for validation case 5 – oxygen.*

11.7 Acknowledgments

This work was performed in support of the US Department of Energy's Fossil Energy Crosscutting Technology Research Program. The Research was executed through the NETL Research and Innovation Center's Transformational Technologies for New and Existing Plants FWP. Research performed by Leidos Research Support Team staff was conducted under the RSS contract 89243318CFE000003.

11.8 Disclaimer

References

1. Intelligent Solutions, Inc, "Modeling physics with artificial intelligence & machine learning," August 2019. [Online] Available from: http://intelligentsolutionsinc.com/Blogs/Blog07-Physics.shtm. (Accessed 1 August 2019).

2. Mohaghegh S. Data Driven Analytics for the Geologic Storage of CO2, 2018. CRC Press USA united states.

3. Mohaghegh S, Maysami M, Gholami V. Smart proxy modeling of SACROC CO2-EOR. *J Fluids* 2019;**4**(85).

4. Bromhal GS, Birkholzer J, Mohaghegh SD, Sahinidis N, Wainwright H, Zhang Y, Amini S, Gholami V, Zhang Yan, Shahkarami A. Evaluation of rapid performance reservoir models for quantitative risk assessment. *Energy Procedia* 2014;**63**:3425–31. https://doi.org/10.1016/j.egypro.2014.11.371.

5. Shahkarami A, Mohaghegh S, Gholami V, Haghighat A, Moreno D. Modeling pressure and saturation distribution in a CO2 storage project using a surrogate reservoir model (SRM). *Greenh Gases Sci Technol* 2014;**4**:1–27. http://doi.org/10.1002/ghg.

6. Fullmer W, Hrenva C. Quantitative assessment of fine-grid kinetic theory based predictions of mean-slip in unbounded fluidization. *AIChE J* 2016;**62**(1):11–7.

7. Shahnam M, Gel A, Dietiker J-F, Subramaniyan AK, Musser J. The effect of grid resolution and reaction models in simulation of a fluidized bed gasifier through nonintrusive uncertainty quantification techniques. *J Verif Valid Uncertain Quantif* 2016;**4**:45–52.

8. Ansari A, Mohaghegh S, Shahnam M, Dietiker JF and Li T, "Data Driven Smart Proxy for CFD Application of Big Data Analytics & Machine Learning in Computational Fluid Dynamics, Part Two: Model Building at the Cell Level," NETL-PUB-21634, NETL Technical Report Series; U.S. Department of Energy, National Energy Technology Laboratory, Morgantown, WV, 2017.

9. ANSYS Inc., ANSYS Fluent User's Guide, Release 16.1, Southpointe 2600 ANSYS Drive, Canonsburg, PA, 2015.

10. Gubba S, Ingham D, Larsen K, Ma L, Pourkashanian M. Numerical modelling of the co-firing of pulverised coal and straw in a 300MWe tangentially fired boiler. *Fuel Process Technol* 2012;**104**:181–8.

11. Zhou H, Yang Y, Liu H, Hang Qi. Numerical simulation of the combustion characteristics of a low NOx swirl burner: influence of the primary air pipe. *Fuel* 2014;**130**:168–76.

12. Modliński N, Madejski P, Janda T, Szczepanek K, Kordylewski W. A validation of computational fluid dynamics temperature distribution prediction in a pulverized coal boiler with acoustic temperature measurement. *Energy* 2015;**92**:77–86.

13. Sun W, Zhong W, Yu A, Liu L, Qian Y. Numerical investigation on the flow, combustion, and NOX emission characteristics in a 660 MWe tangential firing ultra-supercritical boiler. *Adv Mech Eng* 2016;**8**.

14. Chen S, He B, He D, Cao Y, Ding G, Liu X et al. Numerical investigations on different tangential arrangements of burners for a 600 MW utility boiler. *Energy* 2017;**122**:287–300.

15. Ge X, Dong J, Fan H, Zhang Z, Shang X, Hu X, Zhang J. Numerical investigation of oxy-fuel combustion in 700°C-ultra-supercritical boiler. *Fuel* 2017;**207**:602–14.

16. Choi C, Kim C. Numerical investigation on the flow, combustion and NOx emission characteristics in a 500 MWe tangentially fired pulverized-coal boiler. *Fuel* 2009;**88**:1720–31.

17. Edge P, Heggs P, Pourkashanian M, Williams A. An integrated computational fluid dynamics–process model of natural circulation steam generation in a coal-fired power plant. *Comput Chem Eng* 2011;**35**(12):2618–31.

18. Shi L, Fu Z, Duan X, Cheng C, Shen Y, Liu B et al. Influence of combustion system retrofit on NOx formation characteristics in a 300 MW tangentially fired furnace. *Appl Therm Eng* 2016;**98**:766–77.

19. Smith TF, Shen ZF, Friedman JN. Evaluation of coefficients for the weighted sum of gray gases model. *Heat Transfer* 1982;**104**(4):602–8.

20. Yin C. Refined weighted sum of gray gases model for air-fuel combustion and its impacts. *Energy Fuels* 2013;**27**:6287–94.

21. Tan P-N, Steinbach M, Karpatne A, Kumar V. Cluster analysis: basic concepts and algorithms. Introduction to Data Mining, Pearson. USA united states, 2018:487–93.

22. Sharma S. "Artificial neural network (ANN)," 2017. [Online]. Available from: https://www.datascience-central.com/profiles/blogs/artificial-neural-network-ann-in-machine-learning. (Accessed 1 August 2019).

23. Chollet F. "Keras," https://keras.io, 2015. (Accessed 1 August 2019).

24. Brownlee J. "How to Control the Stability of Training Neural Networks With the Batch Size," Machine Learning Mastery Pty. Ltd, 2019. [Online]. Availables: from: https://machinelearning-mastery.com/how-to-control-the-speed-and-stability-of-training-neural-networks-with-gradient-descent-batch-size/. (Accessed 10 March 2020).

25. Oliphant TE. Guide to NumPy. Travis E. Oliphant 2006:34–46.

26. Mohaghegh S, Ameri S. Artificial neural network as a valuable tool for petroleum engineers. *SPE* 1995;*29220*:25–31.

27. Mohaghegh S, Arefi R, Ameri S, Rose D. Design and development of an artificial neural network for estimation of formation permeability. *SPE Comput Appl* 1995;**7**(6):151–5. http://doi.org/10.2118/28237-PA.

28. Mohaghegh S. Top-down, intelligent reservoir modeling of oil and gas producing shale reservoirs: case studies. *Int J Oil Gas Coal Technol* 2012;**15**:12.

29. Mohaghegh S. "Smart proxy modeling for numerical reservoir simulations–big data analytics in E&P," 2015. [Online]. Available from: https://webevents.spe.org/products/smart-proxy-modeling-for-numerical-reservoir-simulations-big-data-analytics-in-ep. (Accessed 1 August 2019).

30. Alenezi F, Mohaghegh S. A data-driven smart proxy model for a comprehensive reservoir simulation-Riyadh, 2016. IEEE.

31. Boosari S. Predicting the dynamic parameters of multiphase flow in CFD (dam-break simulation) using artificial intelligence–(cascading deployment). *Fluids* 2019;**4**:45–52.

32. Richards G, Straub DL, Ferguson DH, Robey EH. "LNG Interchangeability/Gas Quality: Results of the National Energy Technology Laboratory's Research for the FERC on Natural Gas Quality and Interchangeability," U.S. Department of Energy, National Energy Technology Laboratory (NETL), 2007.

33. Carvalho MG, Coelho PJ. Heat transfer in gas turbine combustors. *J. Thermophysics* 1989;**3**(2).

34. Klobucar L. Thermal Radiation Heat Transfer between Surfaces. University of Ljubljana, USA united states, 2016.

Index

Page numbers followed by "*f*" and "*t*" indicate, figures and tables respectively.